Animal Anomalies

What Abnormal Anatomies Reveal about Normal Development

Among the offspring of humans and other animals are occasional individuals that are malformed in whole or in part. The most grossly abnormal of these have been referred to from ancient times as monsters, because their birth was thought to foretell doom; the less severely affected are usually known as anomalies. This volume digs deeply into the cellular and molecular processes of embryonic development that go awry in such exceptional situations. It focuses on the physical mechanisms of how genes instruct cells to build anatomy, as well as the underlying forces of evolution that shaped these mechanisms over eons of geologic time. The narrative is framed in a historical perspective that should help students trying to make sense of these complex subjects. Each chapter is written in the style of a Sherlock Holmes story, starting with the clues and ending with a solution to the mystery.

Lewis I. Held, Jr. is Associate Professor of Developmental Genetics in the Department of Biological Sciences at Texas Tech University. He is a fly geneticist who has taught human embryology for 35 years. He studied molecular biology at the Massachusetts Institute of Technology (BS, 1973), investigated bristle patterning under John Gerhart at the University of California, Berkeley (PhD, 1977), and conducted postdoctoral research with Peter Bryant and Howard Schneiderman at the University of California, Irvine (1977–86). This is his sixth scholarly monograph, following *Models for Embryonic Periodicity* (Karger, 1992), *Imaginal Discs* (Cambridge, 2002), *Quirks of Human Anatomy* (Cambridge, 2009), *How the Snake Lost its Legs* (Cambridge, 2014), and *Deep Homology?* (Cambridge, 2017).

"Orthodoxy is so rife in science these days it is strangling originality. The spread of 'best practice' as well as the related belief that there is only one ideal way to understand and explain things has stifled both diversity and imagination. It is increasingly commonplace to explain a natural phenomenon by confusing the consensus view with the truth. There has been a growing dependence on a few model systems such as the mouse, *Drosophila*, and *C. elegans*; this pragmatic approach has proved successful, but it is also confining and a bit dull.

Lewis Held shows us there is another way – to look at the natural world open-mouthed and open-minded. We are taken on a lively ramble through myriad natural phenomena in countless species and the attempts of scientists to understand them. There is an infectious sense of the wonder and complexity of everything. There are innumerable nuggets to be found and it is fun. As Professor Held says in his introduction: 'Tedium is the bane of textbooks, and I wanted to write a more whimsical opus.' He has succeeded: his book sings like the descant in a choral rendition of a familiar hymn. I recommend it, but don't try and read it all in one go!"

Peter A. Lawrence
University of Cambridge, UK

"With rigorous arguments presented in captivating prose and crystal-clear drawings so rich in information, this new masterpiece by Lewis Held is a unique introduction to the genetics of development. Here, monstrous and normal illuminate each other, as products of the same developmental logic. This book is full of inspiring insights, on a par with the works of the great developmental biologists highlighted in its pages."

Alessandro Minelli, University of Padova, Italy, and author of
Understanding Development

Cyclopic male goat from a Boer goat breed stock, which was born in 2012 on a farm in Saint Lawrence (near Midland), Texas [1325]. This kid, which had a normal twin sister, is being held by Garry Batla, a son of Delmer Batla, the farm's owner. Delmer's granddaughter Kendra sent this picture to *Ripley's Believe It or Not!*, where it was published in the 10th edition of their annual book series. The Batlas fed the kid with an eyedropper because its mouth was too deformed to suckle, but despite their loving care it lived for only two weeks. They never gave it a name. Cyclopia, which affects humans also, is normally a rare condition [221,309,1153], but on one occasion it occurred as an epidemic in a whole flock of sheep when they happened upon a meadow of toxic plants [112,686]. The plants contained a chemical that inhibits the Hedgehog pathway, which governs the midline of the vertebrate body [207]. That pathway is discussed in this book along with other modalities of intercellular communication.

In Homer's *Odyssey* a giant Cyclops held Odysseus and his men captive in his cave, and they seemed doomed until a strategy occurred to the great general that ultimately allowed him to defeat the Cyclops and liberate his soldiers. Another goal of this book is to recount the exploits of researchers in the field of developmental biology who have been as clever as Odysseus in the experiments they have designed and executed over the years. Their feats are no less heroic than the mythical deeds which Homer lauded so lyrically. Photo courtesy of Delmer Batla.

(A black and white version of this figure will appear in some formats. For the color version, please refer to the plate section.)

Animal Anomalies

What Abnormal Anatomies Reveal about Normal Development

LEWIS I. HELD, JR.

Texas Tech University

CAMBRIDGE
UNIVERSITY PRESS

University Printing House, Cambridge CB2 8BS, United Kingdom

One Liberty Plaza, 20th Floor, New York, NY 10006, USA

477 Williamstown Road, Port Melbourne, VIC 3207, Australia

314–321, 3rd Floor, Plot 3, Splendor Forum, Jasola District Centre, New Delhi – 110025, India

79 Anson Road, #06-04/06, Singapore 079906

Cambridge University Press is part of the University of Cambridge.

It furthers the University's mission by disseminating knowledge in the pursuit of education, learning, and research at the highest international levels of excellence.

www.cambridge.org
Information on this title: www.cambridge.org/9781108834704
DOI: 10.1017/9781108876612

First published 2021

Printed in the United Kingdom by TJ Books Limited, Padstow Cornwall

A catalogue record for this publication is available from the British Library.

ISBN 978-1-108-83470-4 Hardback
ISBN 978-1-108-81974-9 Paperback

Contents

Color plates can be found between pages 138 and 139.

Foreword

Developmental biology can be seen as going through a golden age: our understanding of thirty years of work on developmental molecular genetics is enabling us to explore in ever greater detail three important areas. These are the developmental basis of evolutionary change (evo-devo), the regulation of stem-cell development, and the mode of operation of the complex networks that drive developmental change. The result is, on the one hand, a stream of rather technical papers and reviews that developmental biologists are duty-bound to read and, on the other, a range of necessarily simplified books that are aimed at the general public.

What is missing are books for working biologists who want to stand back from the bench and sit down with a good read about their subject, one that reminds them of why they enjoy being biologists. Lewis Held's *Animal Anomalies* may be what they have been looking for! It brings together information on developmental anomalies from across the animal world and provides the molecular and developmental context in which to appreciate them. Although some are well known, many are not, but all pose intriguing problems, many of which remain to be solved.

Lewis Held has spent many months investigating the obscurer corners of the biological literature and has brought back to life an area of developmental biology that today seems mainly to have slipped through the cracks. The result is a fascinating investigation of how embryogenesis can take wrong turnings. Every developmental biologist will know some of these examples, but may not have appreciated how they are now explained; very few, however, will be aware of most of the abnormalities that Held has uncovered.

These heritable curiosities have an additional significance: because they show us the outer reaches of developmental potential, they provide tests for standard theory. Perhaps more important, they can force us to go beyond our current thinking to produce novel explanations and experimental tests for what is going on. The classic example here is the set of homeotic mutants in *Drosophila*, first discovered by Bridges in 1915. It took more than sixty years for the underlying homeobox genes and their mutations to be identified, but only a few more years before their role in anterior–posterior patterning in embryos across the phyla was revealed. Who knows what further Nobel-Prize-winning discoveries lurk in the unexplained molecular thickets of those anomalies yet to be explored?

The book should have a wider resonance. Molecular biologists who focus on simple eukaryotes and the prokaryotic world will find this book to be an intriguing way into the molecular complexities that keep developmental biologists awake at night. Geneticists should enjoy this exploration of a molecular complexity that the founding fathers of their subject could never have imagined. Perhaps more important, however, is that everyone who teaches in the general area of development will find here a host of fascinating topics and anecdotes that will engage their student audiences, even on a hot Friday afternoon in summer.

Jonathan Bard
Oxford

Preface

Each of us started life as a single cell. Inside that egg's DNA there must have existed some sort of instruction manual for how to assemble our body, but the exact nature of that guidebook has eluded embryologists for centuries. Only recently have we succeeded in deciphering those plans by researching how simpler animals develop, and most of our understanding has come from aberrant individuals that manifest heritable defects.

> Although we seek to understand the events occurring in normal embryonic development, it is the bizarre mistakes made by the developing organism, either spontaneously or in response to experimental manipulations, that are most instructive. [163]

One of the most informative such mutants was the four-winged "bithorax" fly, the investigation of which led to the discovery of the Hox Complex and its widespread role in constructing both vertebrates and invertebrates. The bithorax phenotype exemplifies the phenomenon of "homeosis" wherein one body part transforms into another. That term was coined by William Bateson, who cataloged hundreds of naturally occurring anatomical abnormalities in his 1894 tome *Materials for the Study of Variation* [80]. Many of the unusual specimens he collected were likely due to mutations, while others must have arisen through congenital (non-genetic) errors.

Bateson pursued his collection not as an end in itself, but to promote a hypothesis explicitly enunciated in the remainder of his book's title: ... *Treated with Especial Regard to Discontinuity in the Origin of Species.* In contrast to Darwin, who thought that new species diverge from old ones by a gradual process of incremental change, Bateson supposed that new species could arise *discontinuously*, via the sorts of anomalous "sports" in his catalog [1017]. Richard Goldschmidt later extended Bateson's argument in a monograph of his own whose title – *The Material Basis of Evolution* – pays homage to Bateson's classic [447].

Goldschmidt thought that deviants could found new species under the right environmental conditions – whence the term "hopeful monsters" (see Fig. 11.2) [305,307], but few, if any, of the variants in *Animal Anomalies* meet that expectation. The focus of this book is more on development than on evolution [1346]. The field of developmental biology is experiencing a renaissance, thanks in part to the burgeoning of "–omics" and stem-cell innovations [1194].

The most exciting time in the history of developmental biology is right now. Fueled both by new technologies and by new thought from other fields, we are exploding old notions and opening fantastic new horizons in embryology. The only problem we face is a problem of perception.

How do we – as a *community* – convey this excitement to others? Since storytelling is the most fundamentally human activity, I think we start by telling our stories. [1346]

Animal Anomalies celebrates some major triumphs of this field by recounting key scientific experiments that yielded pithy insights. It uses a case-study approach where each chapter follows a simple rubric: a puzzling anomaly leads to a search for the phenotype's etiology. As the detective story unfolds, take-home lessons emerge about how animal development works, with further examples ("tangents") added to explore those general principles ("GPs") in detail. The GPs are numbered to enable cross-referencing.

More comprehensive treatments of these same axioms can be found in the Embryo Project Encyclopedia (https://embryo.asu.edu), Lewis Wolpert *et al.*'s *Principles of Development* [1405], Gilbert and Barresi's *Developmental Biology* [433], and Alberts *et al.*'s *Molecular Biology of the Cell* (4th ed., Chapter 21) [17]. The best primer in this genre remains Sean Carroll's *Endless Forms Most Beautiful* [187].

Ideally, Carroll's book should be read before trying to digest mine because I dive right into the narrative instead of ramping up slowly with a lot of introductory material. Tedium is the bane of textbooks, and I wanted to write a more whimsical opus. Yes, there is a lot of jargon to absorb, but I've opted to define the technical terms when they're needed to grasp the stories at hand, rather than risk losing readers by force-feeding them preambles about the history, theory, or vocabulary of the embryological concepts.

A similar approach was used in a different venue by another Carroll. *Lewis* Carroll exalted the wonders of what has been called "recreational mathematics" through his Alice books [46,184]. Like those books, *Animal Anomalies* was written more for amusement than edification, in a spirit of "recreational embryology" (e.g., trying to decipher zebra stripes and leopard spots in Chapter 10). Despite my attempts to soft-pedal the pedagogy, the text remains regrettably ponderous, especially in the early chapters. However, for readers who stick with it there is one reward awaiting them near the end: the chapters progressively get shorter or, as the Gryphon said to Alice, "That's the reason they're called *lessons* – because they lessen from day to day."

Many outstanding treatises discuss anomalous humans [1039], including a volu-minous compendium called *Anomalies and Curiosities of Medicine* published two years after Bateson's classic volume [463], Armand Leroi's *Mutants* [777], Mark Blumberg's *Freaks of Nature* [127], and a host of medical genetics textbooks. Such works extend the didactic tradition of the *Wunderkammer* cabinets of curiosities, popular in the seventeenth century [475], but which devolved into the voyeurism of P. T. Barnum's freak shows in the nineteenth century and the hucksterism of *Ripley's Believe It or Not!* in the twentieth [737]. The latter exploitations led me to feature *non*-human anomalies instead, but it would be silly to try to avoid human topics entirely, because animal models have helped to solve clinical syndromes [654,1389,1398]. Hence, human cases do appear here as well, albeit sparingly.

Why focus on frogs, flies, dogs, and cats? Frogs and flies have been two of the model organisms employed by embryologists for the past century, and much of our understanding of developmental mechanics has come from these two groups. Dogs and cats have been subjected to artificial selection for millennia, based on fortuitous peculiarities that happened to appeal to their human handlers, so they also offer a rich reservoir of traits that depart from the norm of wild ancestors. Now is a good time to revisit those traits, because many have recently been demystified through genomic surveys.

The internet abounds with pictures of freaky animals, but many are fakes. All of the pictures in this book have been fully authenticated, and all of the sources have been thoroughly investigated.

For example, the notion of a frog with eyes in its mouth (Chapter 1) seems incredible, but I corresponded with the herpetologist, Jim Bogart, who personally examined this specimen. This frog was pictured on the cover of the 1995 book *Phenotypes* by David Rollo [1078], and David was helpful in referring me to Jim, as well as to Scott Gardner, who shot the picture. Although I never inspected this individual myself, I was involved, directly or indirectly, with the other oddities discussed in Part I. Chapter 1 explores two key phenomena: induction and morphogenesis.

Chapter 2 recounts the research done by John Gerhart [413], my PhD advisor at the University of California, Berkeley. Before I graduated in 1977, John had voiced skepticism about an old experiment on the "gray crescent." His hunch turned out to be correct [416], and his revised interpretation, based on work done with students who came after me, revealed an unsuspected mechanism of cytoplasmic reorganization [665]. This chapter also delves into Falkenberg's rule, a puzzle posed in 1919, which concerns the orientation of the left–right axis in conjoined twins. When such twins share a thorax, the left member of the pair looks normal, while the right one has its internal organs reversed half of the time. Why? The mystery was solved in 1996 by a team of researchers studying chick embryos [783], with further insights added in 2017 by a different team studying frogs [1285].

Chapter 3 discusses Bateson's rule – a riddle even older than Falkenberg's rule – about the symmetries of extra legs in a variety of insects, crustaceans, and vertebrates. After brainstorming with Stan Sessions, the world expert on such defects in frogs, the two of us submitted our conjecture to the *Journal of Experimental Zoology*, which published it [564]. Hence, there is much overlap between this chapter and that article. Whether or not our hypothesis turns out to be right, this project closed a circle for me, since my first book, published in 1992, was based on Bateson's book [553].

Flies have provided the preferred model system in genetics for more than a century, so they offer an opportunity to explore the underworld of gene regulation which drives development, and that is an understatement. Little did I realize when I chose the fly's second-leg basitarsus for my doctoral research in 1974 that it would turn out to be a microcosm of animal embryology in general [547]. The ~1800 cells on that leg segment employ all of the cardinal signaling pathways of metazoans to build a cuticular pattern [555]: the Dpp, Wg, and Hh pathways fix the axes (Fig. 3.10) [554], the Notch pathway pinpoints the bristles (Fig. 6.10) [550,562], the PCP pathway orients them [566], and the EGFR pathway induces the bracts (Fig. 1.4) [556]. To paraphrase William Blake, it feels in hindsight like I was seeing the world in a 300-μm-long cylinder of prickly cuticle. The patterning mechanisms employed by flies are surveyed in Part II (Chapters 4–6) [561].

Regrettably, those devices cannot be explained without a host of gene names as alien as the toves and borogoves of *Jabberwocky*, but fear not! There is a frabjous way around this frumious impasse. The names of fly genes turn out to be quite charming once you know how they were coined, and the etymologies are all available at Tom Brody's *Interactive Fly* website. Google it, and you'll see.

Chapter 4 continues the theme of obdurate old conundrums. As in Chapters 2 and 3, Chapter 4 presents a mystery that was first uncovered by my own research team in 1986 [566]. It concerns a polarity-modifying mutation that causes fly tarsi to have twice as many joints as normal, with each of the extra joints being upside-down. Why? This riddle has basically been solved, but many questions remain.

Chapter 5 begins with the most famous mutant ever described – the four-winged "bithorax" fly. Its discovery in 1915 was as fortuitous as Howard Carter's finding of Tut's tomb seven years later. By delving into the etiology of the mutant's four-wing phenotype, Ed Lewis (1918–2004) stumbled upon a cluster of genes that has been organizing the head-to-tail axis of animals for the past 600 million years [774,804]. This "Hox Complex" has furnished a goldmine of insights into how genes construct anatomy. One tangent that I discuss is the frog leg, whose extra tibia and fibula arose by the tweaking of Hox gene expression. This same trait arose in two jumping mammals (tarsiers and galagos) [560]. Given the billions of people on earth, it is surprising that this anomaly has never shown up in a human. Basketball teams would surely love to have players with this sort of catapulting superpower.

Chapter 6 investigates a gene cluster – the Achaete-Scute Complex – that offers some useful contrasts with the Hox Complex. Disabling its main genes causes virtually all bristles to vanish, making the fly virtually bald. This nude syndrome resembles our own species insofar as humans are what Desmond Morris aptly called the only "naked ape." Some years ago I wrote a review on the genetics of human hairlessness [558], and I adapt that essay here by permission of Springer Nature (2010). Also adapted for this chapter is material from *Imaginal Discs* (2002) [555], whose contents needed updating based on insights unearthed in the interim. The goal of that book – and this chapter – was to probe how genes represent anatomy in DNA language [198,840].

Dogs have been guinea pigs, so to speak, in a global experiment spanning millennia. The original goal was to domesticate wolves to do the bidding of their human masters, but as the descendants of those progenitors proliferated, mutants emerged that people found useful or beautiful or comical. By mating those exceptions together and subjecting their offspring to selective breeding, generation after generation, dog fanciers sculpted the sundry varieties that we know today [11]. Darwin realized the power of artificial selection to modify morphology, and he used it as an analogy for the process of natural selection in his *Origin of Species*, though he employed pigeons as his cardinal example, rather than dogs [68,359]. The mottled coat patterns we see in dogs are thought to have arisen as a result of the domestication process itself [1392,1393], but if so, then it seems strange that humans do not show similar splotches [1004], since we are thought to have *self*-domesticated over the eons [864]. Geneticists have recently been digging into the dog genome [951,982,995,1124], and some choice nuggets are featured in Part III (Chapters 7–9) of the present book.

Shaggy dog stories that were omitted but are still worth mentioning [1212] include how the Dachshund got its short legs (answer: an *Fgf4* retrogene that causes chondrodysplasia) [996], how the Borzoi got its long snout (answer: an odd ratio of tandem-repeat glutamines to alanines in the *Runx2* gene) [1312], how the Dalmatian got its spots (answer: a deviant allele at the *Mitf* color locus plus a second-site mutation) [675], how the bulldog got its underbite (answer: a missense mutation in the bone-promoting *BMP3* gene) [841,1125], how the Siberian Husky got its blue eyes (answer: a 99 kb duplication upstream of the homeobox gene *Alx4*) [287], how the Bearded Collie got its beard (answer: a combination of *Fgf5* and *R-spondin2* alleles) [170], how the Mexican Hairless lost its hair (answer: a frameshift mutation in the *Fox13* gene) [324], why Labrador retrievers collapse when they exercise (answer: an error in the *dynamin* gene *DNM1*) [891], and why dogs are so affectionate in general (answer: they carry the hypersociality alleles of Williams syndrome "leprechauns") [775,1341,1425].

It is worth pausing to note that readers who are not geneticists may have still been able to get the gist of the last paragraph despite not being familiar with the arcane verbiage, in the same way that Catholics and Jews can benefit from their respective services without knowing Latin or Hebrew. The reason for bringing this point up is that this whole book had to face this dilemma. It ended up being written for an ecumenical audience without being "dumbed down," so it is up to readers to judge whether its technical terms are a deal-breaker.

Chapter 7 examines a Chinese dog breed with rumpled skin called the Shar-Pei. This deformity has been traced to a mutation that causes a polysaccharide – hyaluronic acid – to perfuse the dermis so as to loosen the skin. Similar phenotypes exist in cats, humans, and flies, and some of them can be understood on the same basis. Surprisingly, hyaluronic acid also helps carve the convolutions of the human brain, and we delve into how our big brain evolved – an issue I've covered before [557]. Fingerprints likewise exhibit labyrinthine patterns, and we assess their genetic basis by reviewing some old forensic studies of the Dionne quintuplets. The take-home lesson from this excursion is that some patterns organize themselves "under the radar" of the genome, based on physical forces that are subject to stochastic vagaries – like a herd of cattle that is spooked to stampede in unpredictable directions. The idea that physics shapes anatomy was the central thesis of D'Arcy Thompson's classic monograph *On Growth and Form* [1280]. The question of determinate growth remains one of the most profound mysteries in developmental biology [893,1336,1338], and it figures prominently in the next two chapters.

Chapter 8 begins with the Bully Whippet, a stocky breed whose double-muscling trait is due to a mutation in the *myostatin* gene. Strikingly similar phenotypes and etiologies have arisen in various species of livestock [9], and a single human case has been documented as well [1129]. Other organs aside from muscles do not appear to use dedicated inhibitors like myostatin: there is simply no equivalent "osteostatin," "neurostatin," or "colostatin" to limit the growth of bones, nerves, or intestines for example. Rather, most other tissues (e.g., the liver) rely upon a generic (Hippo) signaling pathway for this function. A separate (insulin) pathway is utilized by organs to amplify their growth beyond that of the body as a whole (e.g., deer antlers). Several of these case studies could lead to breakthroughs in solving clinical problems.

Chapter 9 offers a history lesson for the younger crop of developmental biologists, who may be a bit too glib about all the glitzy genomic tools at their disposal [1205]. They should heed the haunting homily of the elf queen Galadriel in the prologue of the first *Lord of the Rings* movie: "Much that once was is lost, for none now live who remember it." Some of us *are* old enough to remember the pioneers who blazed our trails, not with technical trickery but with mental wizardry. They were keen thinkers who tackled daunting phenomena and solved knotty mysteries [611] – for example, John Gerhart's exegesis of cortical rotation (Fig. 2.2), William Bateson's synopsis of limb symmetries (Fig. 3.3), and Curt Stern's analysis of genetic mosaics (Fig. 6.9). Here I highlight the work of Pere Alberch and Jonathan Cooke, who are relatively unsung heroes from that earlier epoch, though Alberch's ideas have recently been revived by Rui Diogo [306,307,309]. Alberch had the acuity to see subtle constraints in the hind paws of his parent's pet dog, and Cooke had the temerity to claim that Wolpert's popular model of morphogen gradients was full of holes, which he proceeded to plug with Alan Turing's reaction–diffusion tenets [232]. The chapter's theme is polydactyly – a symptom in more than 300 human syndromes [109,1309,1330].

Only one mammal group other than ours has had a Broadway musical all to itself. *Cats* celebrates cats in a way that is inconceivable for any counterpart *Dogs* production. There is something ineffable about how cats behave that gives them a unique "purrview." They are clearly too smart for their own good, and they know better than to look up to us. Dogs obey us, but cats exploit us. They even got the Egyptians to worship them as gods! Supercilious behavior and superior intelligence aside, however, cats cannot measure up to dogs in two respects. They are all about the same size, and they are all about the same shape [473]. Where they tend to vary is in their coat patterns – at least as much, if not more so than dogs [340,663]. Consequently, cats portray how mammal skin gets "painted" while dogs exemplify how mammal bodies get sculpted [323]. Part IV (Chapters 10–12) takes advantage of this fact to probe how three odd coat patterns arise.

Chapter 10 starts with a tale of two cats – one little and one big. The little one is the domestic tabby, and the big one is the cheetah, known for its speed but also for its spots. There is a mutation in cheetahs that turns their spots into stripes and irregular blotches, and it so happens that virtually the same mutation makes tabbies blotched as well. In one case (the tabby) we go from tiger-like *stripes* to blotches, and in the other (the cheetah) from *spots* to blotches. How curious! This riddle prompts us to ask deep questions, like the languid Caterpillar interrogating Alice, about the nature of spots and stripes in general. Here Turing's model helps us realize (with apologies to Joni Mitchell) that we never really knew stripes or spots at all. They are both visible manifestations of invisible machinations at the cellular level that we are only now beginning to grasp. Along the way we will ponder the coat patterns of zebras, giraffes, and leopards. Our tour of the zoo will be informed by clever analyses of insects and fish, including a classic experiment by another titan from an earlier age [758]: Sir Vincent Wigglesworth was a superb researcher whose feats deserve wider appreciation [337,338,761]. He was the insect physiologist who trained Peter Lawrence [757], the patriarch of developmental genetics who in turn trained Gary Struhl [1331], whose contributions to the field are legendary. The chapter ends with a fish story from the lab of Shigeru Kondo, another luminary, who solved the riddle of how the fish gets its stripes. Kondo apprenticed with Walter Gehring, a co-discoverer of the homeobox (Chapter 5) [1388], who in turn was trained by Ernst Hadorn, the discoverer of transdetermination [503]. The field of developmental biology boasts an illustrious genealogy indeed.

Chapter 11 concerns the Siamese cat, which, like many breeds, is named for the part of the world where the originating mutation arose – in this case Siam, the country we now call Thailand. The same mutation that causes its distinctive coloration also causes it to cross its eyes [851,1362], and the story of how that correlation was deciphered is told in a charming essay entitled "Serendipity and the Siamese cat: the discovery that genes for coat and eye pigment affect the brain" [662]. Bateson was the first to notice the similarity between Siamese cats and Himalayan rabbits [81,617], and since that time (1909) analogous mutations have been found in other mammals, including humans. In all of these cases the outside temperature affects the mutated tyrosinase enzyme, resulting in dark extremities. Some species of butterflies and mammals have evolved a reliance on outside temperature as a means of giving them different phenotypes in different seasons of the year, and some species of reptiles have incorporated it into their sex-determining mechanisms. Other species use other environmental influences such as nutrition, crowding, or social interactions as a way of creating alternative anatomies, especially various insect castes. How evolution wired animal genomes to utilize these external cues remains a mystery.

The narrative closes with Chapter 12, which uses the calico cat as a springboard for diving into the phenomenon of mosaicism in general. The potential of genetic mosaics for investigating gene function (in lieu of tissue transplantation) was first realized by Curt Stern (1902–1981), one of whose experiments was discussed in

Chapter 6. His *Genetic Mosaics and Other Essays* [1203] was one of my favorite books in graduate school at Berkeley, where I was privileged to meet the great man himself before Parkinson's disease forced him to retire [1291]. Stern was an illustrious member of Morgan's fly lab [717], but there were so many stars in that firmament that it is hard to rank their luminosities. Even so, no story about mosaics would be complete without discussing gynandromorphs, and no account of those half-male, half-female creatures would be complete without recounting how they were harnessed to construct fate maps by another of Morgan's disciples, Alfred Sturtevant. It was not my intention to write a history book, but as my own career nears its end, it is hard to resist one last glance at the totem pole of giants on whose shoulders I have been honored to stand.

Abbreviations include ECM (extracellular matrix), EGFR (epidermal growth factor receptor), Fgf (fibroblast growth factor), GOF (gain of function), GP (general principle), GTPase (guanosine triphosphatase), LOF (loss of function), and PCP (planar cell polarity). Gene names are italicized, as per convention in fly genetics, and protein names are capitalized. Capitals are used for genes when the originally isolated mutation was dominant – e.g., *Serrate* (gene) and Serrate (protein) – vs. cases in which the originally isolated mutation was recessive – e.g., *achaete* (gene) and Achaete (protein). Names can denote what a gene does, as for growth factors, but often they reflect the phenotype when the gene malfunctions – e.g., a serrated wing or a bristleless ("a-chaete") cuticle. Numbers in brackets are citations of sources listed in the *References*, which can steer readers to related research in the literature.

Constructive critiques of the book proposal and of draft chapters were kindly provided by Richard Campbell, Jane Maienschein, Cliff Tabin, and Adam Wilkins. Many colleagues provided pictures, and they are acknowledged in the relevant figure legends. Dominic Lewis, my editor at Cambridge, was steadfast in his support of the project; Hugh Brazier, my copy-editor for this third book in a row, wielded his scalpel as deftly as a neurosurgeon; and my "coaches" all along were my sister Linda Wren and my friend Sam Braudt. Even Scott Gilbert, author of the most widely used developmental textbook of all time [433], gave his blessing to this more casual approach when I discussed it with him at the outset.

The educational strategy used here was modeled after my favorite professor at MIT, Seymour Papert, who exhorted all of us undergraduates to look for big ideas, which he called "general powerful principles" (distilled here to "GPs"), that we could use as heuristic tools to build conceptual frameworks in whatever field we chose to pursue [991]. In keeping with Papert's didactics, GPs should be applied to other cases as soon as they are abstracted from a particular case so as to affirm their generality. That means going off on tangents, at least briefly, after each GP is formulated, which is the rubric I've adopted.

In an essay for *Newsweek* entitled "How life begins," Sharon Begley extolled the grandeur of human embryogenesis from egg to baby: "It is as if a single dab of white paint turned into the multicolored splendor of the Sistine Ceiling" [91]. The majesty of development has long been shrouded in mystery, but many of the tricks behind the magic have now been deciphered, thanks largely to the analyses of anomalies. The underlying mechanisms possess an esthetic all their own [966], but readers will have to plow through a lot of complexities before they can reap the full bounty of the beauty.

Lewis I. Held, Jr.
Lubbock, Texas
July 2020

Part I

Frogs

1 The Introspective Frog

In 1992 a high school student in Ontario, Canada, found a toad in her yard. Deidre knew something was wrong when it didn't open its eyes as she picked it up. When she looked inside its mouth, she was shocked to see a pair of eyes looking back at her (Fig. 1.1). At first, she thought the eyes might belong to another toad that it had swallowed, but she soon realized that they were attached to the roof of its mouth. She took the toad inside and tried feeding it some worms, but it would only eat them if she placed them directly into its mouth [142].

Deidre contacted the local newspaper, and they sent a crew to her home to see the frog for themselves. When the staff photographer, Scott Gardner, got the call over his two-way radio, he rolled his eyes in disbelief, suspecting that the dispatcher was just playing a prank on him, but upon his arrival he saw that the introspective

Fig. 1.1. A frog whose eyes developed inside its mouth. The toad (*Bufo americanus*) reacted to motions only when it gaped [276]. The only known similar case was a leopard frog (*Lithobates pipiens*) found in Minnesota in 1996, with one normal external eye and one internal eye that hung down from the roof of the mouth on a stalk of flesh [1183]. Photo (used with permission) by Scott Gardner, staff photographer for *The Hamilton Spectator*. (A black and white version of this figure will appear in some formats. For the color version, please refer to the plate section.)

amphibian was quite real after all [402]. By this time Deidre had named it Gollum after the semi-aquatic creature in Tolkien's *Lord of the Rings*. The crew got him to open his mouth by tapping his lips with some tasty insects.

The next day Deidre took Gollum to a herpetologist at the University of Guelph. Professor Bogart had seen a lot of amphibian abnormalities over the years [524], but never one like this. Like a doctor at an emergency clinic, he recorded the essential facts of the patient's presentation [1268]: "male, *Bufo americanus* (common in Ontario), two inches long, and at least two years old." He was surprised that a nearly blind toad had survived in the wild for so long.

As for how Gollum got this way in the first place, Dr. Bogart could not be sure. He surmised that the eyes had developed upside-down, but he could not tell whether the cause was genetic or environmental. He wanted to mate Gollum to see whether the trait was heritable, but Deidre adamantly refused to loan her pet to him for that purpose. Indeed, she even declined to donate Gollum's body for an autopsy after he died. Only one other similar frog was subsequently found in Minnesota [1183], which makes an external agent (e.g., a pesticide) less likely.

The aquatic habitat of tadpoles exposes them to potential damage by parasites and predators [653], though the bilateral symmetry of Gollum's eye trait would seem to argue against any such targeted external injury. Unlike other vertebrates, frogs have muscles that can depress the eyes toward the oral cavity to aid in swallowing, but Gollum's phenotype cannot be explained by eye rotation alone because the skin atop the head was unbroken where his eyelids should have been, and there was no sheath of palatal skin covering the lenses of his eyes.

In theory, Gollum's palatal skin could have been somehow pierced by fully formed eyes, but a more plausible explanation, based on what we know about vertebrate eye development (see below), is that his palatal skin was incorporated into the eyes themselves when his retinas accidentally grew down toward his mouth.

GP-1: Inductive signaling can enhance precision

Nobel laureate Sydney Brenner (1927–2019) was famous for his wit [759], and one of his cleverest sayings was that embryonic cells acquire their fates based either on who their parents were – the "British Plan" – or on who they happen to know – the "American Plan" [1060]. Or, to paraphrase, embryonic cells adopt distinct roles based on (1) instructions they inherit via cell lineage or (2) signals they receive from their neighbors. The nematode embryos studied by Brenner's lab primarily use the first kind of source [1207], whereas vertebrate embryos routinely rely on the second one.

Proof of intercellular signaling during vertebrate eye development was adduced by Hans Spemann (1869–1941), the first embryologist to win a Nobel Prize. Spemann recounted his experiments in the 1938 book *Embryonic Development and Induction* [1187]. His key conclusion was that the optic cup induces the lens – the first kind of induction ever documented [1100]. The optic cup grows out from the developing brain (neural tube) to form the retina, and the lens arises wherever the cup contacts the overlying ectoderm (prospective epidermis). In some frog species a lens can be elicited virtually anywhere in the head or trunk region by transplanting an optic cup (or antecedent vesicle) beneath the ectoderm [974], thus implying a causal relationship between cup and lens (boldface added):

> The nature of these potencies [of skin regions outside the normal site of lens formation] especially can be ascertained only if, as I suggested, the optic cup is brought into contact with foreign parts of the epidermis, either by transplantation of the optic cup itself or of the epidermis covering it ... The most incontestable results are obtained by the first-mentioned method, in which the optic vesicle is exposed, cut off, and pushed backward under the [trunk] epidermis. This experiment was first made by W. H. Lewis ... on *Rana sylvatica* and *palustris*, with the result that, in numerous cases, **a lens formed above the transplanted optic vesicle**. After a short development the lens was still in connection with the skin and thus indicated its [trunk skin] origin. [1187]

It is therefore possible that Gollum's optic cups took a wrong turn and wound up inducing lenses in the roof of his mouth (Fig. 1.2). Such a detour would explain why his eyes looked normal despite being displaced, as well as why they had no cloudy patina of palatal skin, as that skin would have become transparent lens tissue. One obvious way to test this hypothesis would be to see whether an artificially transplanted optic cup can induce a lens in the palatal ectoderm of this species. It is a shame that no offspring were obtained to see if Gollum's anomaly was genetic. The defect remains enigmatic, partly because we don't yet know what factors dictate the normal trajectories of the optic cups.

A priori one might have imagined that evolution could have evoked the vertebrate lens and retina from separate sites within the embryo and then fitted them together, but that would have run the risk of misalignment. As anyone who wears glasses realizes, clear vision requires fine precision, and induction of the lens by the optic cup guarantees fidelity of fit. If lens–retina coordination is so useful for acuity, then we should see it in the eyes of non-vertebrates as well. Indeed, lens and retina development are intertwined in the similar but non-homologous eyes of cephalopods [715].

Based on genetic studies in the mouse, the chief intercellular signals that induce the vertebrate lens are Bone Morphogenetic Proteins (BMPs) [500], with other signaling pathways playing supporting roles [262]. Recent research shows that ectodermal cells cannot respond to these inductive BMPs unless they are primed to do so in advance [647]. The main priming agent that makes them "competent" is the conserved transcription factor Pax6, though other regulatory proteins interact with Pax6 in a complicated genetic network [262].

GP-1 tangent: Bract induction

A much simpler instance of induction concerns the "bract" – a tiny cuticular structure in flies. The inducer in this case is a ligand of the Epidermal Growth Factor Receptor (EGFR) pathway, and the competence agent is Distal-less

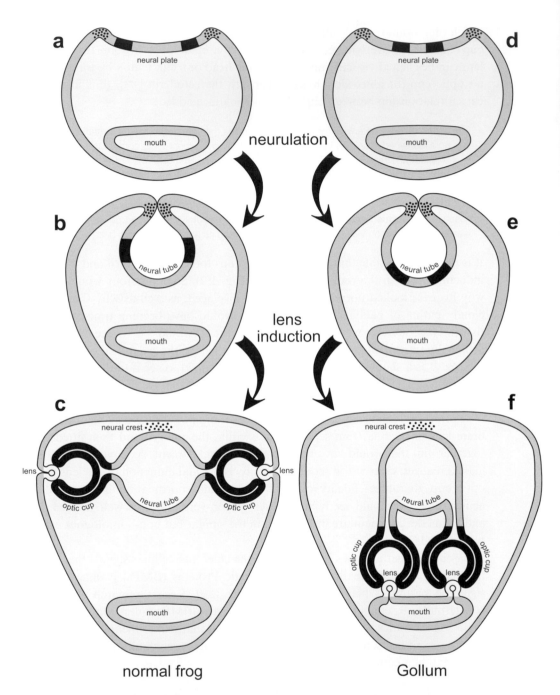

Fig. 1.2. Hypothetical etiology of Gollum's eyes. The development of a normal frog's eyes (**a**, **b**, **c**) is compared to that of Gollum's eyes (**d**, **e**, **f**), assuming that Gollum's optic cups suffered a ventral detour from the orthodox lateral trajectory. Stages are depicted schematically: **a**, **d**: early neurula; **b**, **e**: late neurula; and **c**, **f**: lens induction, showing invagination of the lens placode to form a vesicle. (A similar induction elicits the auditory vesicle that becomes our inner ear [742].) All panels are coronal cross-sections of the head

(Dll) – a protein, which, like Pax6, contains an ancient homeodomain (DNA-binding) motif. The rationale for delving into this vignette is to briefly show how induction can ensure precision (GP-1) on a *cellular* scale, not just at the tissue or organ levels as in the lens-induction case.

Bracts are thorn-like protrusions that are secreted by single cells in fruit flies. They never occur alone but are always found adjacent to bristles on the legs and wings where the gene *Distal-less* (*Dll*) is expressed, and indeed, loss-of-function (LOF) mutations in *Dll* block bract – but not bristle – development [174]. For many years circumstantial evidence continued to mount, arguing that bristles induce bracts, but definitive proof only came in 2002, when the inductive signal was identified as Spitz, a ligand of the EGFR pathway [289,556].

Mechanosensory bristles in *Drosophila melanogaster* are formed by a cluster of cells, all of which descend from a common ancestor called the "sensory organ precursor" (SOP). SOPs arise at consistent sites within the fly epidermis during metamorphosis (Fig. 1.3). They undergo three mitoses to yield five cells [420,1053], each of which acquires a unique identity via instructions that it inherits from the SOP, obeying Brenner's British Plan, where lineage dictates destiny.

In contrast, the bract cell is recruited into the bristle complex via the American Plan, where your fate depends on who you know. No one knew *which* member of the bristle clan induces the bystander until 2012, when the mystery was solved by Ying Peng and Jeff Axelrod. Their paper not only indicted the socket cell beyond any shadow of a doubt, but also uncovered a novel mode of close-range induction [1011]. Instead of "spitting" Spitz willy-nilly in its vicinity, which is the norm for paracrine inducers [1016], the socket cell reaches under the epidermis to tickle the unsuspecting neighbor with a Spitz-laden lamellipodium.

The revelation of this rude gesture on the part of the socket cell made perfect sense to aficionados of the bract world, because it neatly solved another nagging riddle. Why do bracts only develop on the proximal side of bristle sockets? Now we know. They do so because socket cells only extend their subterranean feelers in a proximal direction (toward the body). How do they determine which way is proximal (vs. distal)? They use the equivalent of a compass to tell which direction is which. It is called the Planar Cell Polarity (PCP) pathway – an evolutionarily ancient "app" that animal cells rely upon to navigate all sorts of challenges during development [561].

←

Fig. 1.2. (*cont.*)

(dorsal above), with ectoderm in gray. NB: Neurulation is induced by the underlying notochord (not shown) [236]. Most of the gut is endodermal [208], but the buccal cavity ("mouth") is colonized by ectoderm [205,1184]. Black zones in **a** and **d** are optic cup primordia, which might have been more medial in Gollum as a result of erroneous patterning. Cups in **c** and **f** are black, lenses are white, and neural crest cells are black dots. Lenses are larger relative to optic cups than shown here. For further details see Figure 1.5 and [262,882]. After [557,1187].

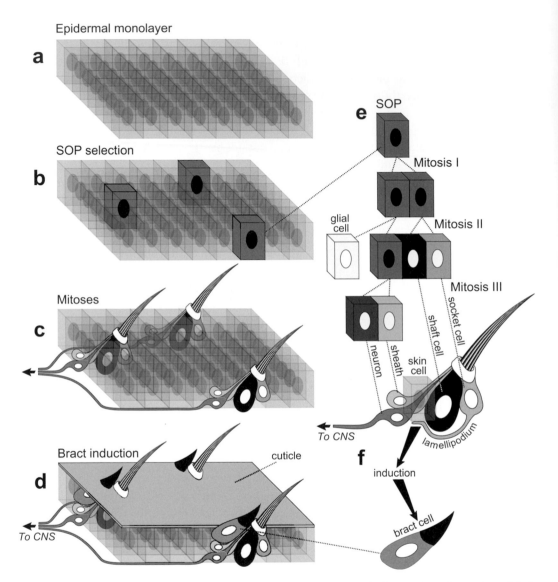

Fig. 1.3. Induction of bract cells during fly leg metamorphosis. In panels **a–d**, a region of fly leg epidermis is drawn schematically, showing the emergence of mechanosensory bristles from single sensory organ precursor (SOP) cells. In all panels, proximal is to the left and distal to the right. **a.** Cells are depicted as translucent boxes with a nucleus (oval) inside. Their actual packing, however, is not nearly so regular as shown in this array. **b.** Three SOPs (darker boxes with black nuclei) are drawn as examples. **c.** Completion of differentiative mitoses. **d.** Bract cell formation. **e.** Pedigree of the bristle lineage, with glial cell to the side because it will migrate away [420,1053]. **f.** Induction of a bract cell by the socket cell, which extends a lamellipodium proximally to reach the nearest neighboring (ordinary skin) cell. Induction actually occurs before terminal differentiation of the bristle. Axons of neurons coalesce into bundles as they leave the skin, headed for the central nervous system (CNS). The size of the bract cell is exaggerated. Redrawn with modification from [552,555].

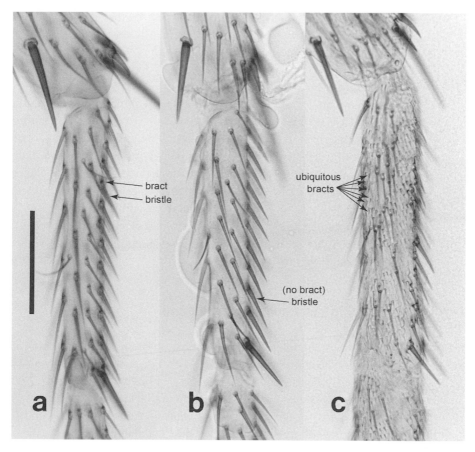

Fig. 1.4. Modulation of bract induction by manipulating the EGFR pathway. Each panel shows the anterior face of the basitarsal segment of a right *D. melanogaster* second leg (proximal at top; scale bar = 100 μm). **a**. Normal phenotype. Bracts are the tiny triangular structures above most bristle sockets. **b**. Fly whose EGFR protein was inactivated (total LOF). No bracts develop. A similar bractless phenotype is seen for *Distal-less*-LOF [174]. **c**. Fly whose EGFR pathway was hyperactivated (extreme GOF) by overexpressing the *Ras1* gene. Most of the skin cells have formed thin bracts at the expense of bristles. See [289] for a similar analysis that yielded comparable phenotypes. From [556]; used with permission from *Elsevier/RightsLink*.

It is possible to override this directional preference by artificially driving the EGFR pathway to higher levels within the entire bristle complex or within the epidermis as a whole, which results in many more bracts than normal being elicited. An example of such a hypermorphic phenotype is shown in Figure 1.4c.

As tidy as this tale might seem, there are still loose ends that remain to be tied up. Chief among them is the function of the bract itself. It lacks any connection to a nerve, nor does it restrict the motion of the bristle shaft, so it can't be acting in touch sensation. Indeed, its removal by mutation does not appear to leave the fly worse off than before. Such silly trifles (think of the muscles that wiggle your ears [557]) are often explicable as vestiges of antecedent devices that did serve a purpose in an ancient ancestor, but no such explanation can rescue us here. Some such baubles help animals entice mates (e.g., the gaudy fan of the peacock), but bracts are not sexually dimorphic, nor does fly vision seem good enough to discern bracts from a distance anyway.

Hence, we are left with the puzzle of why evolution has gone to the trouble of deploying an intricate inductive mechanism to situate a seemingly needless structure (the bract) next to a functional one (the bristle). Conceivably, the answer lies buried in the fly genome somewhere, but there are likely to be few Don Quixotes willing to devote much time trying to find it. Nevertheless, the military precision with which this induction is executed illustrates the power of intercellular signals to build multicellular ensembles. The epitome of architectural accuracy is the ommatidium of the insect compound eye [201], where an elaborate cascade of sequential inductions assembles the cell types within each modular conglomerate [57,736].

GP-2: Embryos tend to build anatomy by origami

Animals are three-dimensional organisms, but many of their organs begin as two-dimensional sheets that fold extensively to attain their final shapes [436,1311,1428]. This 2D → 3D "origami" strategy [377,658] is obvious in Figure 1.2, where (1) the central nervous system emerges by rolling the surface into a tube [225], (2) the optic cups originate by outgrowth from the walls of that tube [84,183], and (3) the lenses arise by inpocketing of the surface wherever the cups encounter the ectoderm on their outward journey [262]. All three of these examples involve ectoderm, but the endoderm undergoes a comparable contortion called "gastrulation" to form the digestive tract [1181].

A brief aside on terminology may be useful here. Ectoderm, mesoderm, and endoderm are the primary germ layers (outer, middle, and inner) of animal embryos as defined at the gastrula stage, and each has its own talents and limitations [506]. Mesoderm employs a mesenchymal type of tissue plan (loose 3D network) more often than an epithelial one (2D sheets), though somites are a blatant exception [667], as is our heart [242]. Then there is the neural crest, which behaves, *sui generis*, as a fourth germ layer. It starts within an ectodermal sheet but dissolves into a mesenchyme when the neural tube involutes (Fig. 1.2b,c) [668,1276], and the cells that are thus liberated from their epithelial bonds migrate all over the embryo to adopt various fates [156] but remain mostly mesenchymal [859].

Gastrulation and neurulation are such integral aspects of development that we take them for granted, yet there is no obvious reason for animals to have settled upon the 2D gimmickry of origami versus other ways of making tubes [1237], such as the excavation of solid 3D cylinders [29,53] (that is how neurulation proceeds in actinopterygians [3]). Presumably, evolution is to blame. The first animals are thought to have had their digestive and nervous tissues on their surfaces [595], while their descendants at some point tucked these tissues inside, probably for protection [506,1114].

The 2D maneuvers we witness at the tissue level must be driven by shape changes at the cellular level [293,604,659], and ultimately by cytoskeletal motors at the molecular level [846,853]. Conveniently, we can travel down into that mechanized underworld by following the next steps in the story of lens development [633], which begins with lens induction (Fig. 1.2c) [202]. Along the way we will encounter many evolutionarily conserved gadgets that govern the development of a wide variety of other organs across the entire animal spectrum.

The first overt sign of lens formation is a thickening of the ectoderm into a placode that then invaginates. Indeed, virtually all epithelial invaginations follow this same recipe, with cells changing from a cuboidal to a columnar shape as a prerequisite for involution [1006]. Why must cells be primed in this geometrical way? The answer might be that the longer intercellular junctions of tall (vs. short) cells can serve as flexible hinges [1123], or that the greater stiffness of the cytoskeleton gives them added leverage [1006], or both.

After the placode cells thicken, these cylindrical cells constrict their apices to adopt a conical shape [271], which causes the placode to buckle into a pit (Fig. 1.5a) [724]. The pit's concavity is adjusted by a contrivance that is nearly universal in the animal kingdom: the actin-remodeling GTPases RhoA and Rac1 act as "tuning knobs" to increase or decrease the bowl's curvature to whatever "Goldilocks" setting the genome happens to specify [203].

This dimpled pit then detaches from the overlying ectoderm to form a vesicle (Fig. 1.5b), but the vesicle cannot remain hollow if it is to focus images as accurately as the glass lens of a camera [75]. To accomplish that, it must become a solid ball of cytoplasm so that its refractive index is homogenized. That final state could theoretically be achieved in a multitude of ways, but vertebrates tend to follow the same procedure [262]. The protocol entails filling the vesicle cavity with cells that elongate from the side of the wall that is closest to the future retina (Fig. 1.5b,c) [44].

Are the cells there *intrinsically* prone to elongate? No. They are induced to do so by an *external* signal (Fgf) that diffuses toward the lens from the optic cup [860]. The existence of such a signal was inferred in 1963 when researchers excised a lens from a chicken embryo and returned it upside-down. If this experiment was performed just before the stage depicted in Figure 1.5b, then the cells that had been stretching came to an abrupt halt, and the cells on the opposite side (now facing the future retina) began lengthening instead [243].

Regardless of their source, the elongating cells eventually fill the cavity to create the "primary" lens – a solid ball topped by a remnant layer of cuboidal cells (Fig. 1.5c). Theoretically, this lens should be able to grow as the newborn animal gets bigger, but it cannot do so because its cells have forfeited the ability to undergo mitosis: their nuclei disintegrate during differentiation to enhance acuity even more since DNA and cytoplasm differ in refractive index [76]. Given the clear benefit of this tactic, it is not surprising that cephalopod lens cells also lose their nuclei [199],

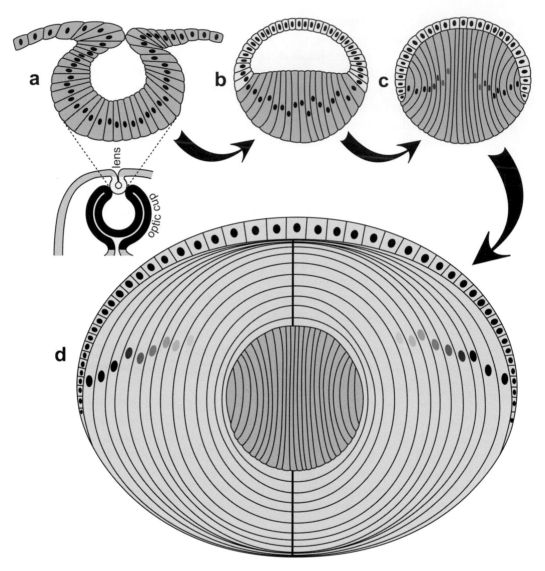

Fig. 1.5. Maturation of the lens after induction. A mouse lens is cross-sectioned at four stages of development (rotated 90° from Fig. 1.2c). **a**. Lens pit, whose cells constrict apically in response to BMP [646]. **b**. Lens vesicle, whose cavity is being filled with cells that cease mitosis and elongate (dark gray) [44], while remaining cells (light gray) continue to divide [227,794]. **c**. Primary lens, whose cell nuclei (ovals) disintegrate (black → gray) from the middle outward. **d**. Secondary lens (light gray), which encircles the primary lens (dark gray). Mitoses push cells toward the equator, where they stretch around the primary lens to meet a counterpart at the suture [261]. Nuclei vanish (as in **c**), which ensures transparency [76]. NB: The ectodermal surface layer (future cornea) left over after the vesicle detaches is omitted from **b–d**. The Pax6-dependent process whereby the crests in **a** merge to liberate the lens malfunctions in Peters' anomaly – a rare disease where a connecting stalk persists [261]. The lens (inset in **a**) is much larger relative to the optic cup than depicted here. The zebrafish lens arises by an alternative (3D → 3D) pathway [487], and lens regeneration entails budding

and insect lenses are made of extracellular cuticle where the issue of nuclear sacrifice is moot [1195].

Vertebrates solve the problem of restricted proliferation in a peculiar way (Fig. 1.5d) [262]. Cuboidal cells continue to divide, and when daughter cells get pushed to the equator, they increase in length 100- to 1000-fold to encircle the primary lens from both sides [1145]. Hence, an onion-like "secondary" lens comes to surround the primary lens, letting the overall lens retain a spheroidal shape as it grows [860]. Nuclei vanish, like those of the primary lens, but before they do so, both cell types purify their cytoplasm by reducing it to a single type of protein called a crystallin. Cephalopods do this too but use non-homologous proteins [546]. In order to avoid spherical aberration, vertebrates and cephalopods both vary the concentration of crystallin proteins (and hence refractive index) as a function of radial distance from the lens center [75,171].

GP-2 tangent: Imaginal discs

The epitome of origami is embodied by the imaginal discs of flies and other insects that undergo total metamorphosis [379]. As an added bonus, the cellular mechanics of their epithelial acrobatics are well understood [659]. Imaginal discs get their name from the fact that they form parts of the imago (the Latin term for an adult insect) and from their resemblance to collapsed balloons (i.e., round and flat). Their key properties are nicely demonstrated by the leg disc (Fig. 1.6).

Each leg of the fruit fly begins as a 2D cluster of 10–20 ectodermal cells that are set aside from the surrounding surface cells, which form the larval skin [78]. The cluster withdraws into the body cavity during the embryonic period [830] and grows inside the larva until metamorphosis [222]. Initially it is a solid mass, but it delaminates to yield a hollow sac [830]. As it grows, the thicker (columnar) side of the sac acquires concentric folds [861] like a Danish pastry [1339]. These folds telescope out as the disc everts to create a hollow cylinder that becomes the adult leg.

Eversion of the leg disc takes a few hours [1272], but it can be sped up to just a few minutes by exposing discs to the enzyme trypsin [228]. Apparently, the columnar cells are spring-loaded to unleash their coils at the slightest provocation, like a Jack-in-the-box toy. Trypsin's power to trigger this release has been traced to its ability to simulate endogenous proteases [300]. Those proteases degrade extracellular-matrix (ECM) proteins (e.g., collagen) that restrain columnar-to-cuboidal transitions.

Fig. 1.5. (*cont.*)

from the iris or cornea [1328]! (Lens regeneration may have evolved in response to lens-eating parasites [575,974].) Curiously, the fly lens shares a number of features with the vertebrate lens [119,201]. See [89,1144] for geometrical perspective and [261] for unsolved puzzles. The gold medal for weirdest origami stunt must go to the hollow, spherical embryos of the colonial alga *Volvox*, which turn themselves inside out [856], and an analogous inversion occurs in a colonial choanoflagellate [161]. Redrawn from [262]; used with permission.

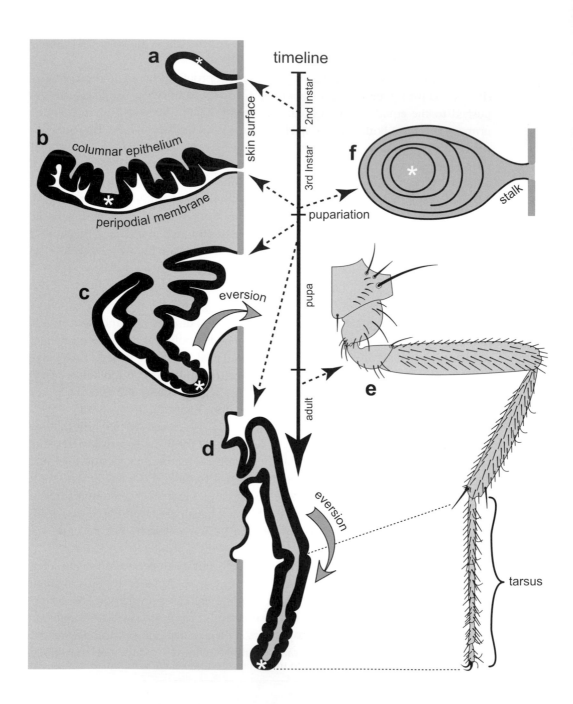

Fig. 1.6. Development of a second-leg disc in *D. melanogaster*. Black shapes (**a–d**) are cross-sections at successive stages (see timeline), with an asterisk marking the future tip of the leg. Cell boundaries and nuclei are omitted, as are the basement membrane, adepithelial cells, outgoing nerve, and overlying cuticle [555]. **a.** Mid-2nd instar, when the solid disc acquires a lumen to become a hollow sac [45,830,865]. All cells remain cuboidal until early 3rd instar, when one side thickens into a columnar epithelium, and the other side flattens into a

The columnar state itself is enforced by Dpp and Wingless – the two main signaling agents that govern the disc coordinate system [1384,1385].

How does the nascent leg elongate without widening to an equal extent, as should occur by cell flattening alone? "Convergent-extension" couplings are common in animal embryos [1148], and they typically involve cell intercalation [1149]. It turns out that choreographed cell rearrangements are also instrumental in both leg [381,1272] and wing [300,659] eversion. In both cases the anisotropy stems from a polarized localization of Myosin-II, but amazingly, none of the known PCP pathways [570] directs this localization [300].

Aside from ECM degradation, anisotropic cell flattening, and cell rearrangement, leg disc eversion is driven, at least in part, by a systemic increase in blood pressure [376], which blows the leg up like a balloon [1272]. Anyone who has ever watched an insect cringe after emerging from its pupal case knows how critical this gimmick is for straightening the wing blades. As the imago contracts its body wall musculature, its hemolymph gets pumped along the wing veins (see Fig. 3.5), which act as rigid struts until the cuticle hardens and the insect can finally relax [999].

Beetles are interesting in this regard because they use hydraulics not only to flatten their wings, but also to inflate their nasal horns (if they have them) [855] and to deploy their hindwings for takeoff throughout their adult lives [1234].

Ancestral insects had two pairs of wings (like dragonflies), but only one pair is needed for flight, so evolution could re-purpose one pair any way it saw fit, so to speak. The dipteran order of flies turned the hindwings into gyroscope-like halteres (see Fig. 5.2) [560], while the coleopteran order of beetles turned the forewings into chitin-fortified elytra that protect the fragile hindwings (Fig. 1.7) [1295]. Most of us have seen a ladybird beetle shut its elytra upon landing, but we may not have noticed the gossamer hindwings disappearing beneath them as they are carefully tucked away.

A fuller understanding of this wing retraction "app" had to wait until 2017 when researchers cleverly replaced one elytron with a transparent acrylic replica so that they could observe the entire origami ritual [1101]. The gadgetry involved is too intricate to describe here since it would distract us along yet another tangent.

←——————————————————————————

Fig. 1.6. (*cont.*)
squamous "peripodial membrane" [861] (cf. neural vs. pigment layers of the optic cup [913]). Both remain monolayers. **b.** Late 3rd instar. The columnar side has concentric folds (top view in **f**) [228] and remains attached to the surface by a thin stalk. **c, d.** Early pupal period. Folds telescope out, and the leg everts through the dilating stalk [1000] to form a hollow cylinder that secretes the adult cuticle. Ritualized contortions during eclosion from the pupal case (not shown) are reminiscent of the escape artist Harry Houdini [376]. **e.** Adult leg (anterior view). Creases in the pupal tarsus ultimately make the adult tarsal joints [426,485]. See [380] for details, [1229] for how folds arise, and [384,539] for how discs fuse at the midline. (First-leg discs share a membrane [890].) See [298,300,659] for wing disc eversion, which is driven not only by cell shuffling, cell flattening, and hydraulic pressure, but also by cell division, cell extrusion, shear forces, and cuticle stresses. Adapted from [228,381,555,1028,1272].

Fig. 1.7. Ladybird beetle. The delicate hindwings of ladybird beetles (e.g., the *Harmonia axyridis* drawn here) are deployed during flight, but when the insect lands, an elaborate folding ritual tucks them under decorative covers called elytra. Those elytra, which evolved from forewings, are made of heavily sclerotized cuticle. From [560].

Suffice it to say that the wings are ratcheted by sandwiching them between the upper surface of the squirming abdomen and the lower surface of the overlying elytra, and that bending is enabled by elastic hinges embedded in the vein lacework. Variations on this theme have been documented in earwigs [303], rove beetles [1102], shield bugs [103], and bamboo weevils [796].

By the time that beetles can fly, their wing cells have long since died – having fulfilled their task of secreting the adult cuticle. The elastic and rigid areas within that cuticle dictate how the wing will passively bend under external forces. In contrast, the origami of embryonic epithelia involves active bending in response to internal forces. Aside from the developmental case studies discussed above, other remarkable examples of 2D morphogenesis include our heart [242], brain [1250,1299], intestine [944], pancreas [247], ear [24], thyroid [208], and thymus [208].

2   Two-Headed Tadpoles

In 1924 Hans Spemann and Hilde Mangold published what is arguably the most seminal paper in the history of embryology [512,1109]. They showed that the dorsal lip of the frog blastopore can cause a secondary twin to develop when it is grafted to a host embryo [915]. The lip forms the notochord, which induces the central nervous system (cf. GP-1), and the rest of the twin assembles around this scaffold [57]. They dubbed the lip the "organizer" [411] because of its power to entrain nearby tissue [284].

In 1960 Adam Curtis tested whether this organizing ability already exists at the site on the egg surface where the dorsal lip will eventually arise [258]. The lip comes from a distinctive part of the fertilized egg called the "gray crescent," and when Curtis grafted that piece to the opposite side of a host egg, a secondary twin did indeed develop there, so he inferred that the gray crescent must possess the same inducing potency as the dorsal lip [259].

In 1981 John Gerhart *et al.* reassessed Curtis's data [416]. Earlier studies with other species had shown that rotation of the egg (without any grafting) could yield conjoined twins [598], and Gerhart's team proved the same for *Xenopus* – the species Curtis had used. Curtis had rotated his host eggs before inserting the donor graft but failed to control for this seemingly trivial manipulation. The twinning he saw had been due to the rotation alone!

GP-3: Cartesian axes are established sequentially

How can the mere act of rotating an egg force it to make conjoined twins? Before fertilization the egg is radially symmetric [417]. When the sperm donates a centriole to form an aster, the aster causes the cortex to rotate 30° in the plane of the sperm entry point (SEP), which sets up a dorsal–ventral (D-V) axis [665,1361]. The 30° tilt exposes a crescent of gray cytoplasm atop the heavier yellow yolk mass, which stays at the bottom of the egg due to the pull of gravity. The organizer eventually forms there, but, in contrast to what Curtis had concluded, the crescent itself has no organizer-like powers.

Realizing that it is the motion of the cortex *relative to the yolk mass* that establishes the D-V axis, Gerhart's group tried a clever experiment. Instead of manually turning the cortex, they used centrifugal force to shift the yolk mass in batches of eggs. Fertilized eggs were allowed to undergo their normal rotation, but then they were embedded in gelatin (to fixate the cortexes) and centrifuged to slosh

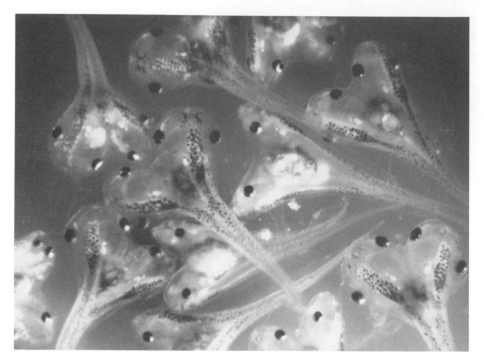

Fig. 2.1. Two-headed *Xenopus laevis* tadpoles. These conjoined twins were created by centrifuging eggs [116,416]. Other two-headed animals that have been reported include lambs [1246], lampreys [1241], lobsters [531], snakes [88], spiders [972], turtles [88], zebrafish [1], and humans [1190]. Photo courtesy of Steve Black. (A black and white version of this figure will appear in some formats. For the color version, please refer to the plate section.)

the yolk in an opposite direction relative to the cortex. Most of these treated eggs (60–70%) yielded conjoined twin tadpoles (Fig. 2.1) [416].

The 30° cortical rotation is able to distribute dorsal determinants along a much wider arc (Fig. 2.2b) [608] by actively transporting them along parallel tracks of microtubules [1361]. Those microtubules assemble de novo at the start of the rotation [976].

This scenario of axis formation is characteristic of amphibia [608], but some other animal taxa reverse the order of axis initiation – creating the D-V axis before fertilization and the A-P axis afterward – while still others set up both axes during oogenesis [450]. Regardless of the order or the timing, however, it is striking that evolution implemented Cartesian coordinates eons before Descartes himself proposed the idea [554].

If the *Xenopus* tadpole were to possess only A-P and D-V axes, then its body would be perfectly symmetric [1104], but its internal organs are in fact asymmetric – e.g., heart, gall bladder, and intestine (Fig. 2.2d). Hence, it must have a left–right (L-R) axis as well. Indeed, most animals manifest L-R asymmetry to some extent [123,125,711], which raises the question of how this axis arises [782]. For most vertebrates it is initiated by motile cilia [126,1345], as will be explained shortly.

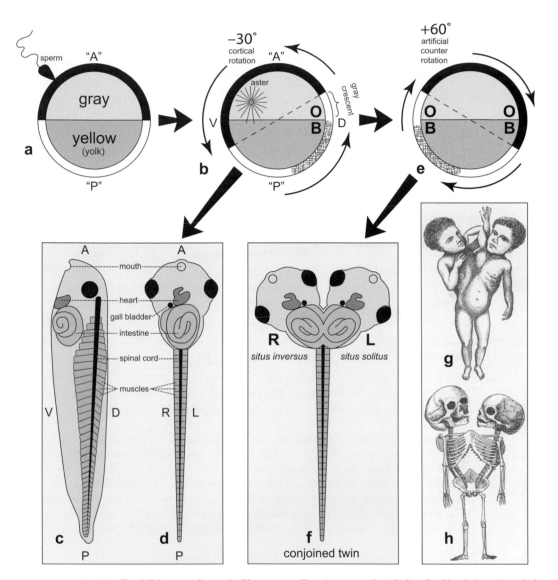

Fig. 2.2. Establishment of axes in *Xenopus*. a. Egg (cross-section) being fertilized. Its sole axis is labeled A-P, but this is an oversimplification [412]. The cortex is actually much thinner (4–8 μm) than the egg (1–1.3 mm diameter) [626]. Inside, the lower half has dense yellow yolk that settles by gravity. **b**. Egg after its cortex has rotated as a rigid shell by 30° (arrows). The yolk stays put, so a "gray crescent" appears opposite the sperm entry point. The rotation spreads dorsalizing signals (hatched area) [665] that antagonize the ventralizing agent (BMP) [477] to establish the D-V axis – a duality that is conserved among animals [284,561]. The future organizer and blastopore are denoted by O and B [848]. **c, d**. Tadpole, side view and ventral view. The heart, gall bladder, and intestine are asymmetric. **e**. Artificial "back-rotation" of the egg effected manually or by centrifugation, which elicits a second D-V axis (O and B) and conjoined twins (**f**). **f**. Twin tadpole, ventral view (cf. Fig. 2.1), where the organs of the left twin are normal (*situs solitus*), but those of the right one are L-R reversed (*situs inversus*). The intestines are actually more chaotic than shown here [1284]. **g, h**. Drawings of surface anatomy and skeleton of the conjoined twins Ritta and Christina, who were born in 1829 and lived for 8 months [472,777] (from [463]). For a survey of symmetry-breaking devices see [147]. A-P, anterior–posterior; D-V, dorsal–ventral; R-L, right–left. After [431,626,665] (cf. [127]).

In 1919, five years before his celebrated paper with Mangold, Spemann described some lesser-known studies of the L-R axis that were conducted by his student Hermann Falkenberg [1188,1284]. Falkenberg used a fine strand of baby's hair to pinch newt embryos in half at the two-cell stage [511] while still maintaining a bridge between the halves. About half of the time, this bisection procedure yielded conjoined twins like the tadpoles in Figure 2.2f [1187], where the left twin had its organs arranged normally (*situs solitus*), while the right one had its organs reversed (*situs inversus*). In the other half of the cases, both twins exhibited *situs solitus*.

This trend of *situs solitus* for the left conjoined twin but random asymmetry (50% *solitus* : 50% *inversus*) for the right twin will here be called "Falkenberg's rule" [1284]. The rule applies with equal validity to humans, but only to that subset of conjoined twins who share a torso (Fig. 2.3e–h): their A-P axes must meet obliquely at the chest or abdomen in order for the right twin to lose its L-R bias completely and adopt its asymmetry arbitrarily [783,904].

One sensational case was the conjoined pair of girls, Ritta and Christina Parodi, who were born in Sardinia in March 1829 and died within a few hours of each other in November of that year. Their side-by-side torsos branched from a single pair of legs (Fig. 2.2g, h), and the internal organs of Ritta, the right twin, displayed *situs inversus* [777]. Their parents had moved to Paris in the hope of exhibiting the twins for profit, but local officials blocked them for the sake of public decency. They may not have become rich, but their two-headed baby did become famous after an autopsy performed by the leading anatomists of that period:

> In the vast amphitheatre of the Muséum d'Histoire Naturelle at the Jardin des Plantes in Paris, Ritta and Christina were laid out in state on a wooden trestle table. The anatomists jostled for space around them. Baron Georges Cuvier, France's greatest anatomist – "the French Aristotle" – was there. So was Isidore Geoffroy Saint-Hilaire, connoisseur of abnormality, who in a few years would lay the foundation of teratology. And then there was Étienne Reynaud Augustin Serres, the brilliant young physician from the Hôpital de la Pitié, who would make his reputation by anatomising the girls in a three-hundred-page monograph. [777]

The reasons behind Falkenberg's rule remained a mystery for more than a century [935]. It was essentially solved by Michael Levin *et al.* in 1996 using chick embryos [783], with further insights provided by Matthias Tisler *et al.* in 2017 through studies of frog embryos [1285]. Both groups focused their arguments on the juxtaposed axes of adjacent twins. Interestingly, the scenario that Levin's group used to explain Falkenberg's rule relied on the same premise employed in Chapter 3 to solve Bateson's rule – namely, a leakage of signals across a boundary that fails to contain their diffusion (boldface added):

> When twins arise from two parallel primitive streaks, there is the possibility of the **activin produced on the right side of the left embryo inhibiting the expression of *Shh* in the left side of the right embryo**. This would result in a normal left embryo, but the right embryo would have no expression of *Shh* in the node, and therefore no expression of *nodal*, and hence would have random heart situs. [783]

The scheme devised by Tisler's team (Fig. 2.4) also invokes the disabling of a limiting boundary, but it involves the flow of fluid over a patch of cells where the L-R axis is imprinted: the left–right organizer (LRO) [1284]. The LRO consists of a few hundred cells, each of which has a posterior cilium [490] that spins clockwise like a tiny propeller. Because the cilium is tilted, the lower half of its stroke cycle is too close to the surface to push the liquid, owing to drag [1176], so the cycle is only effective in its upper half [963], yielding leftward flow. On both flanks of the LRO are cells with fixed cilia. The ciliated cells downstream of the LRO respond to some aspect of the flow by triggering expression of *nodal* to dictate "leftness" [269].

For a single embryo (Fig. 2.4a) the geometry can be distilled to FC_R-LRO-FC_L, where FC denotes the flanking cells on the right (FC_R) or left (FC_L) side of the LRO. For a conjoined embryo (Fig. 2.4e) the geometry would be FC_R-LRO-FC_C-LRO-FC_L, where FC_C marks the central boundary where the edges of the LROs fuse together. Once the cilia start spinning, the fluid flows would be $FC_R \rightarrow FC_L$ and $FC_R \rightarrow FC_C \rightarrow FC_L$, respectively. The lack of L-R bias in the right twin would hence be due to cells in the FC_C failing to sense the cue that should cause them to activate *nodal*, depriving the right twin of any *nodal* expression, and resulting in it adopting L-R polarity randomly.

There is no dispute about how the motile cilia operate, but there are two schools of thought about the nature of the cue that the immotile cilia are sensing [363]. Tisler *et al.* argue that the cilia sense the flow by being deflected [1285]. In that case, the flow at the FC_C must not attain the minimum needed to trigger *nodal*. However, there is scant evidence to support such a claim.

The alternative paradigm makes more sense here – namely, that the cilia of the flanking cells are chemosensory, not mechanosensory. In that case, the LRO would secrete signaling molecules known as "morphogens" (m) [1324] uniformly (Fig. 2.4a, e), and those molecules would be swept left by ciliary propellers across the entire LRO. Regardless of whether the LRO is unitary (Fig. 2.4b) or fused (Fig. 2.4f), the morphogens would wind up at the FC_L since the FC_C could not halt their flow. In conjoined twins, *nodal* would be activated on the left side of the left twin (Fig. 2.4g) but not at all in the right twin. The predicament of the right twin is analogous to a seesaw deprived of any weight on either side, leaving it to the vagaries of the wind to tip it to one side or the other [489].

GP-3 tangent: The twisted maggot

The ciliary propellers that launch the L-R axis in vertebrates have something in common with the cortex swivel that sets up the D-V axis in amphibia – namely, a reliance on microtubules [616]. Microtubules and their associated motors (kinesin and dynein) are a key contraption that animal cells use to move materials. Their other favorite device, of course, consists of actin fibers and their myosin motors.

It was therefore of considerable interest in 2006 when a mutation in a myosin gene – *myo1D* – was found to cause *situs inversus* in fruit flies [409,1186]. Because

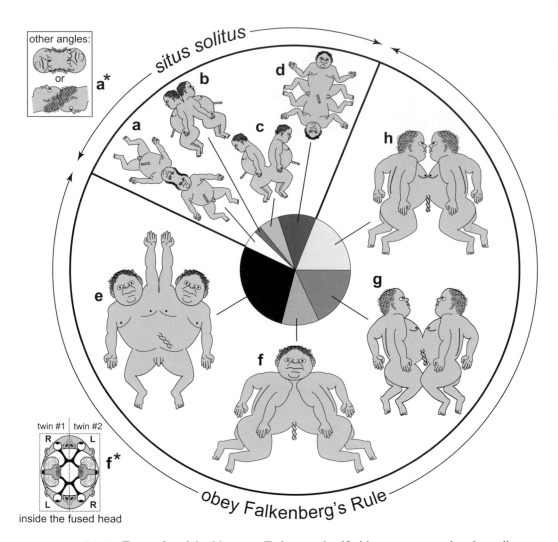

Fig. 2.3. Types of conjoined humans. Twins are classified into two categories, depending on whether both members show *situs solitus* (**a–d**) or the right member obeys Falkenberg's rule (i.e., *situs inversus* ~50% of the time, **e–h**) [1285]. The pie chart indicates frequencies of the eight types of geometries [1190], three of which (**a–c**) have separate umbilical cords (cropped helices). Twins can be joined at the head (**a**, craniopagus, 5%), back (**b**, rachipagus, 2%), sacrum (**c**, pygopagus, 6%), caudal end (**d**, ischiopagus, 11%; note diverted genitals), side of the thorax (**e**, parapagus, 28%; cf. Fig. 2.2g, h), head to umbilicus (**f**, cephalopagus, 11%), thorax with separate hearts (**g**, omphalopagus, 18%), or thorax with conjoined heart (**h**, thoracopagus, 19%) [1190]. Conjoined twins occur at ~1 per 100,000 total births or ~0.3% of monozygotic twin births [681]. The most illustrious pair – Chang and Eng Bunker (type **h**) – were born in 1811 near Bangkok [129,472], whence the antiquated term *Siamese* twins [322]. Most such twins look symmetric, suggesting they are simply identical twins who never finished splitting apart. However, this "fission theory" is disproven by craniopagus cases where the heads point in different directions (**a***) [157,970,1189,1191,1215] and by cephalopagus heads where each of the Janus faces comes half from one twin and half from the other (**f***) [1191]. Rather, such cases support a "fusion theory" where separated twins collide and fuse

arthropods (the phylum to which flies belong) do not use *nodal* as the master gene for the left side of their body [125], the fly's reliance on myosin bolstered the consensus that arthropods and vertebrates must have evolved different ways of setting up their L-R axes [246]. However, that consensus was shattered in 2018 when Melanie Tingler *et al.* knocked out *myo1d* gene function in *Xenopus* and found that 25% of the resulting tadpoles displayed *situs inversus* [1422], with another 25% showing heterotaxia, where the internal organs are heterogeneous in their L-R asymmetry [1283].

Another seminal experiment was reported in 2018. Stéphane Noselli's team, which had discovered the *situs inversus* phenotype of null *myo1D* mutants, decided to see what would happen if they activated *myo1D* where it is not normally expressed. When they forced *myo1D* to be transcribed in skin cells, the body twisted into a corkscrew, and the larva started using a rolling mode of locomotion to move forward (Fig. 2.5). An opposite sort of phenotype was obtained by mis-expressing the gene *myo1C*, and these reciprocal polarities could be partly replicated in vitro using only actin fibers and the respective myosin molecules [767].

This ability to impose asymmetry on a symmetric structure (here the larval skin) by engaging myosin motors suggests that L-R asymmetries could have evolved more easily than we might think [6,767,1352]. Possible instances of such evolutionary tinkering include the spiral coils of snail shells [2,641,918] and the spiral cleavages that precede them [1257], the spiral loops of tadpole intestines (Fig. 2.2d) [1283], the helical scroll valves of lungfish and sharks [534], and the spiral shapes of snake embryos [33,497], not to mention the plethora of other spiral and helical structures throughout the animal and plant worlds [231].

GP-4: Organs adjust robustly if anatomy changes

Attentive readers will have noticed that no mention has been made of *adult* frogs with two heads. If such animals did exist, this chapter would have featured them in its title and its frontispiece. Among the dozens of two-headed tadpoles spawned in the centrifuges of Gerhart's lab, none survived to the adult stage [116], despite the fact that two-headed reptiles (snakes, lizards, and turtles) can live for years, and even two-headed newts (another amphibian) can reach sexual maturity [1188]. I asked him why he thought this was so, and he replied (boldface added):

Fig. 2.3. (*cont.*)

back together [129,134]. Unsolved mysteries include: (1) why conjoined twins are three times more likely to be female than male [879] and (2) why conjoined twinning occurs so often in ducks (2% of fertilized eggs) [1308]. Four cases of conjoined *triplets* have been reported [42]. For lively appraisals of why certain anatomies are less likely than others, consult Mark Olson's review "The developmental renaissance …" [977] and Pere Alberch's classic essay "The logic of monsters" [15]. Adapted from [557,1192] (cf. [127]).

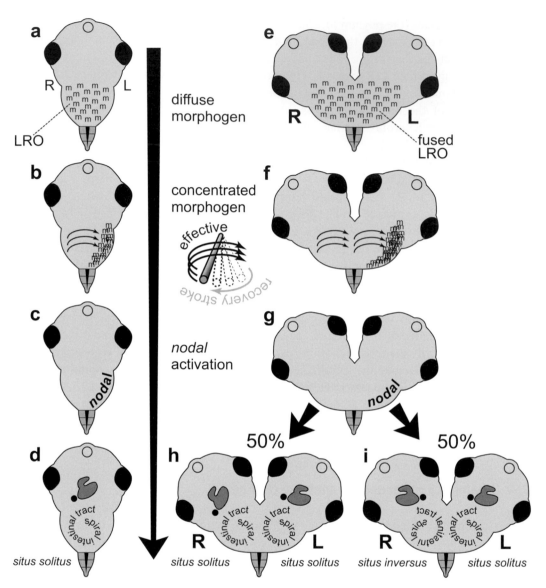

Fig. 2.4. Hypothetical mechanism of L-R axis initiation. Inception of the L-R axis is sketched for single (**a–d**) versus conjoined (**e–i**) *Xenopus* embryos. In each panel the embryo is represented as a truncated tadpole in ventral view. However, the depicted events actually occur on the archenteron roof of the neurula [1284] in an area called the left–right organizer (LRO). The LRO arises from the dorsal-lip organizer in the gastrula [124]. **a, e.** "Morphogen" molecules (m; see text) are secreted by cells throughout the single (**a**) or fused (**e**) LRO (see text). Those molecules could either be freely diffusing or contained in vesicles [1121]. Candidate morphogens include Sonic hedgehog (Shh) and retinoic acid [1264]. **b, f.** The morphogens get swept to the left by rotating cilia (cartoon between **b** and **f**) that are attached to LRO cells. **c, g.** The concentration of morphogen molecules at the left edge of the LRO rises above a certain threshold, whereupon it evokes transcription of *nodal*. **d.** Expression of *nodal* on the left side establishes the L-R axis and assigns locations of asymmetric organs such as the heart, gall bladder, and intestine (cf. Fig. 2.2d). **h, i.** The right twin, which fails to express *nodal*, randomly orients its L-R axis. Modified from [1285] (see text).

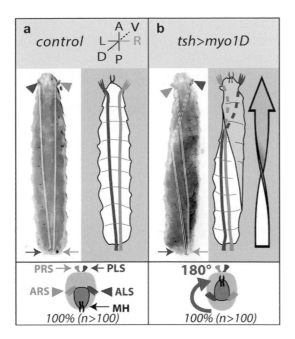

Fig. 2.5. Effect of mis-expressing *myo1D* in larval epidermis. a. Normal anatomy of a fly larva, dorsal side up. Axes at top: D-V, dorsal–ventral; A-P, anterior–posterior; L-R, left–right. Lines along the A-P axis are tracheal tubes ending in anterior (triangles) and posterior (arrows) spiracles, with the left tube and spiracles (ALS, PLS) darkly shaded and the right ones (ARS, PRS) lightly shaded. At bottom is a schematized larva seen from the front, with silhouette in gray and mouth hooks (MH) pointing down. **b**. Expressing *myo1D* with a *teashirt* enhancer (*tsh>myo1D*) causes the body to undergo a 180° twist that tucks ALS and ARS underneath (dashed lines) and points the MH up, without disturbing the PLS or PRS. This drastic change has been deemed a possible exemplar of a "hopeful monster" [307] à la Goldschmidt (cf. Fig. 11.2) [767]. Simplified from [767]; used with permission from *Science/RightsLink*.

> At metamorphosis the gut goes through considerable remodeling as the tadpole/froglet switches from vegetarian to carnivore. **I assumed the twins had problems with this remodeling.** But we didn't do any analysis of the ailing twins. These were complete or nearly complete twins – two full axes or nearly so. Maybe the two-headed forms of snakes and turtles have a gut that is only duplicated anteriorly. I'd guess *Xenopus* tads with only anterior duplication might well make it through metamorphosis. Those can be made by joining two half gastrulae (stage 9+) cut to one side of the organizer, fused to have two organizers separated by 30–45 degrees or so (Ron Stewart did such manipulations). – J. C. Gerhart (email, 4 December 2019).

In fact, the shared intestines of conjoined tadpoles are unusually convoluted [1285], so they might indeed be getting tangled during metamorphosis [542,1126]. Thus, it seems safe to conclude that they represent an unexpected exception to an otherwise quotidian rule.

However, it is possible that we have got it backward. Perhaps we should instead be surprised that *any* conjoined twins survive beyond hatching or birth. (The frequency is ~20% for humans [829].) After all, no manufactured machine (car, TV, computer, etc.)

would work after being partly amputated and then joined to a doppelgänger in mirror image. When looked at from this perspective, it is remarkable that human twins can merge together at so many odd angles (Fig. 2.3) and still manage to live for years in many cases [1364]. Evidently, the body's organs are capable of adjusting robustly to radically altered geometries [707,1322].

John Gerhart is at least as well known for his theory of robustness as he is for his two-headed tadpoles or his work on the cortical rotation mechanism [413]. That theory was set forth in two influential books that he co-authored with Marc Kirschner: *Cells, Embryos, and Evolution* (1997) [414] and *The Plausibility of Life* (2005) [699]. Both books argue (1) that animals can withstand large changes in anatomy because so many of their developmental devices are inherently self-correcting, and (2) that this robustness gives them flexibility in the face of evolutionary challenges. The authors define these related concepts thus:

> The flexibility and robustness of development have been considered under many names – developmental adaptability, plasticity, regulativeness, accommodation, self-adjustment, self-organization, tolerance, buffering, developmental homeostasis, or canalization. ... Robustness connotes something different from flexibility, namely a capacity *not* to change when conditions change, a capacity for homeostasis. Flexibility, by contrast, connotes a capacity to change in relation to changing conditions, to accommodate to change. All these terms, though, reflect the capacity of a process to perform its function despite changes in conditions, materials, or organization. [414]

GP-4 tangent: The three-eyed frog

The nervous system is a case in point. It overproduces neurons to innervate its targets and then prunes the excess, so that parity between pre- and post-synaptic populations is achieved regardless of whether target organs increase or decrease in size by experimental intervention or evolutionary alteration [561].

The three-eyed frog in Figure 2.6 was discussed in *Cells, Embryos, and Evolution* (their Figure 4.20) as an illustration of how robustness operates [414]. The extra eye was implanted during the embryonic period from a same-aged donor before any retinal axons had penetrated the brain. Normally, axons from left and right eyes would innervate separate halves of the tectum with no overlap, but here axons from the third eye have converged onto one of the two tectal halves along with the axons that belong there. The intruding axons segregate into stripes that alternate with stripes of the native axons. These sorts of "ocular dominance columns" are typical of animals like cats and monkeys that have binocular vision [594] but were not expected in the frog, which lacks stereopsis, so this result took the investigators by surprise [230].

The power of axons from different eyes to "sort out" in a species that would normally not need such segregation implies that this ability reflects a deep-seated rule for neural wiring [533,1040]. Indeed, it may be a corollary of the exploratory strategy that nerves use to prune excess links. The survival criterion here appears to involve affinity based on firing synchrony [226,739,1054].

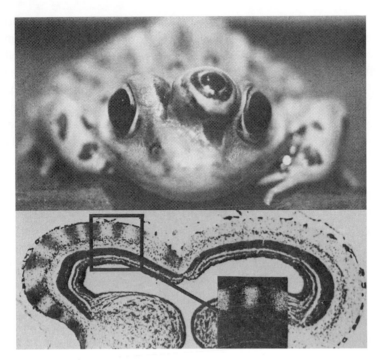

Fig. 2.6. Three-eyed northern leopard frog (*Rana pipiens, above*) and tectal autoradiograph (*below*). The frog's third (middle) eye was implanted during embryonic development from a donor at the same age. The autoradiograph of the optic tectum (coronal section) is from a similarly treated specimen, one of whose native eyes was injected with radioactive proline to trace the path of its retinal axons. Instead of innervating the tectum uniformly, the axons segregate into "ocular dominance" columns [678] that alternate with projections from the third eye, as if they were competing for limited space [594]. The inset (*square*) is a dark-field enlargement. For compensatory adjustments that are equally drastic in the visual systems of albinos, see [594,1370]. The strange case of the missing chiasm is also worth a look [273,975]. From [230]; used with permission from *Science/RightsLink*.

The presence of this ability in amphibians suggests that they possessed a predisposition for binocularity before it ever actually evolved in their descendants [603,1133]. However, this charming notion received a fatal blow when a similar segregation capacity was found for afferent axons in the *auditory* centers of three-*eared* frogs [342]. Moreover, subsequent surveys of the distribution of ocular dominance columns across the vertebrate spectrum have called into question the function, if any, of such modules in general [603,678]:

Ocular dominance columns are absent in the mouse, rat, squirrel, rabbit, possum, sheep, and goat. They are present in the cat, ferret, mink, and more than a dozen primate species. ... Species variation in columnar structure is hard to reconcile with ideas about the functional importance of columns. It is problematic when a system such as ocular dominance columns is present in some species but not in others, without any obvious functional correlate. ... At some point one must abandon the idea that columns are the basic functional entity of the cortex. [603]

Even if the lessons learned from the three-eyed frog have diminished in their significance over the years, many other examples of robustness exist. Among those that were emphasized in *The Plausibility of Life* [699] are (1) the ability of the vascular system to provide an adequate supply of blood vessels to peripheral tissues, regardless of their size, based on local oxygen requirements, (2) the ability of motor neurons to provide adequate innervation for the appendicular muscles of the limbs, regardless of their number or configuration, based on the same oversupply-plus-pruning strategy that is used by the central nervous system, and (3) the ability of the skeletal system of canine mutts to accommodate the wildly divergent bauplan blueprints that are inherited from the respective parental breeds.

To conclude this tangent we can turn to a human example [1347], since there is no more dramatic illustration of robustness than the case of the seemingly normal girl who was born with only one cerebral hemisphere [920]. Her right hemisphere had failed to develop before the seventh week of gestation, and much of her right eye was missing, yet her bilateral coordination and her depth perception were virtually unaffected – so much so that she could even roller-skate and ride a bike [148].

> Here we report on a single case study of a 10-year-old girl (AH) who lacks the entire right cortical hemisphere and most of her right eye (microphthalmia). AH has a modest left hemiplegia for distal movements (hemiparesis) but close to normal vision in both hemifields . . . The fact that the rerouted projections support absolutely normal perceptual functions as well as visuomotor coordination strongly indicates the action of highly efficient adaptation mechanisms beyond early visual cortices that are capable of extracting useful information from maps that, because of rerouted axons and molecular cuing, exhibit markedly abnormal topologies. [920]

Ordinarily, the nasal half of each retina sends its axons to the hemisphere on the opposite (contralateral) side of the body, so AH's left eye should have only been able to see images in its temporal half, but AH's brain managed to compensate in utero. The nasal retinal axons that had started to grow into the vacant half of her head were re-routed at the optic chiasm to the remaining hemisphere, where they competed for space with the temporal axons. In effect, her visual cortex solved the problem of dual projection in the same way as the striped tectum of the three-eyed frog.

3 Extra-Legged Frogs

Extra legs have been reported sporadically in amphibians for centuries [653,1320], having even been preserved in Jurassic fossils [1350], but in the past few decades outbreaks have erupted in ponds across North America [122,1183], yielding freaks like the one in Figure 3.1 and suggesting that pollutants might be to blame [430,1213]. More often than not, however, researchers have traced these plagues to

Fig. 3.1. Extra-legged frog. The extra legs of this Pacific chorus frog (*Pseudacris regilla*) are attributable to invasion of its hindlimb buds by parasitic flatworms during the tadpole stage. On each side there is one normal-looking hindleg plus an extra pair of legs posteriorly. The extra legs on the right are separate distally but enclosed in a common sleeve of skin at the thigh area. The ones on the left are fused in a sleeve along most of their length, and lack some digits. Clearing and staining revealed that the skeletons conform to Bateson's rule (see text). The digit formula (including the normal limb) on the right is 1-2-3-4-5/5-4-3-2-1/1-2-3-4-5, and on the left it is 1-2-3-4-5/5-4-3/3-4-5. Captured at Aptos Pond, Monterey County, California [1134]. Photo courtesy of Stan Sessions.

population blooms of a parasite that infests the nascent limbs of tadpoles. Whether the blooms are the result of ecological disturbances is not yet clear [652].

The parasite that was found to cause the localized epidemics is the flatworm *Ribeiroia ondatrae* [653,820]. *R. ondatrae* has a life cycle with several successive hosts [121]. It reproduces inside herons or egrets, and the resulting eggs enter ponds through the birds' feces. The eggs hatch into first-stage larvae that infect snails and develop into second-stage larvae called cercariae. The cercariae exit the snails and swim to tadpoles, where they congregate in the pelvic folds [1216] – a posterior entry site that explains why hindlimbs are affected more than forelimbs [651]. After burrowing beneath the skin the larvae encase themselves in cysts and remain quiescent until the tadpoles metamorphose [650]. Any resulting leg deformities will disable the frogs, making them easy prey for birds [458], thus repeating the cycle.

Conceivably, *R. ondatrae* could be inducing extra legs by secreting growth-stimulatory signals of some kind. However, the influence of these parasites appears to be purely physical rather than chemical, because uninfected tadpoles can be coaxed to sprout the same kinds of extra legs by just inserting inert beads into their limb buds [1134]. The resin beads used for this experiment were about the same size as the cysts themselves, with the diameter of each one being about a third the width of a hindlimb bud [1216]. Dozens of parasites can wedge themselves into the vicinity of a single bud, and the resulting cysts remain behind in the hip region as the legs grow out, thus preserving unmistakable evidence of the original intrusion at the crime scene itself (Fig. 3.2).

Heavy infestations can shred a bud into pieces like a shotgun blast, leaving cysts embedded throughout the area like lead pellets [1216]. How this mangled mass of debris manages to recover enough to make anything recognizable is remarkable, let alone the fact that the legs which eventually sprout from the affected hip joints look relatively normal in both size and anatomy. A single limb bud can beget as many as 12 legs [1134].

Strangely, the multiple legs that share a common hip socket tend to alternate as mirror images of their nearest neighbors (L/R/L ... or R/L/R ..., where L and R denote left- and right-handed legs), instead of being identical copies of one another (L/L/L ... or R/R/R ...). Thus, the upper frog in Figure 3.2 has a pair of R/L hindlegs on its left side, and the lower frog has a triplex cluster of L/R/L hindlegs on its left side.

GP-5: Cells obey local rules with no global blueprint

These kinds of symmetries among extra limbs intrigued William Bateson (1861–1926), who cataloged them in his 1894 book *Materials for the Study of Variation* [80]. After distilling their invariant features he derived what has come to be called "Bateson's rule": "The long axes of duplex or multiplex appendages lie in one plane, [and] two adjacent members form in structure and position the image of each other, as reflected from a plane mirror bisecting the angle between respective axes" [527].

Most of the appendages that Bateson relied upon in formulating this rule actually belonged to insects, rather than vertebrates (Fig. 3.3a). He even went so far as to build a wooden device with interlocking gears to show how the multiplex legs of insects can subtend a wide range of angles from one specimen to the next without altering the mirror planes between them (Fig. 3.3b).

Bateson discussed a few cases of duplex human arms as well. He was especially perplexed by the fact that a *right* hand could develop on the *left* side of a person's body. What vexed him most was why such a fundamental feature of animal anatomy (bilateral symmetry) should appear unilaterally in a place where it clearly does not belong [79]. Here is Bateson's description of the case in Fig. 3.3c (italics are his; boldface is mine):

> Woman (examined alive) having eight fingers in the left hand arranged as follows ... With the exception of the left arm the body was normal. ... The eight fingers were arranged in two groups of four in each, one of the groups standing as the four normal fingers do, and the other four being articulated where the thumb should be. There was no thumb distinguishable as such, but it is stated that there was a protuberance on the dorsal side of the hand, between the two groups of fingers, and this is considered by Murray to represent the thumbs, for according to his view the limb was composed of a pair of hands compounded by their radial sides. ... **The four radial fingers in size and shape appeared to be four fingers of a *right* hand.** ... Between the two groups of fingers there was a wide space as between the thumb and index of a normal hand, and the two parts of the hand could be opposed to each other and folded upon each other. The power of independent action of the fingers was very limited. [80]

Despite the absurdity of such a geometry, the cells themselves may not sense anything wrong because they reside next to their usual neighbors [553], which appear as doppelgängers across the mirror plane, and the same is true for the multiplex legs in Figure 3.2. Local continuity is preserved in all of these instances, regardless of the globally altered anatomy [790]. These architectural blunders imply that the cells are not consulting a blueprint as they build a leg [781], for if they were, then they would be able to detect deviations and correct them. In general, cells just follow local rules [1373]; they don't seek global goals [533,960,1287]. Cellular myopia also governs anatomy at a larger (body vs. limb) scale, as shown by conjoined twins, which likewise manifest ectopic mirror planes (see Chapter 2).

In his book *Paradoxes of Progress* (1978) [1200], the neurobiologist Gunther Stent considers a local decision that can be phrased in the form of the familiar chicken–road riddle: "Why does the retinal axon cross the chiasm?" The ability of mammals to perceive depth is due to a partial crossing of axons from the left eye to the right side of the brain (and vice versa) so that higher brain centers get input from the same point in visual space as seen from different angles in order to assess the degree of parallax. It so happens that the decision an axon makes as to whether to turn left or right when it reaches the crossing point ("chiasm" is named for the Greek letter X) depends upon the capacity of retinal cells to produce pigment, and that capacity is reduced in Siamese and albino cats (see Fig. 11.1). Hence, the neural wiring in the brains of those cats is shifted laterally

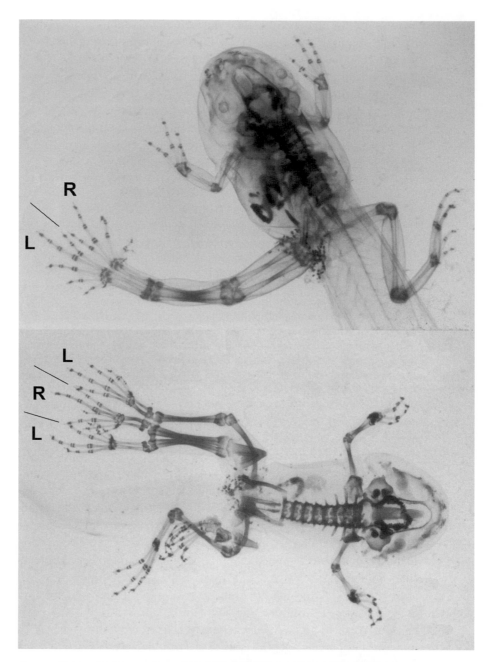

Fig. 3.2. Frogs with planes of symmetry between adjacent legs. Mirror planes are indicated by line segments, with "L" and "R" denoting left- or right-handed legs [1135,1216]. Both specimens are partly metamorphosed *P. regilla* frogs (note the tadpole tails), cleared and stained to reveal bones (alizarin red) and cartilage (Alcian blue). The dark dots peppering the pelvic area are cysts of the flatworm *R. ondatrae*, each of which measures ~150 μm across [1134]. The cysts evidently caused the left leg bud of the upper individual to undergo an anterior–posterior (A-P) duplication, yielding side-by-side legs (note the double femur).

at several echelons [661]. From these facts Stent deduces the same general principle that is codified here as GP-5 (boldface added):

> The probable connection between the absence of retinal epithelial pigment and the misdirection of the optic nerve fibers suggests some testable hypotheses about the **rules** that determine whether an optic nerve fiber grows to the same or to the opposite side of the brain as its retina of origin. The eventual statement of these hypothetical **rules** may contain such terms as enzymes, gradients, growth rates, threshold concentrations, preferential adhesion and nerve impulse frequencies, but **the word "gene" is unlikely to find frequent mention**. [1200]

His larger point here is that the genome does not use a global blueprint to specify the wiring of the brain any more that it does to build the anatomy of the arm. Rather, it programs cells to act in certain ways when they detect certain local cues in a cascade of binary choices – a decision tree, if you will [1435] – that usually "gets it right" but can go off track in weird ways if it's knocked off course (cf. Fig. 11.3). Curiously, different mammals use different rules at the chiasm, and those rules make little sense in terms of phylogeny [948]. For example, humans are more similar to marsupials than to mice (their closer placental relatives) in how the loss of one eye fails to divert the paths of axons from the remaining eye.

GP-5 tangent: Secondary mirror planes

The etiology of mirror planes in the extra limbs governed by Bateson's rule will be explored shortly. Before doing so, however, it may be worth pointing out that vertebrates possess mirror planes aside from the medial one that defines our bilateral symmetry. As shown in Figure 3.4, such secondary planes of symmetry are evident in our own digits and jaw.

A clever attempt to explain the symmetry of our digits (fingers and toes) was made in 1981 by Greg Stock and Susan Bryant, based upon a continuity imperative (reminiscent of Bateson's rule) that they derived from the Polar Coordinate Model [1211]. That model had been proposed five years earlier by Vernon French, Peter Bryant, and Susan [377], but it has largely been disproven by recent molecular evidence [932]. Instead, an even older model of pattern formation, proposed by Alan Turing in 1952 [1306], seems to hold the key to understanding digit formation [1143,1335]. A discussion of how the Turing mechanism works will be deferred until Chapter 9.

←——————————————————————————————

Fig. 3.2. (*cont.*)

Both hindlimb buds of the lower frog sprouted extra legs. The left one made an A-P triplication with one whole leg plus a branched symmetric outgrowth. The mirror planes, which conform to Bateson's rule (see text), are clearly evident from the palindromic A-to-P digital sequences of the R/L legs above (5-4-3-2-1/1-2-3-4-5) and the L/R/L legs below (1-2-3-4-5/5-4-3-2-1/1-2-3-4-5), where numbers denote digits and slash marks are mirror planes. Photos courtesy of Stan Sessions. (A black and white version of this figure will appear in some formats. For the color version, please refer to the plate section.)

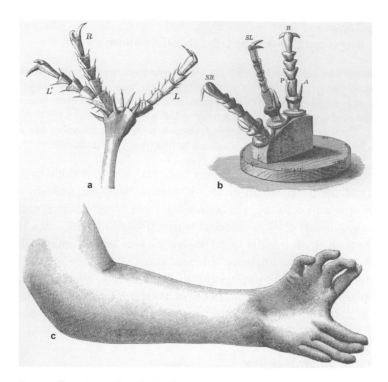

Fig. 3.3. Drawings of a triplication, a duplication, and a contraption that simulates them both. These sketches are from William Bateson's 1894 monograph [80]. **a.** Distal portion of the left middle leg of a ground beetle (family Carabidae) with three tarsi emanating from a single tibia (Case #742): the normal tarsus (L), an extra right tarsus (R), and an extra left tarsus (L′). **b.** Wooden device showing the relations between "supernumerary" legs (SR and SL) and the normal leg (R), with A and P marking anterior and posterior tibial spurs (his Fig. 153). **c.** Unilateral double-hand anomaly (ventral aspect). The latter phenotype (Case #495) conforms to ulnar dimelia syndrome, with 70 cases reported as of 2015 [1042,1292]; feet can be doubled as well as hands [158,535,702]. Bateson cited a similar double-hand deformity in a man (Case #492) who considered his condition actually to be an asset, not only in his job (machinist) but also for playing the piano.

Our jaw manifests an internal mirror plane as well (Fig. 3.4b): the maxilla (upper jaw) displays the same seriation of tooth types as the mandible (lower jaw). Bateson wondered whether the jaw might be inserting a reflection plane by relying on the same kinds of devices that are used at the body's midline:

> In view of the fact that the teeth in the upper and lower jaws may vary simultaneously and similarly, just as the two halves of the body may do, it seems likely that the division of the tissues to form the mouth-slit must be a process in this respect comparable with a cleavage along the future middle line of the body. [80] (p. 197)

In fact, the reflection across the D/V line turns out to be attributable to back-to-back gradients of the secreted signal Fibroblast growth factor-8 (Fgf8), which

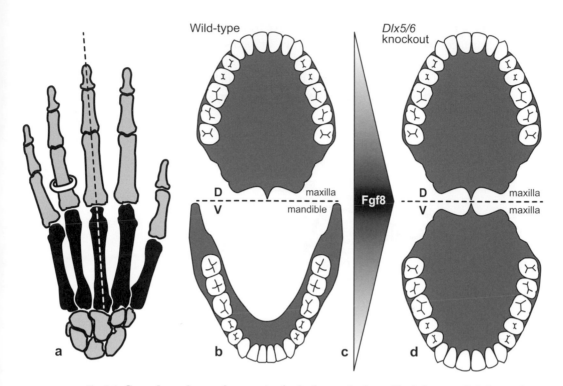

Fig. 3.4. Secondary planes of symmetry in the human body. a. Each human digit is nearly symmetric, as is the metacarpal bone at its base, with the mirror plane denoted by a dashed line along the middle finger. The hand as a whole also has a subtle symmetry. **b.** The tooth patterns of our upper (D, dorsal maxilla) and lower (V, ventral mandible) jaw are quasi-symmetric, with the mirror plane denoted by a horizontal dashed line. **c.** Secretion of the protein Fibroblast growth factor-8 (Fgf8) along the D/V interface establishes back-to-back concentration gradients (triangles) [218,810], which appear to enforce opposite polarities on either side of the D/V mirror plane (see text) [294]. **d.** Transformation of a mandible to a maxilla in a mouse (redrawn with human dentition for comparison) whose *Dlx5* and *Dlx6* genes have been knocked out (null LOF condition) [107,294]. Adapted from [557].

diffuses away in both directions from that line (Fig. 3.4c), while evidently acting as a polarizing agent to give cells their "compass" orientations [294].

In mice the mandible can be transformed into a nearly perfect imitation of the maxilla by disabling the homeobox genes *Dlx5* and *Dlx6* (Fig. 3.4d) [107,294]. Conceivably, this phenotype may be revealing the ancestral (default) condition of the jaw [107], with *Dlx5* and *Dlx6* having been added later to modify the mandible [1258]. However, the story of vertebrate jaw evolution is not nearly as simple as this scenario suggests [738].

Secondary mirror planes are not confined to vertebrates. Flies, for instance, also use them to construct their wings, as do other flying insects. In that case, the signal that is secreted along the mirror plane is Wingless (Wg). This protein diffuses away in both directions to create back-to-back concentration gradients. Based on the positional information that cells receive from that signal, the upper and lower wing surfaces acquire an identical array of veins inside the imaginal disc, despite developing separately from one another (Fig. 3.5a) [555]. Ultimately the two surfaces fuse together to form a flat, two-ply airfoil, whose venation patterns mesh precisely (Fig. 3.5b), like the teeth in our upper and lower jaws.

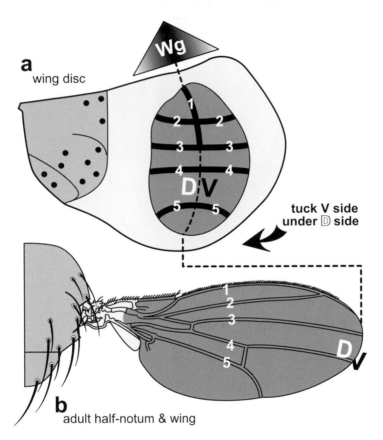

Fig. 3.5. A secondary plane of symmetry in the fly wing. a. Mature right *D. melanogaster* wing imaginal disc (anterior above, posterior below). The dark oval will form the wing, with dorsal (D) and ventral (V) halves coming together. Numbers (1–5) mark future veins, which are symmetric about the vertical dashed line. The signaling molecule Wingless (Wg) is secreted along that line. From there it diffuses laterally to form back-to-back concentration gradients (triangle above) [992]. During metamorphosis the disc folds along the dashed line to tuck the V side under the D side (as in GP-2). Black dots denote future macrochaetes (large bristles). **b.** Adult derivatives of the wing disc, seen from above. The D surface of the wing is visible, but its V surface is hidden below. The D and V veins have merged to form pipes for hemolymph that hydraulically inflates the wing after the fly exits its pupal case. Vein 1 runs along the front edge and is studded with bristles. Given the necessity of a perfect fit between the D and V halves of each vein, it seems odd that they develop independently, rather than one half inducing the other to ensure precision (as in GP-1). Microchaetes (small bristles) are not shown. Adapted from [555].

GP-6: Organs assign cellular positions along axes

One lingering conundrum from the case of the extra-legged frogs is why more than one leg should sprout from a bud after the parasites have chewed it to pieces. The capacity of the remaining cells to regroup into leg-forming islands illustrates the versatility of vertebrate limb primordia in general [285]. During their early stages they behave as self-organizing "embryonic fields" [96,143,285,1367], whose cells are not yet committed to form any specific part of the limb.

The classic experiments that led to this notion of a limb field were performed on salamander embryos by Ross Harrison (1870–1959) and published in 1918 [526]. He found that if he removed the front or back half of an incipient limb bud, then each half could still make an entire leg (Fig. 3.6a–f). He summarized the bud's properties as follows (boldface added):

> Self differentiating as the system is as a whole, the parts within the system do not constitute a developmental mosaic, with the exception of certain portions of the shoulder girdle. The system itself is equipotential, as shown by two tests to which it can be subjected; **a whole will develop out of a part**, and a single normal whole will develop out of two separate rudiments when fused together. [526]

Another experiment that Harrison conducted was to split a limb bud in half with a vertical incision, leaving each section in place. He expected each half to form a whole leg, just as they had done when their complementary half-bud was excised, but a total of only one leg emerged from the bisected bud, apparently due to postoperative fusion of the halves to reconstitute the original field. Harrison's student F. H. Swett realized the problem and devised a way to prevent it. After cutting the bud in half, he pried open the wound and inserted a strip of flank skin that was incapable of participating in limb formation, thereby keeping the two halves a short distance apart.

Like Harrison, Swett expected each of his half-buds to make a whole leg, and the front half-bud invariably did so, but the rear one instead made *duplex* limbs that grew out as mirror images of each other [1243] (Fig. 3.6g–i). Overall, therefore, each of Swett's buds formed three forelimbs in an R/L/R array, where the second slash represents not only a mirror plane but also the flank tissue wedged between the R anterior leg and the R/L duplicate behind it. This outcome mimics the L/R/L hindlegs of the lower frog in Figure 3.2, except that Swett studied *forelimb* buds on the *right* side of *salamanders*.

The strip that Swett inserted was evidently wide enough to block the two halves from merging back together *physically*, but it may have been too thin to stop the half-buds from influencing one another's responses *chemically* via molecular signals diffusing across the isthmus (Fig. 3.6m–o). Each half-bud probably started to make a right limb, but the nearness of the back part of the front limb to the front part of the rear limb could have elicited a left bud de novo between them (R/L/R).

The primary signals that govern salamander limb outgrowth are Fgf8 (see Fig. 3.4) and Sonic hedgehog (Shh) [1261]. As discussed in the previous chapter,

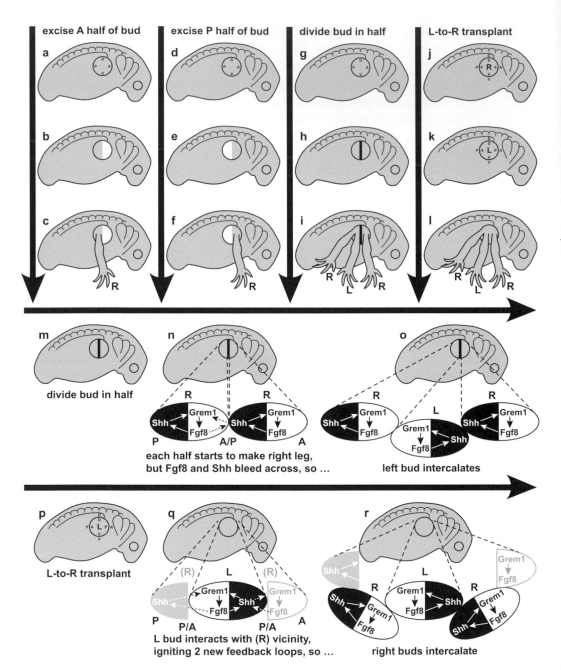

Fig. 3.6. Pioneering experiments on salamander embryos. Panels depict the right side of a salamander embryo at the operative stage, with somites (arches), eye rudiment (tiny circle), and gills (petals) in the neck region. The forelimb bud (large circle) is inscribed with "compass" directions (D, V, A, P = dorsal, ventral, anterior, posterior); the hindlimb bud is not yet visible. Adult forelegs (L and R are left and right) are drawn as if growing from the embryo even though they arise much later. Ovals (lower panels) denote embryonic limb fields, with black and white indicating P and A domains. **a–c.** Removing the front half-bud

such signals are called "morphogens" [1402]. The term was coined by Alan Turing to mean a diffusible substance capable of establishing spatial patterns [1306], but its usage was later narrowed by Lewis Wolpert to denote a molecule whose concentration informs cells about their distance from fixed reference points [1401].

According to Wolpert's theory of "positional information," morphogens (1) emanate from one end of an axis, (2) decrease in concentration as they diffuse away from the source, and thereby (3) allow nearby cells to ascertain their positions along the axis by the intensity of the signal along the concentration gradient [1404]. Put simply, morphogens establish a global positioning system (GPS) that informs cells of their locations within a Cartesian-style coordinate system [1324].

Shh functions as the posterior (P) morphogen for the salamander limb bud, while Fgf8 acts as the anterior (A) one. Morphogens not only encode positional information; they typically also stimulate growth [1062,1357], so it is not surprising that Shh and Fgf8 sustain limb outgrowth. Moreover, the positive feedback loop between Shh and Fgf8 could be instrumental in launching limb growth in the first place.

The diffusibility of Shh and Fgf8 helps explain another set of experiments, which Harrison published in 1921. He excised most of the *right* forelimb bud of a salamander embryo and replaced it with a *left* forelimb bud. This surgery placed the left bud's back side (oozing Shh) at the front end of what had been the right bud (containing residual Fgf8) and its front side (oozing Fgf8) at the rear of the right bud area (containing residual Shh). The result of this confrontation was the outgrowth of two new R forelimbs on either side of the L transplant (Fig. 3.6j–l) [527]. Those extra limbs could have been incited by the proximity of Shh and Fgf8,

---

$\leftarrow$

Fig. 3.6. (*cont.*)

does not result in half a leg. Instead, a whole leg emerges from the back half alone [526]. **d–f**. Similar result when back half-bud is excised [526]. **g–i**. Bisecting the bud and inserting a strip of (inert) flank skin (black rectangle) was expected to yield two legs, but it made three instead [1243]. Why does the middle (L) leg develop from the hind (vs. front) half? Probably because the rear half-bud has fewer limb-competent cells to start with than the front half-bud [526], thus giving the latter a head start [79] and allowing its diffusible signals to exert a greater influence across the gap. **j–l**. Replacing the right forelimb bud with a left one reverses the A-P axis and yields three legs as in **i** [527]. **m–o**. Hypothetical explanation for bud bisection results (**g–i**), based on the positive feedback loop (Fig. 3.7p) between Sonic hedgehog (Shh) and Fibroblast growth factor-8 (Fgf8) that is mediated by Gremlin-1 (Grem1) [291,1261]. Two extra (dashed) arrows have been added to the canonical circuit to indicate signals bleeding across the flank skin "barrier" at the A/P interface (**n**): Shh diffuses posteriorly and Fgf8 diffuses anteriorly. **p–r**. Hypothetical explanation for L-R transplant results (**j–l**), based on new feedback loops that are sparked between adjacent limb fields at each P/A interface. NB: The circuitry here only concerns the A-P axis, which was indelibly recorded by limb cells before the surgeries were performed, unlike the D-V axis, which was not specified or else incorporated later [527]. The latter axis relies on different signals from the A-P axis – namely, the Wnt pathway for the dorsal side and the BMP pathway for the ventral side [291]. The handedness or "chirality" of a limb (i.e., left vs. right) is defined by the relative orientations of these two axes.

and their handedness could have been dictated by the reversed polarity of the Shh–Fgf8 axis at the front and back edges of the transplanted bud (Fig. 3.6p–r).

More recent experiments have proven that Shh and Fgf8 are both necessary and sufficient for the outgrowth of salamander limbs (Fig. 3.7).

Shh serves as the P morphogen in all vertebrate limbs, and the feedback loop between Shh and Fgf8 also appears to be universal [1437]. However, the exact role of Fgf8 varies among taxa: salamanders express Fgf8 in anterior mesenchyme [1261], while birds and mammals express it in apical (distal) ectoderm [1337].

Frogs appear to represent a compromise situation. In both *Xenopus laevis* [1421] and the direct-developing frog *Eleutherodactylus coqui* [493], Fgf8 is expressed apically (as in birds and mammals), but it is transcribed more strongly on the A than the P side (as in salamanders), so it could indeed function as a polarizing agent along the A-P axis. Hence, the (A-P) model that has been proposed for salamander limb development (Figs. 3.6 and 3.7) may also apply to frogs, allowing us to explain the leg anomalies caused by the parasites.

After a limb bud gets tattered, the surviving cells probably coalesce into miniature fields with Fgf8 expressed anteriorly and Shh posteriorly (Fig. 3.8d). Initially, those fields would therefore retain the handedness of the original bud, but as they overgrow the cysts between them, the Fgf8 and Shh from adjacent fields could overlap (Fig. 3.8e, f) and ignite new feedback loops of opposite polarity (i.e., Shh expressed anteriorly and Fgf8 posteriorly) in the tissue between them. The resulting legs (Fig. 3.8g) would alternate in handedness (L/R, etc.) along the A-P axis (cf. Fig. 3.2) in conformity with Bateson's rule.

In conclusion, therefore, more than one limb cannot emerge from the same socket without positional signals leaking across the "limb/limb" interface. This "bleeding" of those signals launches an intervening limb of backward polarity ("bmil"), culminating in a palindromic "limb/bmil/limb" Batesonian sequence of pattern elements [564].

This scenario (Fig. 3.8) is merely an extrapolation of Swett's intervention (Fig. 3.6g–i), where each cyst would correspond to a strip of flank skin, though most buds would be diced repeatedly instead of being bisected once. Frogs are as adept as salamanders at regenerating their limbs during the tadpole stage [501,1146], but they lose this ability as they mature [895].

GP-6 tangent: Extra-legged flies

As mentioned previously, Bateson's rule was mainly founded on the duplicated or triplicated legs of insects, rather than those of frogs or other vertebrates. Hence, we must ask whether any of the foregoing argument applies to insect legs. The answer, in short, is yes. The actors may differ, but the play remains very much the same. The leg disc of the fruit fly *D. melanogaster* was discussed earlier (Chapter 1), so it is a useful place to start.

In the 1970s a mutant strain of flies was found to exhibit the same kinds of duplicated or triplicated legs as those in Bateson's book [1095,1096]. The strain

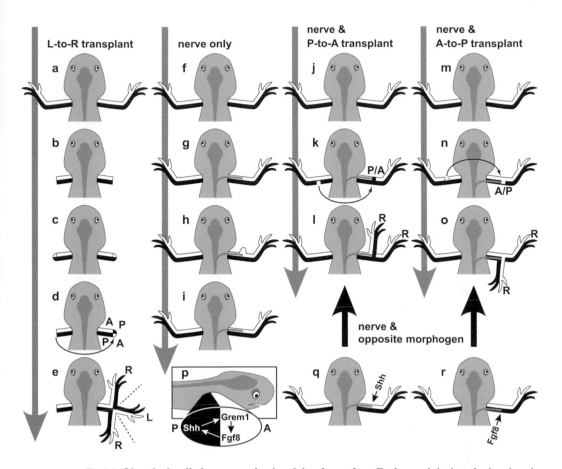

Fig. 3.7. Unorthodox limb regeneration in adult salamanders. Each panel depicts the head and forelegs of a salamander. White and black mark anterior (A) and posterior (P) halves of the leg, with dark gray denoting brain and spinal cord. **a–e.** Salamanders are exceptional among vertebrates for being able to fully regrow their appendages after reaching the adult stage [1298]. The first step after amputation (**b**) is formation of a blastema of dedifferentiated, embryonic-like cells (**c**) [19,501]. If the blastema of a left (L) foreleg is transplanted onto a right (R) stump (**d**), then a new foreleg grows out at each of the two A/P interfaces (**e**), due in theory to the same kinds of interactions that lead to triplications in salamander embryos (Fig. 3.6p–r). **f–i.** Re-routing a nerve into the A part of the forelimb (**g**) elicits a blastema (**h**) that regresses (**i**) [895]. **j–l.** Doing the same plus inserting P tissue into the A region (**k**) causes an extra (R) limb to grow out (**l**). **m–o.** The reciprocal operation – putting A tissue into nerve-supplied P region (**n**) – causes an extra (R) limb to grow out (**o**). **p.** Blastema (embryo-like) limb field on a right stump (cf. **c**) in cross-section (oval) on a salamander viewed from the side, with black and white indicating P and A domains. Outgrowth is governed by a positive feedback loop between Sonic hedgehog (Shh) and Fibroblast growth factor-8 (Fgf8) mediated by Gremlin-1 (Grem1) [1261,1437]. Grem1 belongs to a different (BMP) signaling pathway [1337]. **q–r.** Proof that direct contact between A and P cells is not needed for limb outgrowth [932], as had been assumed in the Polar Coordinate Model [377]. The Shh protein can be used as a proxy for P cells (**q**), and the Fgf8 protein can be used as a proxy for A cells (**r**), leading in each case to limb outgrowth (**l, o**) [932]. NB: Cell interactions leading to limb triplication after L-R transplants in adults (**a–e**) [627] may differ from those in embryos (Fig. 3.6p–r) because signaling networks function differently in limb development vs. regeneration [137,1233]. For a sober assessment of whether humans can ever regrow arms or legs see [20]. Adapted from [932,1261].

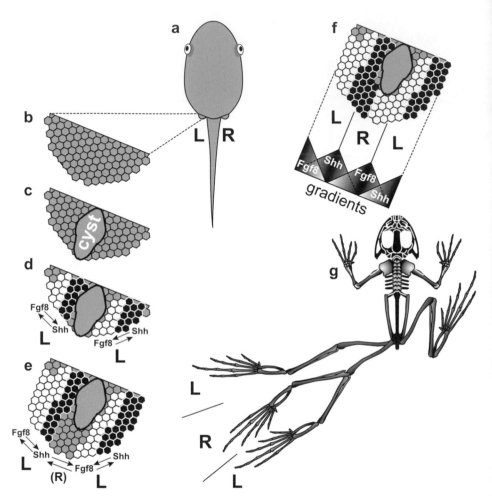

Fig. 3.8. Hypothetical etiology of extra legs in parasite-infested frogs. a. Tadpole, whose hindlimb buds are drawn as bumps that flank the tail (L = left; R = right). **b**. Enlarged view of the L hindlimb bud, with cells shown as hexagons. Gray indicates uncommitted (pluripotent) embryonic cells. **c**. A parasite has wedged itself into the center of the bud and formed an inert cyst. **d**. Coalescence of cells into miniature limb fields on either side of the cyst. Black cells secrete Shh, white cells secrete Fgf8, and the two morphogens reinforce one another's expression by the positive feedback loop (arrows) that governs salamander limb development (Fig. 3.7p; Grem1 has been omitted for clarity). **e**. Outgrowth causes limb tissue to extend beyond the cyst, allowing morphogens from the two fields to bleed across the intervening space and ignite an Shh–Fgf8 loop of opposite (R) polarity. **f**. Continued outgrowth of these fields produces L/R/L legs that obey Bateson's rule. The "mountain range" of morphogens (from A to P) consists of (1) a single Fgf8 gradient, (2) a back-to-back pair of Shh gradients, (3) a back-to-back pair of Fgf8 gradients, and (4) a single Shh gradient, with the back-to-back pairs of gradients resulting from the seeping of morphogens in both directions (A and P) from their source sites.
g. Metamorphosed frog, with a bouquet of hindlegs on its left side. In this imaginary rendition, the two rear members of the cluster are sketched as branching from a shared femur – the sort of phenotype often seen in parasitized frogs (Fig. 3.2). The legs that end up fusing may depend upon the relative rates of mini-field outgrowth and the geometries of their contact angles when they grow beyond the cyst barriers.

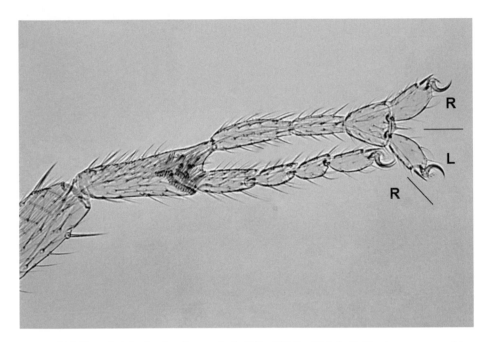

Fig. 3.9. Triplicated right foreleg from a *lethal(1)ts726* **fly.** This individual was exposed to a 2-day pulse of 29 °C in the third larval instar. Only the tarsal region is shown. R = right branch; L = left branch. Note the three pairs of claws, which alternately point up, down, and up, indicating that their legs are R/L/R. This pattern conforms with Bateson's rule like the frog legs in Figure 3.2, but these mirror planes invert their polarities along the dorsal–ventral, instead of the anterior–posterior, axis. Photo courtesy of Jack Girton.

carried a recessive mutant allele of the *suppressor of forked* gene [370] called *lethal (1)ts726*. As its name indicates, the mutation causes lethality (at the cellular and organismal levels [441]), is located on the X (1st) chromosome, and is temperature-sensitive (ts). Mutant larvae that are raised at the lower temperature of 22 °C become normal flies, while those raised at the higher temperature of 29 °C die before the adult stage.

Exposure of mutant individuals to a 2-day pulse of 29 °C early in larval life results in duplicated legs [443,1096], whereas exposure to a 2-day pulse during the second half of the larval period yields triplicated legs like the one in Figure 3.9 [437]. This particular leg happens to be a right foreleg, whose branches display an R/L/R alternation in handedness along the dorsal–ventral (D-V) axis.

The fly's leg disc uses the same morphogen – called Hedgehog (Hh) – as the frog's limb bud (Shh) in its rear compartment, but Hh plays a very different role, so triplicated fly legs cannot be explained in the same way as triplicated frog legs. The main task of Hh is not to specify A-P coordinates directly, but rather to diffuse a short distance into the A region and activate two secondary morphogens [1093]. One of those – Decapentaplegic (Dpp) – is switched ON both dorsally (strong expression) and ventrally (weak expression) [1110], while the other – Wingless (Wg) – is switched ON ventrally but not dorsally (Fig. 3.10).

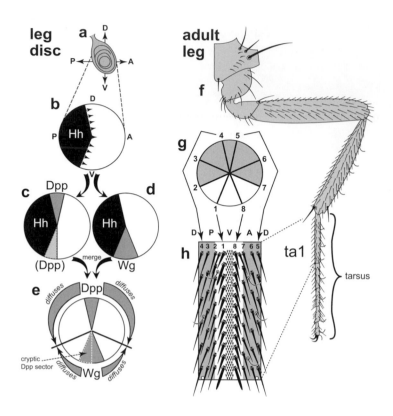

Fig. 3.10. Axes of fly leg geometry. a. Left *D. melanogaster* second-leg disc (cf. Fig. 1.6). Directions are dorsal (D), ventral (V), anterior (A) and posterior (P). **b**. Schematic depicting the disc as a circle. The P compartment (black) expresses the morphogen Hedgehog (Hh), which diffuses anteriorly (arrows) to evoke two secondary morphogens (Dpp and Wg). Hh is activated by *engrailed* (not shown) [1093], which encodes P identity throughout the fly [555]. **c**. Dpp is expressed strongly in a narrow D sector and weakly (dotted outline) in a narrow V sector. **d**. Wg is expressed in a wider V sector [54]. **e**. The complementary "spheres of influence" (diffusion ranges?) for Dpp (~225°) and Wg (~135°) are marked by thick lines that were mapped via null clones [563]. Such clones indicate that the V sector of Dpp (dashed outline) is non-functional ("cryptic"), since disabling Dpp there has no effect. The lower level of Dpp in its V (vs. D) sector is due to suppression by Wg [1274]. **f**. Left second leg of an adult, anterior view. The tarsus has five segments, the most proximal of which is the basitarsus (ta1). **g**. Fate map of tarsal bristle rows in the second-leg disc [549]. This pinwheel becomes a parallel array (**h**) by telescoping of the disc during metamorphosis (cf. Fig. 1.6). To visualize this process, imagine that the wheel in **g** is drawn on a flattened balloon and that you pinch and pull the center upward to form a cone (adult leg) [173]. The realms controlled by Dpp and Wg are shaded gray and white, respectively, in **g** and **h**. The entire Wg-governed area expresses transcription factors H15 and Midline (up to rows 2 and 7), while the Dpp-governed area expresses Omb down to rows 3 and 6 (not shown) [1093]. These pie-shaped sectors imply that Wg and Dpp are not diffusing randomly but rather are being constrained somehow to travel around the circumference (see text). **h**. Panoramic map of the bristle pattern, drawn as if ta1 had been cut along its D midline and flattened to show its whole surface.

The V half of the Dpp expression zone is non-functional [563], allowing Dpp to act as a morphogen for the D portion of the disc, while Wg reigns as the V morphogen. Dpp and Wg inhibit one another [50,718], but they act cooperatively to trigger outgrowth where they overlap in the center of the disc [173,175]. This triggering ability was proven in 1993 when small clones of Wg-expressing cells were randomly initiated [1219]. They were found to cause branching outgrowths (triplications), but only when they happened to overlap the Dpp sector.

Thus, fly leg outgrowth depends directly on its D-V axis, unlike the frog leg bud, which appears to rely more upon its A-P axis. This dependency might explain why the mirror planes of fly leg triplications tend to flip back and forth along the D-V, rather than the A-P, axis [437]. Regardless of this axial difference, flies and frogs both use positive feedback loops to drive appendage outgrowth. In frogs it is Shh and Fgf8 that activate one another. In flies Dpp and Wg each have their own auto-stimulatory loop, and they interact synergistically when high levels merge [172,845]. In both cases, the loops offer robustness in the face of injury-related deviations.

The V wedge of Wg expression is twice as wide as the D wedge of Dpp expression (due to a lower response threshold to Hh?), but it controls only half as much (~38% vs. 62%) of the leg circumference (due to a lower rate of diffusion?) [563]. The complementary shapes of their respective "spheres of influence" (~135° vs. ~225°) areas imply that Wg and Dpp are actively transported along curved paths [457], instead of passively diffusing from their source zones [560]. As they travel along these arcs, their concentrations would decrease – hence forming curved gradients – until they meet at the boundary lines where bristle rows 2 and 7 will later appear.

Considering this geometry, Dpp and Wg could theoretically be encoding the angular coordinate envisioned by the Polar Coordinate Model [377]. However, the pinwheel depicted in Figure 3.10g is only an inferred fate map: bristle rows do not actually arise until after the leg has everted into a cylindrical shape [980]. Hence, it seems more likely that Dpp and Wg are encoding a Cartesian variable – namely, distance around the circumference relative to the D (Dpp) and V (Wg) midlines [173], with the other coordinate being supplied, at least in part, by gene expression zones along the proximal–distal axis [718].

The *lethal(1)ts726* mutation kills arc-shaped clusters of cells in the leg disc when mutant larvae are exposed to the restrictive temperature [215,442], and the frequency of triplex legs varies with the magnitude of cell death – a.k.a. apoptosis [902] – suggesting that the patches are causing the triplications [442]. Two types of triplications were found in the 29 °C-treated mutant flies, both of which obey Bateson's rule. Converging triplications stem from the D side of the disc and taper as they grow, often aborting before making claws. Diverging triplications stem from the V side and bifurcate into branches that get more complete circumferentially as they grow (Fig. 3.11a).

If cell death is occurring in the same sausage-shaped areas that sprout the extra legs (Fig. 3.11b, e), then those legs can be explained by assuming that apoptosis activates Wg in the Dpp sector of some discs (Fig. 3.11c) and Dpp in the Wg sector of others (Fig. 3.11f). The ensuing proximity of Dpp- and Wg-expressing

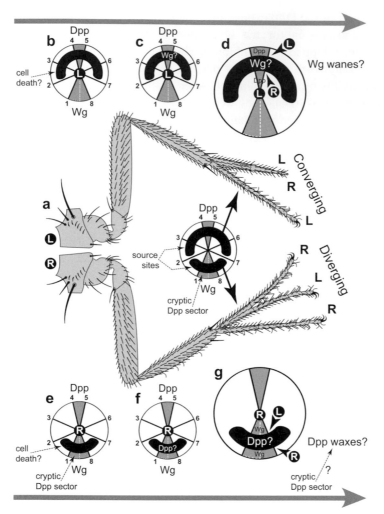

Fig. 3.11. Hypothetical etiology of converging versus diverging triplications. a. Two types of leg triplications arise in *lethal(1)ts726* mutant flies exposed to 29 °C during the third larval instar: converging (above) and diverging (below). They emerge from different parts of the disc that are indicated by black, sausage-shaped areas in the schematic, where gray sectors are expression domains of Dpp and Wg, and pinwheel spokes denote future bristle rows (1–8; cf. Fig. 3.10). Converging outgrowths, which taper distally, emerge from the dorsal arc, while diverging ones, which branch distally, stem from the ventral arc. **b–d.** Converging etiology, starting with a left (L) leg. **b.** Black area is the presumed region of cell death. **c.** Apoptosis supposedly elicits Wg expression. **d.** Two new legs grow out, one from each area of contact between Wg- and Dpp-expressing cells. Given their dorsal–ventral polarities, the more dorsal extra leg will keep the handedness of the disc (L), while the more ventral one flips (R, right). Eventually, the fused legs taper, possibly because the ectopic Wg wanes. **e–g.** Diverging etiology, starting with a right (R) leg. (Circles are drawn to match **b–d** but are inverted relative to the R leg since its ventral side is up.) **e.** Black area is the presumed region of cell death. **f.** Apoptosis supposedly elicits Dpp expression. **g.** Two new legs grow out, one from each area of contact beween Wg- and Dpp-expressing cells. One keeps the disc's handedness (R), while the other flips (L). Each branch gets more complete circumferentially

cells on either side of each cell death zone should then trigger a new leg to arise (Fig. 3.11d, g) [173], and the members of each such pair will be of opposite handedness, in conformity with Bateson's rule. Indeed, Wg and Dpp are both spurred to high levels of expression whenever tissue is ablated (by tweaking genes in the apoptotic pathway) [523,1178], and regeneration can be stopped dead in its tracks by blocking Wg [525].

Because Wg and Dpp antagonize one another so fiercely [645,845], it is hard to see how they could emerge in the heart of "enemy territory" without being snuffed out right away. The solution to this riddle appears to be that apoptosis induces amnesia before delivering its coup de grace: it causes cells to "forget" who they are by wiping the memory markers (Polycomb proteins, etc.) from their identity genes [1408]. Thus, a cluster of cells in the Dpp sector, say, could forget its Dpp identity and then freshly "reboot" in either its old Dpp state or a new Wg one [912]. Patches that return to a Dpp state would blend back into the background, yielding a normal leg after healing, while those that reawaken in a Wg state would elicit extra legs. If the choice between these options is random, then no more than 50% of legs in the *lethal(1)ts726* flies should ever be triplex. Indeed, the actual maximum is ~36% [437].

So far so good, but why should triplex legs that arise ventrally diverge instead of converging like the ones that arise dorsally? The secret "superpower" that enables them to finish the triplication process may be their cryptic Dpp sector. Dpp is transcribed (albeit at a low level) in the Wg sector, so it could easily be launched from that platform to an intensity that could sustain outgrowth. There is no comparable sector of Wg in the Dpp area, though Wg still crops up somehow whenever Dpp is suppressed [563,645]. Wg might be expressed so late in the process that it cannot overcome the prevailing Dpp, hence forcing the outgrowth to wither.

Duplications do not pose the same converging-versus-diverging dilemma as the *triplications* because they typically grow to completion with hardly any tapering [443]. Given that the complete (diverging) triplications come from the ventral (Wg) side of the disc, it should come as no surprise that virtually all of the duplications do also. Their etiology likely mimics that of diverging triplications (Fig. 3.11e–g),

← ───

Fig. 3.11. (*cont.*)

as it grows, possibly due to rising levels of Dpp. NB: The two new legs need not fuse as shown here: the center member of a triplication could theoretically fuse with the original leg instead, depending on their relative proximity (cf. [439] for evidence). This scenario echoes Meinhardt's Boundary Model [873], except for his tenet that outgrowths must include A- and P-compartment cells (and be sustained by Hh), which was later disproven [173,444]. In order for the Polar Coordinate Model to explain such triplications, the dying areas would have to stretch perpendicularly before healing so as to bring distant cells into contact [437], but the time needed for swaths of dying cells to be cleared makes such contact unlikely before the onset of metamorphosis (Jack Girton, personal communication), and the fact that implants can triplicate without fully evaginating [440] seems to rule out a need for reshaping. Blocking the Wnt (Wg) pathway was recently shown to cause extra (mirror-symmetric) fore- and hindlimbs in frogs [1248]. After [555].

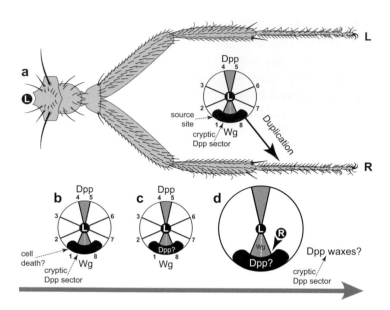

Fig. 3.12. Hypothetical etiology of duplications in *lethal(1)ts726* fly legs. Flies that manifest this phenotype were exposed to 29 °C for 2 days in early larval life. **a**. Left (L) second leg, which has sprouted a mirror-image right (R) duplicate. (The normal R partner is not shown.) The duplicate comes from the sausage-shaped area (black) [443,1096]. Gray sectors are expression domains of Dpp and Wg, and pinwheel spokes denote future bristle rows (1–8; cf. Fig. 3.10). **b–d**. Proposed sequence of events. **b**. Black area marks the presumed region of cell death. **c**. Apoptosis supposedly elicits Dpp expression. **d**. A new leg grows out from the area of contact beween Wg- and Dpp-expressing cells. It is left-handed due to the inverted polarity of its dorsal–ventral axis. NB: In order for the Polar Coordinate Model to explain this sort of duplication, the dying area would have to stretch perpendicularly before healing so as to bring distant cells into contact [438].

with one exception. Based on the elements that vanish at the plane of symmetry, the area of apoptosis must occur along the edge of the disc, thus causing only one Wg/Dpp interface – and hence one extra leg – instead of two (Fig. 3.12).

The basic assumptions behind these etiology arguments were tested by using a laser to kill cells in arc-shaped areas inside *wild-type* (non-mutant) discs. Reassuringly, comparable leg triplications do indeed arise [440]. Moreover, remarkably similar triplications can be induced in wild-type leg discs by using a scalpel to partly bisect the disc on its ventral side – implying that the injured area need not be entirely internal to trigger outgrowth [162].

Hence, the kinds of insects that were depicted in Bateson's book, which possess legs as juveniles (i.e., the hemimetabola), could theoretically be goaded into forming triplications by the kinds of external injuries (cuts or bites) that occur routinely in nature (cf. crustacean battle wounds) [1140]) ... assuming, of course, that they use Dpp and Wg in roughly the same way as flies do. That assumption has been verified in crickets [59].

In the same manner as exhibited by salamanders (Fig. 3.7a–e), triplications can be elicited in cockroaches by transplanting right legs onto left stumps (or vice versa) [377] or by merely nicking the ventral (Wg) side of the leg [130,163,1140]. Overall, therefore, the Dpp–Wg model that explains extra-leg formation in flies seems generalizable to at least some other insect groups [30,543].

Obviously, however, that Dpp–Wg model is not applicable to vertebrates. There the Shh–Fgf8 circuit operates instead – a change of actors, but not of the play. In both taxa the limbs obey Bateson's rule owing to the bipolar nature of their feedback loops. Like the astral microtubules that radiate from the poles of a mitotic spindle, the morphogens at the two ends of the loop diffuse in all directions, so that when extra dipoles intrude into one another's vicinity (due to parasites, injury, apoptosis, etc.), they will bleed across the gap to establish new dipoles of opposite orientation, resulting in the L-R flip-flops of Bateson's rule [564]. Despite the infelicity of these deformities, the cells that construct them are merely executing the cue-driven ("If X, then do Y") instructions provided to them by the genome. Pere Alberch called it "the logic of monsters" [15].

Positive feedback loops are easy to spark but hard to stop (like forest fires). Nevertheless, the extra legs of parasite-infested frogs typically grow to the right size, as do the extra legs of insects. How do limbs, or for that matter organs in general, manage to stop growing so neatly [334,521,1338]? One idea for vertebrate limbs is that the poles of the feedback loop grow so far apart that they can no longer spur one another [1261,1329], but that trick can't work for the Dpp–Wg loop, given their perpetual proximity. Alternatively, timers may be involved [290], and evidence for their usage in vertebrate limbs has been adduced [1084,1103]. For cricket legs (and maybe fly legs as well [762]), the trigger for cessation appears to be the steepness of a proximal–distal gradient within each leg segment [60]. The regulation of growth will be explored in later chapters.

Part II

Flies

4   The Double-Jointed Fly

Flies belong to the phylum Arthropoda, meaning jointed foot. The *Drosophila melanogaster* leg has four tarsal joints, each of which has a beak-shaped ball in a dome-shaped socket (Fig. 4.1). The ball-and-socket joints evolved from simpler articulations in ancestral insects [1256]. They develop as infoldings of the skin, which epitomize the origami strategy (GP-2) so commonly used by animal epithelia [239,485,719,892]. The ball is made first; then the socket is molded to fit snugly [1255]. Cuticle is deposited in laminated sheets [1127] as a 3D printer extrudes plastic [1254]. Tarsal segments are separated by rings of thinner cuticle that allow bending. The tarsus lacks muscles and is operated by a tendon, like a marionette [1180]. Some other insects possess fewer than five tarsal segments, but none has more [719].

In 1982 David Gubb and Antonio García-Bellido described a variety of mutations that disorient the hairs and bristles on fruit flies [496], in the same way that a

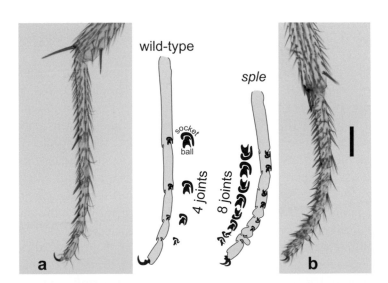

Fig. 4.1. Double-joint syndrome. Left midleg tarsus of a wild-type *D. melanogaster* fly (**a**) compared to that of a fly homozygous for the *spiny legs* (*sple*) allele of the *prickle* (*pk*) gene (**b**). Anterior is up, and ventral is to the left. Scale bar (**b**) = 100 µm. Tracings depict cross-sections with ball-and-socket joints (outlines). The joints have been redrawn alongside enlarged two-fold. The mutant tarsus (**b**) has twice as many joints, and the extra ones are upside-down. The *pk* gene acts in the Planar Cell Polarity (PCP) pathway, and the *pk^{sple}* mutation also affects bristle polarities [566]. Photos by the author.

person might have an unruly cowlick on the back of the head [49]. They mainly studied the effects of the mutations on wings, but my lab was more interested in legs, so we requested their stocks to have a look.

To our amazement the legs of the mutant strains exhibited extra tarsal joints, all of which were upside-down. Why? We did not know. The highest number of extra joints occurs in flies homozygous for the *spiny legs* (*sple*) allele of the *prickle* (*pk*) gene (*pk^{sple}*). Such flies typically have twice the normal number of tarsal joints (eight instead of four), though the basitarsal joint is often incomplete (Fig. 4.1). We published our analysis of the leg phenotypes in 1986, hoping that other researchers could help solve this mystery [566]. It took 26 years for that to happen.

GP-7: The PCP pathway governs cell orientations

In 2012 Amalia Capilla *et al.* pieced together various molecular clues to come up with a clever solution to the double-jointedness puzzle (Fig. 4.2) [180]. Their proposal relies upon an ancient signaling mechanism that encodes spatial vectors, rather than cellular positions [760]. Commonly known as the Planar Cell Polarity (PCP) pathway [459], this toolkit is nearly universal among animals [561].

The PCP pathway gives epithelial cells the equivalent of a magnetic compass. The compass "needle" takes the form of PCP proteins that move to one or the other side of the cell [169] based on morphogen gradients in which the cell is embedded [48,394,1221]. The PCP genes are the same ones that Gubb and García-Bellido had studied under loss-of-function (LOF) conditions. To delve more deeply into this story, some background is required.

The fly leg is a hollow cylinder that inflates from a flat "balloon" during metamorphosis (see GP-2; Fig. 1.6). The proximal–distal (P-D) axis emerges after the anterior–posterior (A-P) and dorsal–ventral (D-V) axes have been specified by Dpp and Wg (see GP-3) [1093]. Given how those morphogens encode cellular positions as graded concentrations (see GP-6; Fig. 3.10), it would be natural to think that the P-D coordinate might use an analogous gradient [554], but that expectation is only partly fulfilled. Positions within the tarsus are specified to some extent by a gradient of the morphogen Vein in the EGFR (epidermal growth factor receptor) pathway [172,390], but leg segments in general rely more on qualitative zones of overlapping gene expression to assess their locations [351,1240], to adopt their identities [4,719], and to pinpoint where to make their joints [937,1049].

Joints are induced to form at precise locations along the P-D axis of the leg by the Notch pathway [114,1256]. That pathway differs from the other modes of cellular communication (e.g., Dpp, Wg, Hh, and EGFR) insofar as its signals – Serrate (Ser) and Delta (Dl) – are not diffusible [72,410]. Rather, Ser and Dl span the membrane of the signaling cells and can only activate Notch by direct cell-to-cell contact. Once the Notch receptor has been thus prodded, it detaches from the membrane of the responding cell and enters that cell's nucleus, where it stimulates the transcription of a cognate subset of genes. In this case, those genes would evoke a joint.

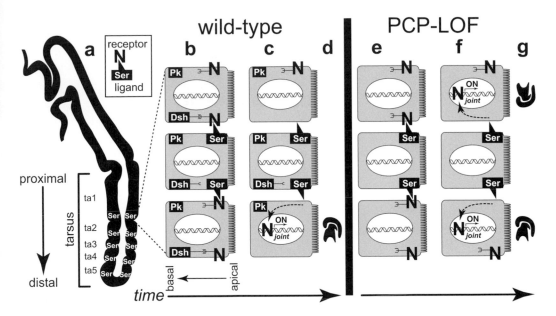

Fig. 4.2. Explanation for why mutations in Planar Cell Polarity (PCP) genes cause extra joints.
a. Partly everted left *D. melanogaster* midleg (cf. Fig. 1.6d). The thick outline represents the
one-cell-thick epidermis in cross-section. Joints form on the dorsal side (to the right). Serrate
(Ser) is expressed at the distal end of each leg segment [1049], as is Delta (Dl; not shown), but
only the tarsal rings (ta1–ta5) are labeled. *Inset:* Abstract ligand–receptor fit between Ser and
Notch (N), both of which are transmembrane proteins. **b–d**, **e–g**. Schematic enlargement of
three adjacent cells (gray rectangles) at the ta1/ta2 juncture (dorsal face), with microvilli on
the apical (cuticle-secreting) face, nucleus (white oval), DNA (double helix), and key
proteins (black rectangles or "N") indicated. **b–d**. Wild-type leg. **b**. Prickle (Pk) and
Dishevelled (Dsh) reside at the proximal and distal sides respectively. Ser binds N in both
abutting cells. **c**. After binding Ser, the inner part of the N protein of the distal neighbor goes
to the nucleus and turns ON joint-creating genes [280], but the N protein of the proximal
neighbor undergoes endocytosis instead due to its link with Dsh. **d**. A joint arises distally but
not proximally. **e–g**. PCP-LOF leg. **e**. LOF mutations in *pk* or other PCP genes disrupt the
asymmetric partitioning of Dsh [1061,1356], thus preventing Dsh from inhibiting N. **f**. Without
Dsh, N proteins of both the proximal and distal cells turn ON joint-creating genes. **g**. A joint
forms proximally as well as distally. This diagram is oversimplified insofar as other PCP
proteins are omitted [459], and distinctions between Pk isoforms are ignored [51]. Adapted
from [114,180,1240].

A ring of Ser and Dl expression arises at the distal end of every leg segment
(Fig. 4.2a), but Ser is more critical for joint induction than Dl [114]. There is no
obvious reason why the Ser-expressing rings should not induce joints above as well
as below, yielding a double-joint syndrome even in wild-type flies [1154]. Indeed, the
ability of Ser to act bidirectionally presents the same geometric dilemma as the
abilities of Shh and Fgf8 to spread bidirectionally in the multi-legged frogs discussed previously (see Fig. 3.8). In that case, the outcome was mirror-image legs (à
la Bateson's rule); here it would be mirror-image joints (Fig. 4.3).

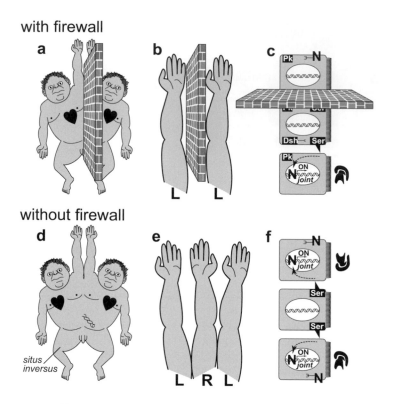

Fig. 4.3. Phenotypic effects of "firewall" failure in various contexts. Barriers to intercellular signaling (cartooned as brick walls) can insulate embryonic fields from one another so as to ensure independent development (**a–c**). Hence, removing those obstacles promotes the emergence of new symmetry planes at unusual sites (**d–f**). Several such symmetry phenomena have remained enigmatic until recently. **a, d.** Falkenberg's rule was formulated in 1919 [1188] and solved in 1996 [783] (cf. Fig. 2.4). It set forth the mystery of why *situs inversus* (reversed asymmetry of the heart, etc.) should arise in 50% of the right twin in conjoined twin pairs where the twins develop side-by-side. **b, e.** Bateson's rule was formulated in 1894 [80] and solved in 2019 [564] (cf. Fig. 3.8). It posed the question of why incipient limbs of the same handedness (e.g., L/L) should evoke an interposed limb of opposite handedness (L/R/L) when they develop side-by-side. **c, f.** Double-joint syndrome was formulated in 1986 [566] and solved in 2012 [180] (cf. Fig. 4.2). It expounded the riddle of why extra, inverted joints should arise near the normal ones when cell polarities are disturbed. NB: It is not barriers per se, but rather the edges of isolated territories that confine the diffusible signals in Falkenberg's rule (perimeter of the embryo) and Bateson's rule (perimeter of the limb-forming domain).

According to Capilla *et al.*, the reason no joint develops above the Ser ring in wild-type flies is that the equivalent of a firewall is erected there by Dishevelled (Dsh) [180,1240]. Dsh adheres to the distal end of each cell, where it binds Notch and removes it from the membrane by endocytosis, thus precluding entry into the nucleus. Under these conditions the Ser ligand may indeed be "yelling" at the cell above it, but that side of the cell is "deaf" because its Notch "ear" has been kidnapped by Dsh. Why does an extra joint arise proximally in pk^{sple} flies? Apparently because PCP complexes cannot remain intact in the absence of Pk [1061,1356]. When Dsh and its compatriots abandon the cell surface, Notch is free to respond to the Ser signal by going to the nucleus, where it activates joint-creating genes [280].

This hypothesis leaves several questions unanswered [561]. Chief among them is why the extra joints should be upside-down, given that the PCP machinery essentially disintegrates in PCP-LOF mutants [1220]. Indeed, one is left to wonder how even the orthodox joints can maintain a normal orientation if they are deprived of the "compass" that lets them orient properly along the P-D axis [1255]. Further questions include:

1. Why don't PCP-LOF mutations elicit an extra joint in ta5, given that there is also a Ser ring at its distal end? Indeed, an extra, inverted joint does arise in ta5 when the homeobox gene *defective proventriculus* is disabled [210,1154].
2. Why don't PCP-LOF mutations cause extra, inverted joints for segments that are proximal to the tarsus? After all, Ser rings are just as instrumental in the formation of those joints [1050].
3. How are PCP-LOF mutations able to elicit extra joints when redundant mechanisms (e.g., EGFR and Fringe) seem to be guarding those sites from just such a possibility [389,719]?

GP-7 tangent: The fly eye

One of the oddest manifestations of the PCP pathway occurs in the fly eye (Fig. 4.4), where groups of cells pivot 90°, but the groups turn in opposite directions on either side of a mirror plane like meshed gears. The eye of a *D. melanogaster* adult has ~750 units called ommatidia – each of which, in turn, contains eight photoreceptor cells (R1–R8) [736]. Ommatidia are hexagonal, while R1–R6 are arranged as trapezoids, with R7 and R8 cradled at their centers. The trapezoids in the dorsal (D) half of the eye point up, while those in the ventral (V) half point down, giving the eye a plane of symmetry along its equator. That plane turns out to be useful for the wiring of the visual system because it allows ommatidia immediately above and below it to combine their inputs to achieve a higher resolution there (in the middle of the visual field) than elsewhere [8,519]. The human eye likewise has a high-resolution fovea near its center, but ours is a spot instead of a stripe [844].

Curiously, however, the fly eye has no such mirror plane until late in development. Photoreceptor clusters are assembled gradually in the wake of a wave called the morphogenetic furrow that sweeps across the eye disc from posterior to anterior (Fig. 4.5) [736]. Receptor cells are added one or two at a time until they reach the full complement of eight cells, though the groups are not yet trapezoidal. The mature clusters in the D and V halves of the eye all look alike at this stage, with their R7 cell pointed toward the P margin. The cells that will become R3 and R4 cells have not yet acquired distinct identities, so they are termed uncommitted "R3/4" cells.

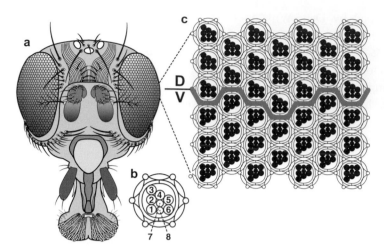

Fig. 4.4. Mirror symmetry in the fly eye. a. Head of an adult *D. melanogaster*. Each eye has ~750 ommatidial facets. **b.** Cross-section of one ommatidium, showing its trapezoid of photoreceptors (numbered) and accessory cells (outlines). **c.** Cross-section of a sample area that straddles the equator (gray line). Note that trapezoids in the dorsal (D) region point up, while those in the ventral (V) region point down. The mirror plane between them defines the equator, which zigzags across the eye in a roughly straight line. After [555,561].

At this stage, the PCP pathway intercedes (Fig. 4.6) on behalf of the gradients of Dachsous and Four-jointed that span the D and V halves of the eye (Fig. 4.5) [25,51,1011]. As in the fly leg (Fig. 4.2), the PCP proteins Dsh and Pk are dispatched to opposite sides of each cell, and Dsh removes adjacent Notch receptors via endocytosis [180]. Depletion of Notch evidently leads to its replacement by the ligand Dl [765], which binds the Notch receptor on the adjacent cell. This inter-action prods the R3/4 cell furthest from the equator to adopt an R4 state, leaving the R3/4 cell nearest the equator to assume the R3 state [713].

Once R3 and R4 identities have been assigned, the trapezoids in the D half turn 90° clockwise, while the trapezoids in the V half turn 90° counterclockwise. We do not yet know how R3 and R4 steer the rotation of the entire R1–R8 cluster, but the EGFR pathway is definitely involved [713], and PCP proteins evidently drive the motion via Rac, Rho, and the actin–myosin cytoskeletal network [1137,1222].

Aside from orienting joints in the fly leg and ommatidia in the fly eye, the PCP pathway also orients bristles on the fly surface [94,297,455], hairs in mammal skin [297,459,869], and stereocilia bundles in the vestibules of mammalian inner ears [857]. Certain areas of the inner ear's utricle and saccule exhibit a mirror plane that is remarkably similar to that of the fly eye (Fig. 4.7).

GP-8: The Notch pathway defines regional boundaries

Not only does the Notch pathway carve joints between leg segments in flies [1240] and other insects [894], it seems to have been generally tasked with drawing

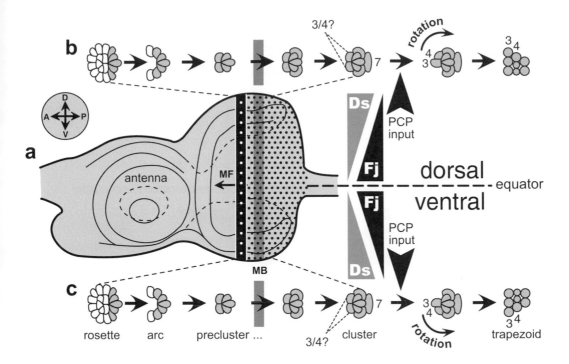

Fig. 4.5. Stages of ommatidial development. a. *D. melanogaster* eye disc near the end of the third larval instar. Inset gives compass directions relative to the body: anterior (A), posterior (P), dorsal (D), ventral (V). The A half of the disc will form the antenna. Ommatidia (dots) arise at regular intervals within the morphogenetic furrow (MF) as it sweeps across the disc from P to A. D and V halves of the eye differ in the expression of regional identity genes within the Iroquois Complex (not shown) [1167]. Reciprocal gradients (at right) of the global PCP regulators Dachsous (Ds) and Four-jointed (Fj) [25,51,677] somehow allocate PCP proteins within each cell (see Fig. 4.6) [854,1278]. **b, c**. Stages in the assembly of the eight-cell photoreceptor cluster [736]. White ovals denote cells that later blend back into the epithelium; gray ovals are incipient photoreceptor cells. Rosette, arc, and precluster stages contain R2, R3, R4, R5, and R8 precursors. The remaining receptor cells (R1, R6, and R7) are born in the mitotic band (MB), which trails the MF. By the time the cluster stage is reached, the entire complement of eight cells is present. However, cells labeled "3/4?" have not yet decided whether to become R3 or R4. That decision is biased by Fj and Ds, which recruit Prickle (Pk) and Dishevelled (Dsh) to opposite sides of each cell in the D and V halves of the eye (see Fig. 4.6). Dsh causes the cell furthest from the equator to activate its Notch receptor so as to become R4 [854]. The polarity of the R3–R4 pair dictates whether the cluster rotates clockwise (**b**) or counterclockwise (**c**) to form a D-pointing or a V-pointing trapezoid, respectively. Modified after [555].

boundaries throughout the animal kingdom [146,623,624]. Thus, it slices body segments in spiders [1214] and defines somites (the precursors of vertebrae and ribs) in vertebrates [97]. It engraves reference lines between the vertebrate midbrain and hindbrain [195], between neighboring rhombomeres [117], and within the thalamic portion of the diencephalon [692]. It separates mouse fingers [988] and marks the margin of both the chick wing [754,1074] and the fly wing (see below).

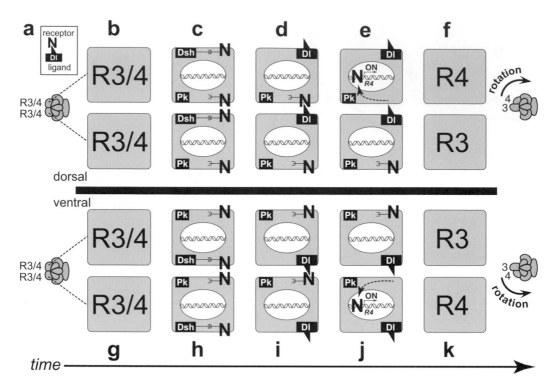

Fig. 4.6. How PCP proteins create mirror symmetry in the fly eye. The horizontal black line is the equator. **a**. Two cells (R3/4) in each cluster of eight photoreceptors (gray ovals; cf. Fig. 4.5) will become R3 and R4 but are indistinguishable at this stage. *Inset:* Abstract ligand–receptor fit between Delta (Dl) and Notch (N), both of which are transmembrane proteins. Dl is used as the N ligand here instead of Ser (cf. Fig. 4.2a). **b–f**. Biasing of Dl-N signaling by PCP proteins in dorsal R3/4 cells. **b**. Naïve R3/4 cells (gray rectangles) in the dorsal half of the eye. Enlarged schematics (from **a**). No apical microvilli are shown (vs. Fig. 4.2) because the apical–basal axis is perpendicular to this plane. **c**. Details of R3/4 cells: nucleus (white oval), DNA (double helix), and key proteins (black rectangles or "N"). Dishevelled (Dsh) resides on the polar (upper) side, while Prickle (Pk) resides on the equatorial (lower) side [25,1222]. Dsh binds Notch [1137], but Pk does not. **d**. Dsh removes Notch from the membrane by endocytosis [180], allowing it to be replaced (via a Dl-N feedback loop [765]) by Dl. **e**. Dl on the lower cell binds N on the upper cell, causing it to be cleaved from the membrane, whereupon it goes to the nucleus and turns ON R4 identity genes. **f**. The N-activated cell adopts R4 identity, the Dl-expressing cell adopts R3 identity, and the cluster rotates in the direction of the 3→4 vector, which in this case is clockwise. **g–k**. Biasing of Dl-N signaling by PCP proteins in ventral R3/4 cells. As in dorsal cells, the N-activated cell adopts R4 identity, the Dl-expressing cell adopts R3 identity, and the cluster spins along the 3→4 vector [655]. However, in this case the direction is counterclockwise. These opposite rotations give the eye its plane of mirror symmetry (see Fig. 4.4). LOF mutations in *dsh*, *pk*, or other PCP genes disorient polarity throughout the eye (not shown) [459]. NB: This hypothetical scenario is an oversimplification insofar as it ignores the distinct properties of Pk isoforms [25] and
the roles of Frizzled dosage [677,713] and Numb expression [315] as biasing agents in the R3-vs.-R4 decision. After [297,854].

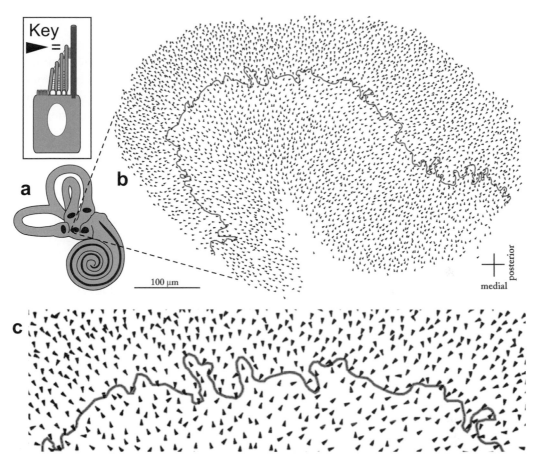

Fig. 4.7. Mirror symmetry in the mammalian utricle. a. Inner ear of a human, which resembles that of a mouse [494]. Black spots are cristae of the semicircular canals plus maculae of the utricle and saccule, all of which have sensory hair cells lining fluid-filled labyrinths. *Inset*: One such hair cell with its staircase of motion-sensitive stereocilia rising to meet a thicker kinocilium. The black spiral is the sensory corridor of the cochlea. **b.** Macula of utricle with ~3600 hair cells. **c.** Enlargement of the central part of the macula. Stereocilia bundles point toward the wiggly line [857], which is known as the LPR (line of polarity reversal). (Saccule vectors point *away* from its LPR [857].) Polarity is controlled by the same PCP pathway that orients ommatidia of the fly eye [857], but we don't yet know whether these hair cells rotate relative to their initial orientations. The vector map is courtesy of Ellengene Peterson [791]. Adapted from [561].

As for why this pathway has been so often conscripted for erecting fences between territories, the reason is not hard to find [36]. As mentioned above, its canonical ligands are transmembrane proteins whose "juxtacrine" range is confined to neighboring cells [263], unlike the diffusible ligands of the other major signaling pathways, whose "paracrine" ranges give them the ability to establish morphogen gradients. Ergo, Notch can etch fine lines. For exceptional situations where an area's perimeter must exceed one cell diameter [114,1240], the Notch pathway can still suffice by having the signaling cells extend ligand-studded filopodia spanning several cells [180,281].

In virtually all of these undertakings the Notch receptor cooperates with a partner protein called Fringe [111]. Fringe glycosylates Notch and prevents it from being activated by Serrate – one of its chief ligands [160]. This post-translational modification attunes Notch exclusively to Delta – its other chief ligand [263]. How these various gears mesh with each other is most clearly seen in the fly wing.

GP-8 tangent: The fly wing

The *Notch* gene was named for incisions that its mutant alleles cause along the fly's wing margin [486], so it came as no surprise when *Notch* was found to be activated there during the larval period [279]. Unlike leg joints, where Serrate (Ser) is the main ligand for Notch (Fig. 4.2), and ommatidia, where Delta (Dl) is the determining factor (Fig. 4.6), both Ser and Dl are involved to equivalent extents at the wing margin, though they stimulate Notch from opposite directions across that line [313].

Figure 4.8 shows where Notch fits into the overall scheme of wing assembly. The wing disc is patterned along two perpendicular axes (cf. GP-3), with a P district expressing Engrailed (En-ON) and a D district expressing Apterous (Ap-ON). The complementary A and V areas are essentially defined by their En-OFF and Ap-OFF states, respectively. The A/P and D/V border zones eventually acquire unique identities that prod them to secrete the morphogens Decapentaplegic (Dpp) and Wingless (Wg), respectively:

1. En causes the morphogen Hedgehog (Hh) to be secreted throughout its P territory. When Hh spreads a few cell diameters into the A area, it turns *dpp* ON along the A/P boundary. Dpp then diffuses to form back-to-back gradients that inform cells of their relative distance from the A/P line.
2. Ap causes Ser to be expressed throughout its D territory, but Ser cannot activate Notch there because Fringe (Fng) is expressed congruently. Only V cells next to the D/V boundary can respond. Cells in the V (Ap-OFF) province express Dl but are somehow deafened; only D cells next to the D/V boundary can react. The stripe of Notch-activated (Notch*) cells turns *wg* ON along the A/P boundary. Wg then diffuses to form back-to-back gradients that tell cells their relative distance from the D/V line.

The overlapping gradients of Dpp and Wg define an oval area that ultimately becomes the wing. The combination of Dpp and Wg inputs causes cells to turn ON the gene *vestigial* (*vg*) [693]. Vg encodes "wingness" identity [705]. Despite the geometrical similarity of the A-P and D-V districts, Notch is only used along one of the two borders. We do not yet know why embryos use Notch to designate certain boundaries but not others.

Thus, the wing is built by a bootstrapping process, starting with regional identities along orthogonal axes (cf. GP-3), followed by boundaries and

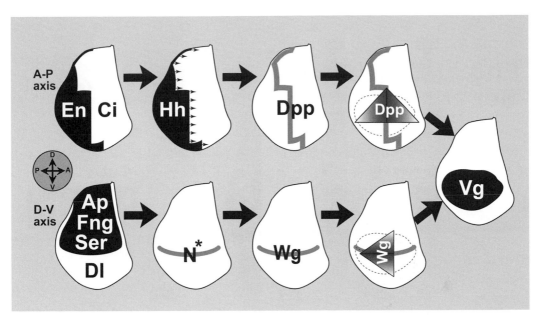

Fig. 4.8. How Notch defines the D/V boundary of the wing disc. Stages in the patterning of a right *D. melanogaster* wing disc (left to right), with A-P steps (above) depicted separately from D-V steps (below). Large arrows indicate causation, small arrows diffusion (cf. Fig. 3.10). The causal chains converge when Dpp and Wg gradients overlap to turn ON the wingness identity gene *vestigial* (*vg*) [693,705]. See text for explanation and additional abbreviations (Ci, Cubitus interruptus; N*, activated Notch). Adapted from [92,561], which offer details of the circuitry.

coordinates (cf. GP-6), with each step causing the next like dominoes. The final pattern is an emergent property of the initial conditions [108]. Self-organizing cascades are used by other discs as well [187,554], and they are widespread among animals in general [57,143,220,460,747].

Figure 4.9 goes into the details of how cells interact along the D/V boundary, as well as showing how the wing margin acquires parallel rows of sensory bristles. Our understanding of the circuitry is incomplete because we do not yet know what agent – here labeled factor "X" – prevents the Notch receptor from responding to Dl signals in the V region of the disc, analogous to how Fng blocks Notch in the D half.

The sensory bristles develop as follows [245]. After Notch is activated by Ser (from the D side) or Dl (from the V side), it turns ON *wg*, and these *wg*-ON cells secrete Wg. Near the peak of its intensity Wg turns ON the gene *achaete* (*ac*) – the licensing agent for bristles – in a band of cells. However, this wide stripe of Ac is split into two thinner stripes due to inhibition from Cut, which is turned ON by Notch. The D stripe goes on to make two rows of bristles, while the V stripe makes one. This triple row is confined to the A half of the wing margin, probably due to inhibition of *ac* by En in the P half of the wing disc.

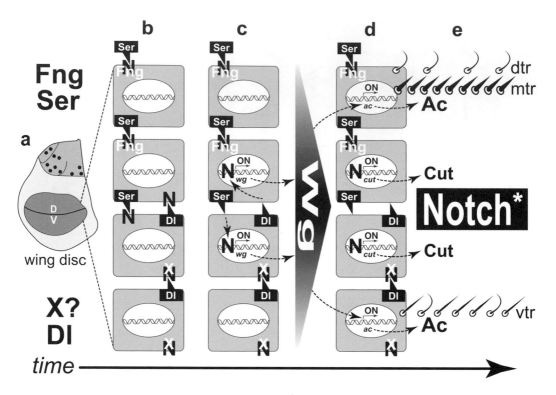

Fig. 4.9. Specification of wing margin by the Notch pathway. a. Mature right *D. melanogaster* wing disc. The future wing (dark oval) and half-notum (light gray) are indicated, as well as the margin (horizontal line) and future macrochaetes (dots). The dorsal (D) half of the disc expresses Fng and Ser (see Fig. 4.8); the ventral (V) half expresses Dl and an unknown agent (X?). **b–d.** Schematic of four cells straddling the D/V line, shown at successive stages. **b.** Fng prevents Notch (N) from reacting to Ser, and X blocks N from reacting to Dl. Hence, the only real signaling occurs in cells where Ser binds N from above or Dl binds N from below. **c.** In those cells the N receptor is cleaved from the membrane, and its inner portion goes to the nucleus, where it turns ON *wg*. Those cells secrete Wg, which forms back-to-back gradients (triangle). **d.** High concentrations of Wg turn ON *ac*, but *ac* is inhibited in cells that abut the D/V line by Cut; *cut* is turned ON by N. This inhibition splits the Ac band into two thin stripes. **e.** Ac enables cells to make sensory bristles along those stripes. A triple row (tr) of bristles ensues: the dorsal tr (dtr) and medial tr (mtr) arise from the upper stripe, while the ventral tr (vtr) arises from the lower stripe [245]. NB: Curiously, the N-activated zone straddling the D/V line initially spans ≥ 10 cells, instead of the eventual two shown here, with the shrinkage in breadth being caused by *nubbin* [245,947]. How can Dl activate non-adjacent cells? Conceivably, the signaling cells use filopodial extensions studded with Dl protein [281]. Redrawn from [561].

The mystery of the upside-down leg joints has led us down a rabbit hole to a Wonderland where Notch signaling erects barriers that balkanize anatomy. In the next chapter we will see how the regions thus created can acquire unique identities by relying on an equally ancient device in the metazoan toolkit: the Hox Complex.

5   The Four-Winged Fly

Flies belong to the insect order Diptera, which literally means two wings. Because two-wingedness defines what it means to be a fly, it must have come as quite a surprise in 1915 when Calvin Bridges found an anomalous *Drosophila melanogaster*

Fig. 5.1. Four-winged fly compared with a normal fly. The *D. melanogaster* mutant (*above*) has four wings and two thoraxes. It is homozygous for three alleles of the *Ultrabithorax* (*Ubx*) gene: *anterobithorax* (*abx*), *bithorax³* (*bx³*), and *postbithorax* (*pbx*). Collectively, they convert the third thoracic segment (T3) into a second thoracic segment (T2) and, in so doing, convert halteres – the bulbous protrusions behind the wings of a normal fly (*below*) – into wings [253]. This phenotype harks back to a dragonfly-like ancestor. Curiously, bithorax traits can be induced in non-mutant (wild-type) embryos by exposing them to ether vapor [445] or heat treatment [572,826,1111] at the blastoderm stage [141,178,588], for reasons we don't fully grasp [425,589,1270,1369]. Photos courtesy of Ian Duncan [327].

with four wings [788]. Its extra wings replaced a pair of club-shaped organs called halteres that flies use as gyroscopes to keep their balance during flight (Fig. 5.1) [301]. The four-wing phenotype epitomizes the phenomenon that Bateson termed "homeosis" – namely, the transformation of one body part into another [787]:

> For the word "metamorphy" I therefore propose to substitute the term **homœosis**, which is also more correct; for the essential phenomenon is not that there has merely been a change, but that something has been changed into the likeness of something else. [80]

Bridges called the mutant *bithorax* (*bx*) after its doubled midsection. The *bx* mutation was later found to be an allele of the gene *Ultrabithorax* (*Ubx*) [831]. By itself *bx* only causes a partial homeosis of halteres into wings, but in combination with other *Ubx* alleles, it converts the entire third thoracic segment (T3) into a second thoracic segment (T2) – or at least the exoskeletal parts thereof [327]. The extra thorax lacks some of the muscles expected for a T2 segment [360,1089], so these four-winged flies cannot fly.

Because halteres evolved from wings, these pairs of appendages are deemed to be serially homologous [1251] in the same way that our arms and legs are serially homologous to one another [1098]. Thus, the human equivalent of a *Ubx*-LOF fly would have arms in place of legs [308,943]. This imaginary phenotype is reminiscent of our chimp-like ancestors, who used all four of their limbs as "arms" for gripping branches. We don't know exactly how the grasping foot of a simian became the walking foot of a hominid [701,797], but we do have a basic understanding of how an insect hindwing became a fly haltere.

There is no better way to understand development than to analyze how it gets modified during evolution, and the reshaping of a wing into a haltere is a case in point. The star of this drama is *Ubx*. *Ubx* contains a "homeobox" motif (named for homeosis) that encodes a DNA-binding "homeodomain." Hence, Ubx can act as a transcription factor to turn any other gene ON or OFF by binding its DNA.

GP-9: Homologous organs diverge via a few key genes

Ubx appears to have been expressed in the T3 segment of primitive insects [1138] as merely a passive bystander [1294]. Then, in the descendant branch that led to dipterans ~225 million years ago, Ubx was able to invade the inner workings of T3's wing-making machinery so as to divert it to a haltere-like state [1055,1206]. The main alterations that resulted from that invasion have been deciphered [560], though we still do not know either the order of their implementation or the rate at which they occurred [555]. They are summarized below:

1. Downsizing. Ubx directly inhibits *dpp* (the main driver of wing growth), thus shrinking the wing down to a mere vestige of its former glory.
2. Inflating the shrunken blade to form a balloon. Ubx directly blocks *blistered* (the "glue" that keeps the surfaces of the wing blade together).

3. Elimination of veins. Ubx directly obstructs *blistered* (a prerequisite agent for vein delineation because it sticks inter-vein areas together).
4. Elimination of bristles. Ubx directly inhibits *scute* (one of two proneural genes that licenses bristle formation, the other being *achaete*).
5. Multiplying sensilla. Ubx enlarges old sensilla nests – possibly by directly activating *atonal* (the gene that licenses sensilla formation).

Thus, despite the stark anatomical differences between a wing and a haltere (Fig. 5.2) the number of genes that had to be altered to convert the former into the latter is surprisingly small. A similar analysis of elytra (rigid covers) versus hindwings in beetles (Coleoptera) confirms that those organs likewise diverged by tweaking few genes [1294,1295,1297]. Our comprehension of how homologous organs in vertebrates evolve is less extensive, but based on what we know so far it appears that the same conclusion applies there with regard to limbs [250,308,1018], digits [1116,1288], faces [1312], jaws [897], teeth [1266], etc. [1355].

Overall, *Ubx* converted hindwings into halteres by becoming a micromanager [10,299,600]. It gives direct orders to genes that operate at all levels of the regulatory hierarchy [577,1005], instead of delegating duties to subordinates in a sequential chain of command, as a chief executive officer might be expected to do [194,1360]. The links between *Ubx* and its target genes were presumably forged gradually as random mutations created sites near the latter genes where the Ubx protein could bind [188,555]. Those sites are called *cis*-regulatory modules (CRMs) because they control the transcription of linked genes on the same chromosome [1130].

A comparable conclusion has been reached for T2 versus T3 legs, which differ less conspicuously from body segment to body segment in wild-type flies [1090]. Hindlegs depart from midlegs mainly insofar as they bear a series of transversely oriented bristle rows ("t-rows") on their tibial and basitarsal segments. Disabling *Ubx* erases the t-rows on T3 legs so that they resemble T2 legs [763,1091], and turning *Ubx* on in T2 legs elicits t-rows so that they look like T3 legs [568,1091,1159]. *Ubx* evidently acquired a direct link to the genes that control t-rows, because *Ubx*-transformed T2 legs also make extra t-rows outside their expected locations [568]. Consistent with its role as a micromanager, *Ubx* is expressed most strongly in those regions of the hindleg that diverge from the midleg, and the same is true for the haltere versus the wing [555].

Hindleg t-rows are used as brushes to clean dust from the wings, and an analogous set of t-rows on the forelegs are used to remove dust from the eyes [1247,1323]. Forelegs must bend forward to scrub the eyes, just as hindlegs must bend backward to scrub the wings, so it is not surprising that the t-rows reside on opposite sides of the forelegs (anterior) and hindlegs (posterior), precisely where they rub against their grooming targets (Fig. 5.3) [1431]. For some unknown reason, t-row bristles in both cases are lighter than normal (yellow vs. brown), and most t-row bristles on the tibia lack bracts.

As the term "transverse" indicates, t-rows are aligned perpendicular to the main axis of the leg, but their bristles still point distally, as do virtually all bristles on the

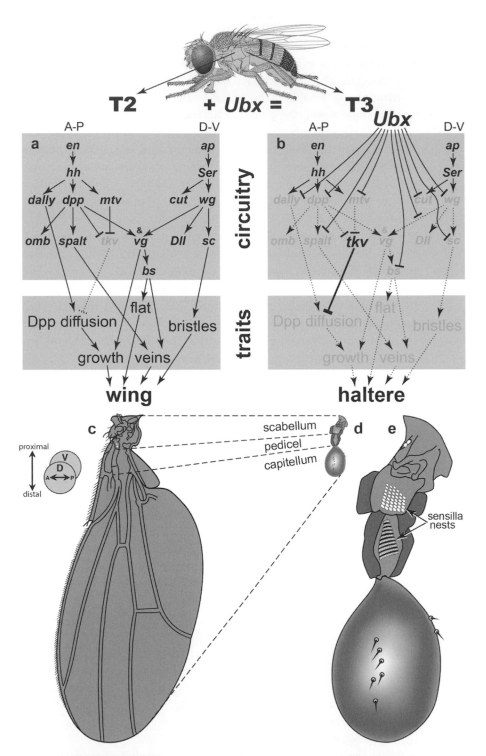

Fig. 5.2. Differences between wing and haltere development. a, b. Genetic circuitry (*above*) responsible for traits (*below*) in dorsal appendages of the second (T2) and third (T3) thoracic

legs. This conformity is attributable to the fact that incipient bristle cells orient their differentiative mitoses parallel to the PCP vectors of the P-D axis, and a similar constraint accounts for why bristles induce bracts on their proximal side (cf. GP-7; Fig. 1.3) [1011].

One glaring exception to this polarity rule is the sex comb – a row of dark, thick, curved, and blunt bristles found only on the male foreleg. This row is parallel (vs. perpendicular) to the P-D axis, and its bracts reside on the dorsal (vs. proximal) side of their respective bristles. This deviant polarity turns out to be due to the fact that the sex comb arises as an ordinary t-row but rotates 90° as it develops [40,567,1289].

The midleg has the simplest pattern of all the legs (no sex comb or t-rows), and it is thought to be the evolutionary ground state for leg identity [190,408]. Given what we have learned about how *Ubx* functions in T3 to create the haltere, it is safe to say that *Ubx* must have steered the hindlegs away from the midleg state to their current condition in T3 by tweaking various genes in leg-making circuits of the genome [568,1091].

GP-9 tangent: The fly foreleg

If *Ubx*'s role is confined to T3, then how did the bristle pattern of the T1 leg arise? The answer is another homeotic gene. Like *Ubx*, *Sex combs reduced* (*Scr*) has a homeobox and serves as a DNA-binding transcription factor, but it acts in the T1 body segment. Disabling *Scr* makes the foreleg look like a midleg, and forcing *Scr* to be expressed in the midleg causes it to develop like a foreleg [559,565]. Interestingly, all bristles on the T1-like midlegs of *Scr*-GOF flies are yellow and bractless like the t-row bristles on the T1 tibia – suggesting that *Scr* enforces the T1 state by affecting cuticle pigmentation and bract induction. Moreover, *Scr*-GOF

←——

Fig. 5.2. (*cont.*)
segments in *D. melanogaster*. A-P and D-V axes start with *engrailed* (*en*) and *apterous* (*ap*), respectively (see Fig. 4.8). Arrows denote activation; T-bars denote inhibition. Faded letters and dotted lines indicate absence or impotence. **a**. Wing development. **b**. Haltere development. *Ubx* inhibits genes at various levels of the hierarchy [388] to transform the wing into a haltere. If *Ubx* is forcibly turned ON in T2, then it usurps the wing-making circuitry of that segment as well to turn the forewings into halteres [1380]. **c**. Upper (dorsal, D) surface of a left wing; ventral (V) surface beneath. Dashed lines denote homologous regions of the wing and haltere. **d, e**. Upper surface of a left haltere, drawn at the same scale as the wing (**d**) or enlarged 5× (**e**). The capitellum serves as an inertial mass [1271], causing the stalk to bend during flight [302], with the resulting torque being sensed by arrays of sensilla in the stalk [446,1035]. Those arrays bear a striking resemblance to the transverse rows of bristles on the legs (see Fig. 5.3). Amazingly, halteres evolved independently in the T2 segment of strepsipterans [866,952]. No drosophilid or strepsipteran species has ever forfeited its halteres and returned to the ancestral four-wing state, but drastic atavisms have occurred for other structures [1132,1381]. Based on [900] and modified from [555,560], which contain gene abbreviations.

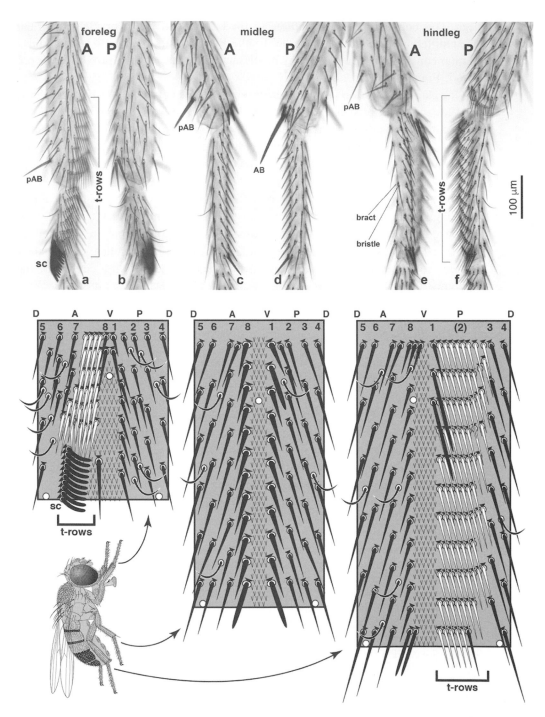

Fig. 5.3. Differences among forelegs, midlegs, and hindlegs. a–f. Right legs (tibial and basitarsal segments only) of wild-type *D. melanogaster* males as viewed from the anterior (A) and posterior (P) side. Apical (AB) and pre-apical (pAB) bristles on the tibia are labeled, and transverse rows of bristles (t-rows) are indicated, along with the specialized t-row known as

causes extra t-rows to develop ectopically – implying that *Scr* induces t-rows directly, just like *Ubx* [1159]. One of those rows was conscripted by *Scr* (acting in concert with the gene *doublesex* [1065]) to become the sex comb (Fig. 5.4) [70,560,728].

Vertical orientation of the sex comb makes sense insofar as it enables a male to grasp the female abdomen more effectively during courtship and mating, like the curled fingers of a human hand [727,852,1277]. Males that are deprived of a comb either surgically or genetically are handicapped relative to their fellow suitors in the competition for mates [949]. Though the adaptive significance of the ultimate pattern may be clear, we do not yet know how the ancestors of *D. melanogaster* stumbled upon the peculiar strategy of cellular rearrangements that they now use to attain that state (i.e., convergent extension) [40,835].

Other *Drosophila* species employ a wide range of odd schemes to construct similar-looking combs (Fig. 5.5) [41,560,727], and the elaborate tricks that some of them use offer an untapped trove of insights into how genes program cells to form patterns during development [835]. Indeed, the intricate maneuvers performed on the tarsi of certain species are as impressive as those of a marching band on parade [835]. They include pivots through various angles, interdigitations of parallel rows, and end-to-end joining of rotated row segments. Given the byzantine complexity of the cellular choreography, the precise alignment of the comb bristles in virtually all of these species is remarkable [587].

GP-10: Hox genes encode regional identities along axes

Ubx and *Scr* belong to an elite group of ~9 homeobox genes that composed a tandem array ~600 million years ago in the ancestor of bilaterally symmetric animals [188,561]. This cluster is called the Hox Complex (abbreviated Hox-C; contraction of homeobox) [296]. Humans have four copies of the Hox-C due to two genome duplications that occurred near the base of the vertebrate clade [177],

←
Fig. 5.3. (*cont.*)
the sex comb (sc). All photos are at the same magnification (see scale bar in **f**). **a, b.** Foreleg. **c, d.** Midleg. The midleg pattern is the simplest and presumed to be the most primitive. Note the thick, dark AB and pAB, which are examples of macrochaetes (big bristles), while the basitarsal bristles are examples of microchaetes (small bristles). **e, f.** Hindleg. Diagrams below are panoramic maps of bristle patterns characteristic of the basitarsal types above, drawn to the same scale, as if segments were slit along the dorsal (D) midline, unrolled, and flattened. In each longitudinal row (1–8), the bristle lengths and bristle intervals increase with distance from the ventral (V) midline. The foreleg has t-rows on its A side, the hindleg on its P side [1159]. In both cases the t-row bristles tend to be lighter (yellow vs. brown), and they abut a stripe of (non-sensory) hairs ("v"s) between rows 1 and 8. Black triangles above bristles are bracts. Mechanosensory (straight) bristles all have bracts; chemosensory (curved) bristles lack them. The three white circles per segment are sensilla campaniformia (stretch detectors). Drawings are idealized insofar as (1) basitarsi are not perfect cylinders, (2) trichoid hairs are not arranged so neatly, and (3) segment widths are exaggerated. From [555,568,1128].

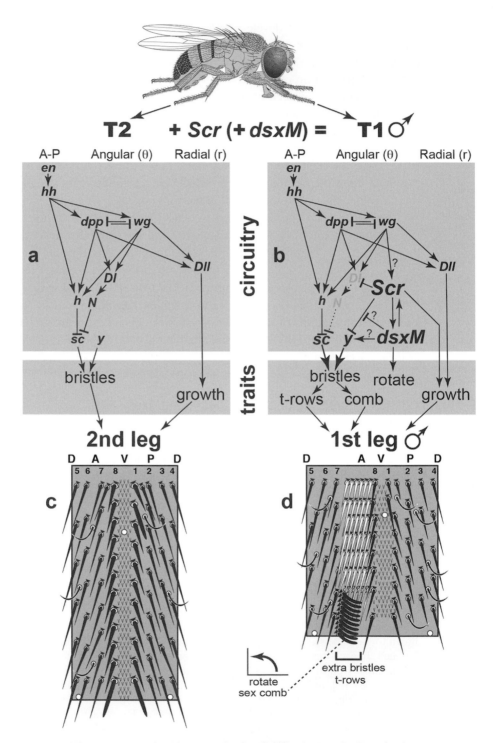

Fig. 5.4. Differences between midleg and foreleg development. a, b. Genetic circuitry (*above*) responsible for traits (*below*) in ventral appendages of the second (T2) and first (T1) thoracic

while *D. melanogaster* has two fragments from a single complex that broke apart in its recent dipteran ancestors (Fig. 5.6) [789,940].

Ubx resides in the fly's Bithorax Complex (BX-C) [832], whereas *Scr* lies in the Antennapedia Complex (ANT-C) [682]. The three genes of the BX-C control T3 and the identity of segments posterior to it, while the five Hox genes of the ANT-C regulate T2 and the segments anterior to it. Indeed, the seriation of genes within the Hox Complexes of most animals matches the order of zones along the A-P body axis. For example, the Hox genes of humans specify the various types of vertebrae along our spinal column (cervical, thoracic, lumbar, sacral, and coccygeal).

This property of "colinearity" was codified in a seminal review of the BX-C in 1978 by Ed Lewis (1918–2004) [328,786,831], who later won the Nobel Prize for his astute analysis [804]. How Hox arrays acquired such congruence in the first place is poorly understood [39,326], but we have at least partly solved the mystery of why it has persisted for so many eons [906]: (1) the genes share control elements whose dispersal would cripple their coordination [622,1130], (2) the complex is associated with a transcriptionally active domain of the genome [837,1073], and (3) vertebrate Hox genes cannot function temporally if they are separated spatially [266,709].

Given how different a fly is from a human (e.g., wings vs. no wings; compound eye vs. simple eye; six legs vs. two), the Hox Complexes must be operating abstractly, above the level of particular structures. Evidently these clusters provide a scaffold of "area codes" that lets the districts along the A-P axis adopt different identities, regardless of the type of metamere (segment vs. somite), germ layer (ectoderm vs. mesoderm), or anatomical incarnation. The same kind of "individuation" occurs in the vertebrate hindbrain, where neural units called rhombomeres are also governed by colinear Hox genes [733].

← _____

Fig. 5.4. (*cont.*)

segments in *D. melanogaster*. As in the wing disc (Fig. 4.8) the leg disc uses *engrailed* (*en*) to launch its A-P axis, but its D-V axis does not use *apterous*. Instead, it employs *wg* and *dpp*, both of which are activated by Hh diffusion across the A/P boundary. Together, *wg* and *dpp* endow cells with angular (θ) coordinates and turn on *Distal-less* (*Dll*), which gives them radial (r) coordinates within a polar coordinate system (cf. Fig. 3.10). Arrows denote activation; T-bars indicate inhibition. Faded letters and dotted lines signify absence or impotence. **a.** Midleg development. **b.** Foreleg development. *Scr* inhibits signaling of *Delta* (*Dl*) to *Notch* (*N*) between rows 7 and 8 so as to boost *scute* (*sc*), spur growth, and sprout bristles [1159]. It also inhibits *yellow* (*y*), thereby turning the novel t-row bristles yellow. Larger font sizes and bigger arrowheads denote increased expression. A male-specific splice variant of the sex-determining gene *doublesex* (*dsxM*) is co-expressed with *Scr* distally, and together they sculpt the last t-row into a sex comb and cause it to rotate [1263]. At that spot *dsx* overrides Scr's inhibition of *y* and evidently stimulates *y* directly as well to make the comb bristles black. Like *Ubx* (Fig. 5.2), *Scr* and *dsx* operate as micromanagers [727]. **c.** Panoramic map of bristles on the midleg basitarsus (cf. Fig. 5.3). **d.** Panoramic map of bristles on the foreleg basitarsus. Changes relative to **c** include: (1) insertion of t-rows, (2) reduction in t-row pigment, (3) conversion of one t-row into a sex comb, (4) increase in comb pigment, and (5) rotation of the comb. We do not yet know how *Scr* shortens the basitarsus. Based on [727] and modified from [560], which contains gene abbreviations.

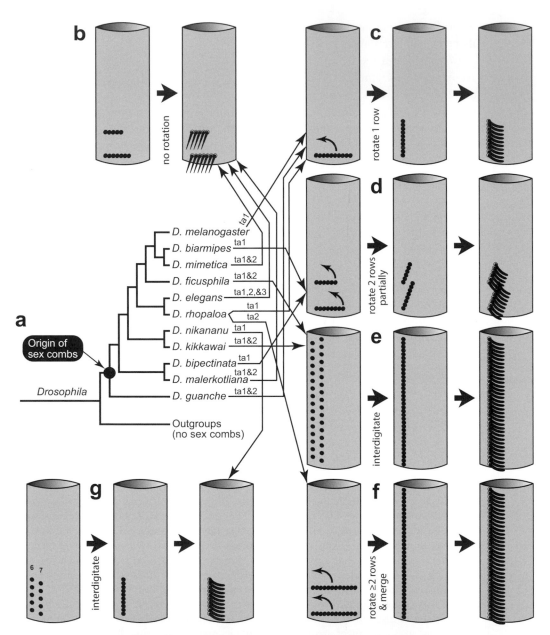

Fig. 5.5. Diverse strategies of sex comb development in *Drosophila*. a. Phylogeny of species in the genus *Drosophila* (subgenus *Sophophora*) that possess sex combs on the male foreleg [727,729]. Arrows connect species (**a**) with the processes used to craft their sex combs (**b**–**g**). Labels above arrows denote tarsal segments (ta1, ta2, etc.) that exhibit the designated style of development [1262]. **b**–**g**. Modes of sex comb formation. Each cylinder represents a tarsal segment from a right foreleg, with black dots marking bristle cell precursors. Bracts are omitted, as are bristles outside the sex comb proper. Curved arrows describe the arc of rotation. **b**. Combs that do not rotate have less distinctive, albeit pigmented, bristles – i.e., tapered and straight, instead of blunt and curved. **c**. The style of rotation in *D. melanogaster*

GP-10 tangent: The frog leg

Evolution has even deployed the Hox-C apparatus well beyond its original head-to-tail domicile – most notably in the tetrapod limb [188], where Hox genes encode bone identities along the proximal–distal axis, as well as specifying finger identities across the hand. The frog leg offers a case in point.

The arms and legs of tetrapods develop from limb buds that grow out from the embryonic flank and install bones in a proximal–distal sequence (see Fig. 3.8). The final limb is organized into three sections: the stylopod (humerus or femur) arises first, followed by the zeugopod (radius–ulna or tibia–fibula), and finally the autopod (wrist–hand or ankle–foot) [1059]. The identities of those sections are dictated by five contiguous genes (9–13) in the Hoxd complex, and to a lesser extent by paralogous genes in the Hoxa complex (Fig. 5.7) [1423]. The fact that the Hoxb and Hoxc complexes are not included suggests that the Hox mode of limb development did not evolve before the vertebrate Hox cluster quadrupled [906], since otherwise all four clusters should have retained some residual involvement.

The mode of action of the Hox Complexes in the limb differs significantly from its function along the body's A-P axis [194]:

1. No homeosis. Null mutations cause reduction or loss of skeletal elements, but no transformations [1423] – e.g., mice that are mutant for both *Hoxa11* and *Hoxd11* lack a radius and ulna but have no other bones there instead [274].
2. Reverse colinearity. The gene at the posterior end of the Hoxd complex (*d13*) is solely expressed in the most anterior digit (Fig. 5.7b), while its more anterior neighbors (*d10–d12*) are expressed more posteriorly [907,941].

During limb development, the lowest level of Hox expression occurs in the wrist and ankle [1059], and this paucity of Hox gene products may be keeping these bones in a stunted state [1406]. Interestingly, a few mammals elongated one of their wrist bones to produce a sixth finger [391], to aid in digging (moles) [706,1108] or in stripping bamboo (pandas) [275,469], but we don't yet know how they did so [1406]. Likewise, a few mammals elongated two of their ankle bones to make an extra tibia and fibula to aid in jumping (tarsiers and galagos) [344,1208], but again we don't know how [406,1085]. Fortunately, frogs also accomplished the latter feat

---

Fig. 5.5. (*cont.*)

is also used by a few distantly related species. **d**. Rotation of more than one transverse row. **e**. Bristle cells from rows 6 and 7 interdigitate to form a comb that spans the whole segment (cf. **g**). **f**. Rotation of parallel rows that then merge to form a segment-spanning comb. An extreme version of this style is used by a species outside the *Drosophila* genus: *Lordiphosa magnipectinata* rotates and merges ~6 t-rows to make a single ta1 comb [41]. **g**. Same interdigitation mode as in **e**, but here the comb is confined to the distal part of the segment. The final comb looks remarkably like the sex comb of *D. melanogaster*, but it is created by an entirely different process. Adapted from [560], which contains gene abbreviations.

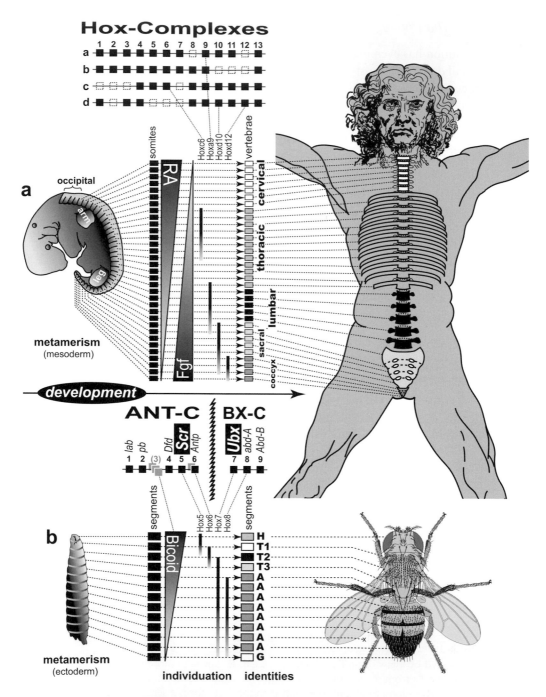

Fig. 5.6. Hox Complexes of humans and flies. a, b. Stages of development (left to right) are metamerism (rectangles), regionalization (gradients), individuation (Hox zones), and differentiation (shades of gray), starting with a 31-day human embryo (**a**) or a fly larva (**b**). The A-P axis is divided into mesodermal somites (**a**) or ectodermal segments (**b**). Positional information is specified by retinoic acid (RA) and Fibroblast growth factor (Fgf) (**a**) or

in an even more spectacular way, and the genetic basis for that transformation has been analyzed in some detail.

In 1998 researchers reported that *Hoxa11*, which specifies the identity of the tibia and fibula (along with *Hoxd11*), is expressed for a longer time in the frog's leg bud than in its arm bud [120]. This delay evidently allows that gene to remain active when the ankle bones start to be specified after the tibia and fibula arise [1068], thereby extending its influence down the P-D axis so as to alter the identity of those bones (Fig. 5.7c) [185,1305].

Based on the various venues where Hox genes operate (body axis, nervous system, or vertebrate limb), an apt analogy for the Hox-C might be a train of colored boxcars that can carry virtually any contents (metameres, rhombomeres, or bones). However, given what we've learned about the roles of *Ubx* and *Scr* in fly anatomy, we cannot view the boxcars as merely inert containers. Rather, each color would come to permeate its contents and enable them to diverge (i.e., individuate) over time by manipulating downstream genes that affect differentiation at the cellular level [1128,1159]. The following quotes are from James Castelli-Gair [193] and his mentor Michael Akam [10], who showed how Hox genes affect various cellular activities (mitosis, apoptosis, etc.) that have nothing to do with segmental (≈ boxcar) identity per se (boldface added):

> **Hox genes are not controlling segment identity, but cellular behavior** that will result in a certain segment morphology. This is what Hox genes do in unsegmented organisms like *C. elegans* [1106] and is probably what they did in the common ancestor of all metazoans. Segment identity is a subjective concept that originates from the observation that in a particular species, a number of cell characteristics are always associated in a given segment. [193]
>
> Some transcription factors appear to be specialists: they specify a particular fate or behavior whenever they are expressed in a cell; the myogenic factors might approximate this role, for example. The *Hox* gene products lie at the other extreme: they are versatile generalists. They operate in many different cell and tissue types, where they modulate, sometimes dramatically but more often subtly, a wide range of developmental processes. **In each of these cell types, expression of a *Hox* gene means something different – to divide or not to divide, to make or not to make a bristle, to die or not to die.** In any given lineage, that meaning probably changes several times during development, in response to hormonal and other developmental cues. [10]

Fig. 5.6. (*cont.*)

Bicoid (**b**). Hox genes provide "area codes" for tagmata (groups of metameres) by overlapping zones of transcription (vertical bars that fade posteriorly) in the same order as their sequence along the chromosome; a few expression zones are sketched to illustrate this colinearity. **a**. Humans have four Hox Complexes, though some genes (dotted outlines) have been lost. Tagmata include cervical, thoracic, lumbar, sacral, and coccygeal vertebrae. **b**. The single Hox-C of flies is split into a Bithorax Complex (BX-C) that has *Ubx* (*Hox7*) and an Antennapedia Complex (ANT-C) that has *Scr* (*Hox5*). *Hox3* and *Hox6* begat paralogs (gray squares) that no longer provide area codes, though two of them (*bicoid* and *ftz*) are still involved in A-P patterning. The story of how Ed Lewis discovered the BX-C is as fascinating as the story of how Howard Carter discovered King Tut's tomb [803,804]. Tagmata are head (H), thoracic segments (T1–T3), abdomen (A), and genitalia (G). Adapted from [561].

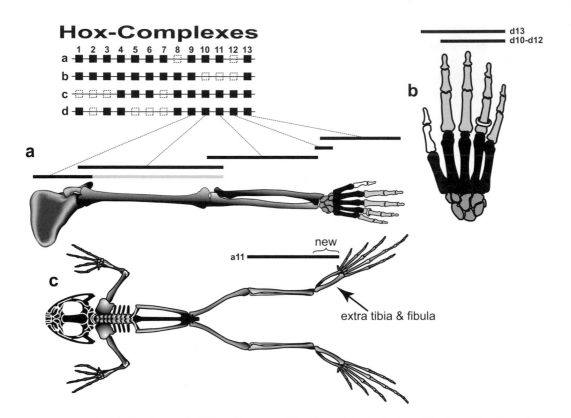

Fig. 5.7. Deployment of Hox-C genes within the vertebrate limb. a. The four Hox Complexes (39 genes) of placental mammals (above). *Hoxd9–13* are expressed in nested domains (lines) along the mouse limb (human left arm shown instead), together with their *Hoxa9–13* paralogs (not shown) [290,1423]. **b.** Hox gene expression in a human left hand. The "Hox code" of the thumb is *d13* ON and *d10–d12* OFF [77,295,557]. Hence, the thumb can be converted into an ordinary finger (i.e., homeosis) by forcing it to express *d12* (with its native *d13*) [712]. Null mutations in *d13* have little effect due to the ability of remaining *Hoxd* genes to substitute for *d13* [710], while partial-LOF mutations in *d13* cause polydactyly, syndactyly, and interdigital webbing instead of homeosis [69,152,1420]. (See [290] for how digit identities get assigned and [292] for technical difficulties in assessing relative roles of adjacent *Hoxd* genes.) **c.** Frog skeleton. Two of the nodular ankle bones evidently evolved to form an extra tibia (tibiale) and fibula (fibulare) by extending the expression of *Hoxa11* distally ("new") beyond its former limit. NB: The frog deviates from the vertebrate blueprint in other respects besides its extra tibia and fibula – notably, in its reduced number of vertebrae [514,689]. See [382,1368] for differences in *Hoxd9* expression between fore- and hindlimbs (shaded extension of the bar in **a**). Based on [557,1015,1076].

This boxcar metaphor leads to the broader concept of a "selector gene" [260], which governs a "compartment" of the body [396,399], regardless of whether the gene or the compartment belongs to any sort of linear series [751].

Just as the ON or OFF states of *Ubx* and *Scr* control the T3 and T1 metameres, respectively, the ON or OFF states of the gene *engrailed* (*en*) subdivide those segments into P vs. A compartments (Fig. 4.8) [1249], the ON or OFF states of *apterous* (*ap*) subdivide the wing into D vs. V compartments (Fig. 4.8) [118], and the ON or OFF states of Iroquois-Complex genes subdivide the eye into D vs. V compartments (Fig. 4.5) [196,1167]. All of these genes qualify as selector genes [840]. Hence, the body of a fly – and to some extent of an animal in general – can be envisioned as a quilt, whose variegated patches are regulated by selector genes acting singly or in combination. The mystery of the four-winged fly has led us to the deep concept of area codes, which is explored more fully – via a different gene complex – in the next chapter.

6 The Naked Fly

In 1916, a year after he found the *bithorax* mutant, Calvin Bridges discovered a *Drosophila melanogaster* that was missing a few bristles from its scutellum (the hind part of the thorax), so he named this new mutant *scute* (*sc*). That same year Alexander Weinstein, another student in Morgan's lab, came across a fly that was missing bristles in front of the scutellum; he called it *achaete* (*ac*), meaning "no bristle" in Latin [149].

Just as the *bithorax* mutation led to a trove of insights into how genes dictate anatomy, the *ac* and *sc* mutants would offer researchers a comparable pirate's map to a similar bounty of buried treasure [397]. The *ac* and *sc* genes reside in the Achaete-Scute Complex (AS-C) along with two other homologs, all of which encode proteins in the basic-helix-loop-helix (bHLH) family of transcription factors [369]. Both *ac* and *sc* are termed "proneural" because they enable ordinary epidermal cells to become bristles, one of whose component cells is a sensory neuron (as seen in Fig. 1.3) [55].

Most *ac* and *sc* mutations only remove subsets of bristles, because these genes are partly redundant [842], but the doubly-null mutant *In(1)ac³ sc¹⁰⁻¹* looks bald (Fig. 6.1) because proneural activity is suppressed throughout the fly epidermis.

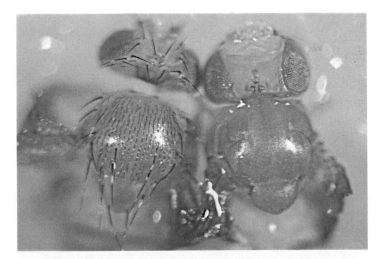

Fig. 6.1. Wild-type fly and *ac³ sc¹⁰⁻¹* mutant fly. The *D. melanogaster* mutant (*right*) is a male whose X chromosome has null alleles for the *achaete* (*ac*) and *scute* (*sc*) "proneural" genes, at least one of which must be functional in order for skin cells to become bristles, so the fly looks naked [1333]. Photo by the author. (A black and white version of this figure will appear in some formats. For the color version, please refer to the plate section.)

"*In(1)*" denotes an inversion (*In*) of the X (*1st*) chromosome, where the AS-C maps. Indeed, most of the lesions recovered from this locus over the years have been inversions or deletions, rather than point mutations [842], and the resulting phenotypes have varied greatly from one allele to the next, depending on the DNA breakpoint sites.

GP-11: Body areas are specified by *cis*-enhancers

In 1995 Juan Modolell and his lab group in Madrid finally figured out why different breakpoints erase unique subsets of bristles in certain regions of the fly surface [176,454]. It so happens that the AS-C is riddled with short stretches of non-coding DNA known as *cis*-enhancers [35,166,815], which cause *ac* and *sc* to be expressed at specific spots within the fly skin [899]. Figure 6.2 shows the "switch-board" of connections between the regional enhancers of the AS-C and the macrochaetes (large bristles) of the head and thorax they designate. Effectively, each enhancer provides an area code for targeting *ac* and *sc* expression.

How do *cis*-enhancers function as area codes? The answer is that they bind upstream transcription factors which define those sites [453]. This sort of logic was discussed before with regard to the triple row of bristles along the wing margin (see Fig. 4.9). In that case, the Notch receptor is activated along the D/V boundary, turning ON *wg* and causing release of the Wg morphogen. The Wg signal diffuses in both directions to activate its own receptor, prompting a down-stream transcription factor to turn ON *ac* and *sc* above a certain level of Wg. This broad Ac-Sc stripe then gets carved into two narrower stripes by the transcription factor Cut after *cut* (like *wg*) is turned ON in a thin band by Notch. Ultimately, the dorsal Ac-Sc stripe makes two rows of bristles, while the ventral stripe makes one. *Voilà!* A triple row!

Each enhancer in the AS-C responds to its own "Venn diagram" of inputs by applying this sort of combinatorial (AND, OR, or NOT) Boolean logic to an overlapping array of regional signals [385,453,1058]. Those signals form a virtual landscape [1067] that Curt Stern, one of Thomas Hunt Morgan's most famous protégés, termed a "prepattern" [1202,1203,1290] (boldface added):

> The epithelium evidently contains precise positional information. This is thought to be embodied in **a "prepattern" constructed by a combination of transcriptional activators and repressors distributed heterogeneously and in different landscapes.** These prepattern factors would be present in domains broader than the *ac-sc* proneural clusters. *as-sc* would be activated only at sites with combinations of prepattern factors appropriate for productive interaction with an AS-C enhancer. Since isolated enhancers direct expression at only one or very few proneural clusters, it follows that each position-specific enhancer is tuned to respond to a different combination(s) of prepattern factors. Thus, according to this model, the AS-C enhancers "read" the positional information laid down by the partially overlapping distributions of prepattern factors ... [899]

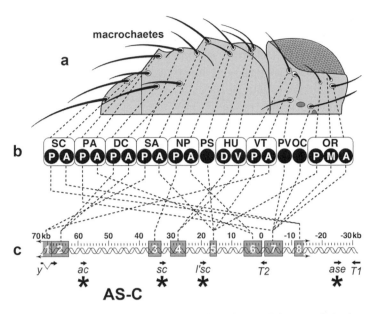

Fig. 6.2. Achaete-Scute Complex (AS-C). The AS-C is one of the best examples known of a genomic "global positioning system" (GPS) device [198,422,423]. **a.** Left half of the head and thorax of a wild-type fly (dorsal view), showing only the 20 macrochaetes (large bristles). **b.** Black circles denote macrochaetes in order from back to front, with relative positions of the members in each pair or triplet marked as P (posterior), A (anterior), D (dorsal), V (ventral), or M (middle) within rounded rectangles. Abbreviations: SC (scutellars), PA (post-alars), DC (dorsocentrals), SA (supra-alars), NP (notopleurals), PS (presutural), HU (humerals), VT (verticals), PV (postvertical), OC (ocellar), OR (orbitals) [802]. **c.** The AS-C spans ~100 kb near the tip of the X chromosome; "0" is an arbitrary reference point. Asterisks mark the four AS-C genes – *achaete* (*ac*), *scute* (*sc*), *lethal of scute* (*l'sc*), and *asense* (*ase*) – that encode homologous bHLH proteins. None of them (arrows) has introns. *T1* and *T2* are unrelated transcription units, and *yellow* (*y*), which has a single intron (bent line), controls bristle pigmentation but appears to have its own set of enhancers [419,451,666,1163]. Dotted lines depict a genotype–phenotype "switchboard" connecting *cis*-enhancers (shaded boxes) along the DNA (double helix) with macrochaetes they control. Enhancers cause *ac* and *sc* to be expressed in "proneural clusters" [254] that give rise to the indicated bristles – evidently via DNA looping that brings each enhancer into direct contact with *ac* and *sc* promoters [1043,1317]. Double arrows indicate unknown limits. Among all of the enhancers the circuitry of the DC enhancer is best understood [386,453,847]. NB: A separate set of prepattern factors acts redundantly with these *cis*-enhancers to ensure an accurate final pattern [1313]. Microchaetes (omitted here) are smaller than macrochaetes (see Fig. 5.3) because they arise later [1164] and have less time to undergo the endoreplication cycles needed for larger shafts [336,684]. Macrochaetes also vary somewhat in length (e.g., pDC vs. aDC), and their sizes are also proportional to the extent of polytenization that they undergo after their sensory organ precursor (SOP) cells finish mitosis [1027]. Higher dipterans evolved bigger bristles by shifting their inception to earlier stages via a novel enhancer [424,1170], and this particular pattern has been conserved for ≥ 50 million years [488,1169,1225]. After [454,555,899].

Why are the enhancers of the AS-C scrambled relative to the sites that they control, in contrast to the Hox-C, where the genes are colinear with the anatomy (as shown in Fig. 5.6) [392]? One obvious reason is that the pattern regulated by the AS-C is *two*-dimensional, not linear, so it would make little sense for its enhancers to conform to only one axis of the pattern. Moreover, these complexes differ in the nature of their modular units: macrochaetes elicited by the AS-C are virtually identical, whereas metameres managed by the Hox-C diverge based on the Hox gene involved [453]. Hence, the various AS-C enhancers only need to tell *ac* and *sc* to "Turn ON here!", so their order along the DNA is moot, but Hox genes govern different compartments, so the enhancers for each gene must be insulated from those for neighboring genes [255]. Most important, the Hox genes (of vertebrates at least) are transcribed sequentially in time, like rosary beads spooled between one's fingers, and their spatial colinearity may have arisen as an "afterthought" [709]. In any event, the chaotic layout of the AS-C is the rule for animal gene complexes in general [1099], while the colinearity of the Hox-C is the exception [34].

Where did gene complexes like the AS-C and Hox-C come from? They began as single genes that underwent tandem duplications over millennia [62,82,939], with the resulting paralogs persisting despite an initial redundancy because they adopted divergent functions [596,939,1092] or synergized robustly [842]. Clusters can resist the kind of fragmentation seen for BX-C/ANT-C if *cis*-enhancers arise so pervasively that they cannot be torn from their cognate genes without disabling their whole complex [622,1130,1218]. The modularity of enhancers allows them to accrue easily [35,730] because they can be inserted harmlessly [165,321,415]. The evolution of the AS-C [649,939,1351] has been traced as extensively as that of the Hox-C [326,938].

GP-11 tangent: The naked ape

Humans are nearly as naked compared to our simian cousins as null mutants of the AS-C locus are relative to wild-type flies. Not surprisingly, anthropologist Desmond Morris called *Homo sapiens* "the naked ape" in the title of his 1967 best-seller [916]. In fact, many areas of our skin that appear to be naked (e.g., our forehead) are not actually devoid of hairs, but rather have a subtle "peach fuzz" called vellus. The only truly hairless parts of our bodies are our palms, soles, lips and parts of our genitalia [629,850].

Why did our hominin ancestors lose so much of their fur? Our best guess is that fur became a liability when they started running long distances and began to suffer from overheating [630]. Under those conditions any mutations that lessened our insulation would have spread through the population. Which genes were impacted? We don't know. No hair-affecting genes have yet been found among the loci that show evidence of selective sweeps [348].

Do we use an AS-C-like system of area codes to specify the hairs in our skin? If the AS-C were as universal among animals as the Hox-C, then we should expect to find one or more clusters of *ac*- and *sc*-like genes somewhere in our genome, and they should operate similarly to the fly's AS-C. We do have *ac* and *sc* homologs [1293], but they have nothing to do with hair formation [1232]. That should come as no surprise because mammal hairs evolved separately from insect bristles and bear little resemblance to them histologically, developmentally, or genetically [611,1411]. Functionally, both kinds of setae act as tactile sensors, but mammal fur provides insulation [629], while insect bristles serve as anti-wetting agents, even going so far as to enable bugs to walk on water [366].

Nevertheless, the question remains as to whether we use the same kind of logic to allocate areas within our skin. What would we expect the imaginary "Hair Headquarters" (HHQ) of our genome to look like, based on what we know about the AS-C? Our HHQ should presumably have at least some of the following features:

1. A single locus with only a few redundant master genes.
2. A collection of eight-or-so region-specific *cis*-enhancers.
3. A scrambling of the enhancers relative to the areas they control.
4. A loss-of-function (LOF) null phenotype involving hairlessness.
5. A gain-of-function (GOF) phenotype involving excess hairiness.
6. A nearby pigment gene that might share a subset of enhancers.

The sixth point is intriguing. Bristles vary in color from yellow (e.g., t-rows) to black (e.g., the sex comb), and the master gene for cuticle coloration (*yellow*) happens to be located within the limits of the AS-C [419,451]. For humans, it would make sense if our master gene for pigmentation – *MC1R* – were embedded in our Hair Headquarters since, in principle, it could then share *cis*-enhancers so as to "paint" our hairs at the same time that they are being "planted" [726]. In this way the HHQ could kill two birds with one stone, so to speak. Men's beards can differ starkly in color from their scalp hair, and other body regions vary independently in color as well [631,886]. Of course, there are still more variables to consider because hair not only varies from site to site in density and color, but it also varies in length, texture, stiffness, curl, and polarity [7,690,1122,1377]. Most of these properties remain enigmatic [104,714,807].

Disappointingly, criterion #6 seems unlikely to be satisfied. No hair-affecting mutations have been mapped to chromosome 16 near *MC1R* (q24.3), which encodes the receptor for melanocortin hormone [591,800] and serves as our (non-homologous) counterpart of the fly's *yellow* gene.

The only genes that overtly satisfy criteria #4 and #5 are members of the Wnt intercellular signaling pathway [383,1120,1433]. Like the fly's *wingless* gene, which belongs to this same family, vertebrate *Wnts* encode diffusible proteins that function as morphogens to establish patterns of structures [1316]. The key evidence for the involvement of *Wnts* in hair development is given below based on data from mice, where the terms "hypertrichosis" and "hypotrichosis" mean excess or missing hair, respectively.

1. *Wnt*-GOF phenotype: hypertrichosis. Overexpression of the stabilized protein β-catenin, which relays the Wnt signal to the nucleus, induces extra hair follicles in embryos [934,1432] and in adults [813].
2. *Wnt*-LOF phenotype: hypotrichosis. Ablation of β-catenin during skin development blocks inception of hair placodes [610], as does inhibition of Wnt signaling via Dickkopf1 [28].

Aside from its ability to increase or decrease the amount of hair, the Wnt pathway also regulates hair spacing [1160], hair differentiation [885,1147], and hair regeneration during wound healing [628]. Moreover, its nuclear effector (Lef-1) binds directly to regulatory DNA sequences at 13 keratin genes that are involved in hair outgrowth [270,1436]. No other signaling pathway comes anywhere close to being so intimately instrumental in the process of hair development [1075].

A central role for Wnts in hair patterning was confirmed by a genome-wide association study of 80 dog breeds. The key gene responsible for bushy mustaches and eyebrows (*R-spondin2*) turned out to be a Wnt regulator [170]. Even the previously mentioned possibility of a link between hair pattern and hair color (criterion #6) finds some support: excess expression of the Wnt transducer β-catenin causes early hair pigmentation [1432].

There are 19 *Wnt* genes in mice [1316], two of which are critical for hair development: *Wnt10a* and *Wnt10b* [1433]. Humans also have 19 *Wnt* genes (Fig. 6.3) [1316], and LOF mutations in our *Wnt10A* gene cause sparse hair in the scalp, body, eyebrows, and eyelashes [132]. A mild phenotype is what we would expect for *Wnt10A*-LOF if *Wnt10A* and its *Wnt10B* doppelgänger act redundantly to compensate partly for one another, as do *ac* and *sc* (criterion #1).

Might *Wnt10A* be serving as our command post for hair patterning? If so, then at least some hair syndromes should map to its *cis*-enhancers (criterion #2). The most pertinent mutations would be those that cause extra or missing hair in certain skin areas (criteria #4 and 5) [400]. The best-characterized hyper- and hypotrichosis disorders are charted in Figure 6.3.

Put simply, the issue here is: How is the two-dimensional "world map" of hairy versus smooth territories in our skin controlled by our one-dimensional genome? How do genes balkanize our surface into islands of hair in an ocean of epidermis? What, for example, is the DNA address for our scalp? Useful clues to this coding riddle are scattered in the literature of human genetics, but they have not yet been pieced together into a coherent model of gene circuitry [198].

Genetic diseases in general are cataloged at the Online Mendelian Inheritance in Man database (OMIM; www.omim.org) and are identified by a unique MIM number. The most common hair-loss affliction is male-pattern baldness (MIM 109200). Nearly 50% of men show hair thinning by age 50 [1097,1165]. The main scalp areas affected are the temples, vertex, and crown [968].

Surprisingly, these same three types of balding are found in certain species of monkeys as well (Fig. 6.4) [150,886]. This commonality implies that our hominin forebears inherited these area codes (*cis*-enhancers?) from their monkey ancestors

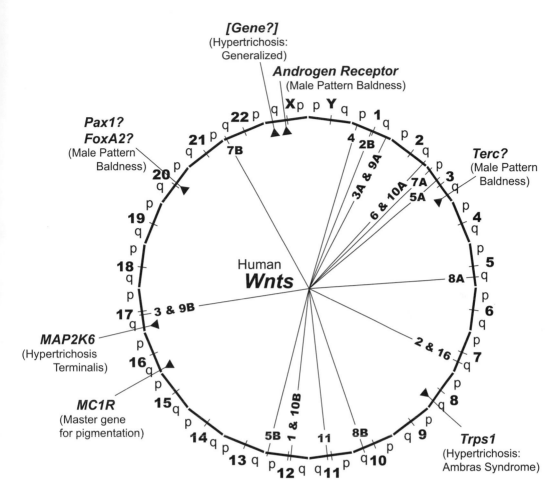

Fig. 6.3. ***Wnts* and other hair-related loci in the human genome.** Our 23 chromosomes are depicted as facets of a wheel, with "p" and "q" arms having equal length and tick marks denoting centromeres. Our 19 *Wnt* paralogs are indicated by spokes, five of which show pairs in close proximity [698]. A few genes whose mutant alleles affect the amount of hair or pigmentation are mapped (arrowheads). None of them coincides with a *Wnt*, but one or more *Wnt* loci might still conceivably harbor a "circuit board" of *cis*-enhancers [136] that delimit areas where terminal hair is allowed to grow within our skin (see text).

long before they began to lose fur from the rest of their body. Interestingly, male-pattern balding entails a reversion of terminal hair back to a vellus state [636] – the very same process that curtailed most of our fur during evolution [1371].

    Three major loci contribute to male-pattern baldness [581,868], none of which, however, is near a *Wnt* gene. One is the androgen receptor on our X chromosome, which explains the sex bias of the baldness trait [1045]. Another is a section of chromosome 3 (q26) containing the gene *Terc* that encodes the RNA moiety of telomerase [1179], which might explain balding's late age of onset, since telomeres shorten as we age. Finally, there is a region of chromosome 20 with *Pax1* and

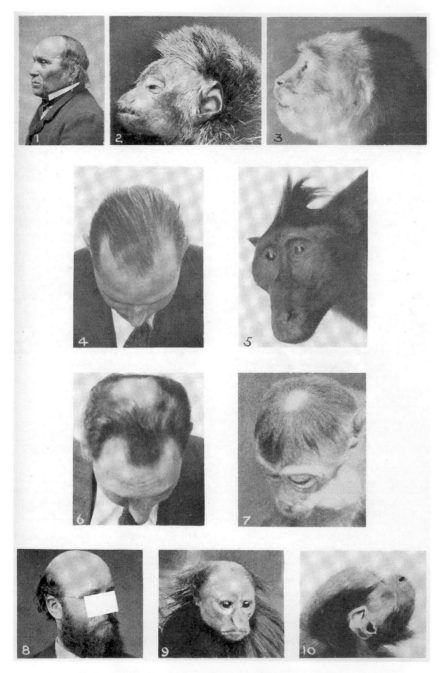

Fig. 6.4. Balding in humans compared with hair loss in monkeys. The left picture in each panel depicts a type of male-pattern balding: (1) raised hairline, (4) receding temples, (6) vertex spot, and (8) bald crown. The picture(s) to the right are of monkeys with similar patterns of hair loss or coloration: (2) monk saki, *Pithecia monachus*; (3) monk saki with lighter hair color in the affected area; (5) Celebes crested macaque, *Macaca nigra*; (7) toque macaque, *Macaca sinica*; (9 and 10) nearly bald crown in a red uakari, *Cacajao calvus*. Remarkably,

FoxA2, which encode transcription factors. Conceivably, they might bind *cis*-enhancers at a *Wnt* site. Indeed, a null *Fox* allele (*Fox13*-LOF) does remove body hair in the Mexican Hairless dog breed [324,994], and *Pax1* is strongly expressed in the scalp. If these factors do regulate hair growth, then their DNA-binding sites could unveil the HHQ "holy grail" [544,989].

Ambras syndrome (MIM 145701) elicits a profusion of hair on the forehead, nose, and ears [83]. The salience of this phenotype, along with its rarity, led to such individuals being exhibited by circuses in the nineteenth century (Fig. 6.5). The trait is caused by a gene(s) in the q23 region of chromosome 8. Working with DNA from three Ambras patients, Angela Christiano and her team [1007] zeroed in on a gene (*Trps1*) that causes a similar "Koala" syndrome in mice (bushy muzzle) [357]. In both the human and mouse mutants, the expression of *Trps1* is lower despite the fact that its coding portion remains intact, implying a position effect of lesions (breakpoints and deletions) nearby. *Trps1* encodes a DNA-binding protein that would be capable of binding a *cis*-enhancer at the hypothetical HHQ.

Hypertrichosis terminalis (MIM 135400) causes extra hair over the entire body (except the palms and soles), though again the head is most affected. Three mutant individuals from different Han Chinese families were found to have slightly incongruent, but overlapping, microdeletions on chromosome 17 (q24) [1235]. DNA from an even hairier man with thick, black, long (> 5 cm) hair over 96% of his body showed a micro*duplication* at exactly the same site. Four genes exist in the area of overlap. Only *MAP2K6* is a reasonable culprit, because deleting the other three genes has no effect. However, enzymes are not typically dosage-sensitive, so the notion that MAP2K6 (an enzyme) could have such effects when present in 0.5 (LOF) or 1.5 (GOF) doses seems odd. The authors conjectured that the lesions exert a position effect on a distant *Sox9* gene, whose LOF mutations cause hair loss [1332]. If *Sox9* is responsible, then we again have a possible lead to pursue, since Sox9 is a transcription factor that could bind at one or more *Wnt* loci.

Congenital generalized hypertrichosis (MIM 307150) evokes hair growth over most of the body, though the face is hairiest by far, the torso less so, and the rest of the body surface smoother still. Males are more strongly affected than females, and the ectopic hair in females sprouts in patches – telltale signs that suggest

←

Fig. 6.4. (*cont.*)

the perimeters of men's bald areas are as sharp as the edges of normally hairy areas (e.g., the eyebrow/forehead boundary) even though there is no hint of such contours in a male's scalp before middle age, so all of these territories might be specified by the same kinds of area codes (see Fig. 6.6). This montage comes from a treatise by Gerrit Miller, Jr. [886], but his vernacular and scientific names have been updated here to reflect standard taxonomy. Simian taxonomy has suffered much tinkering (not to say monkeying) over the intervening 90 years, and during that time Miller's essay has only been cited six times – two of which were in books that I wrote. My interest in this subject is not purely academic, since I started losing my hair at age 40, and I'm still angry about it 30 years on. I'm seeking a gene(s) to blame. Used with permission from the Smithsonian Institution.

Fig. 6.5. Stephan Bibrowsky (1891–1931), age 16. Born in Poland to normal parents, Bibowsky adopted the stage name of "Lionel, the Lion-faced Man" and toured with the Barnum and Bailey Circus [322,737]. His extravagant hair was due to Ambras syndrome. Such people were once thought to be throwbacks to our chimp-like ancestors [102], but none of the great apes actually have furry noses [557]. If such syndromes were truly atavistic, then we might expect to see a return of the kinds of catlike whiskers (vibrissae) that characterize more primitive primates [578,1112]. From [1340].

X inactivation. Indeed, two Mexican families with this syndrome (likely related to each other) were studied, and the trait was traced to a section on the X chromosome (Xq24–q27.1) [364,1253]. Despite the screening of 82 genes within this interval, no underlying mutation has so far been identified [1007].

A few other syndromes are known to cause striking, region-specific hypertrichoses. They display hair on the elbows (MIM 139600), neck (MIM 600457), ears (MIM 139500), nose (MIM 139630), or palms and soles (MIM 139650). In none of these cases has the trait yet been mapped. If the culpable area codes are eventually traced to *cis*-enhancers (at a *Wnt* locus?), then we should be able to confirm their roles by coupling the enhancers to reporter genes and seeing whether the transgenes that are constructed in this way can drive expression in similar parts of the mouse body [1157].

The HHQ holy grail that we are seeking is still hiding in our genome, but Figure 6.6 at least allows us to visualize what it might look like, assuming it fulfils our predictions based upon (1) the AS-C analogy, (2) the likelihood of *Wnt*

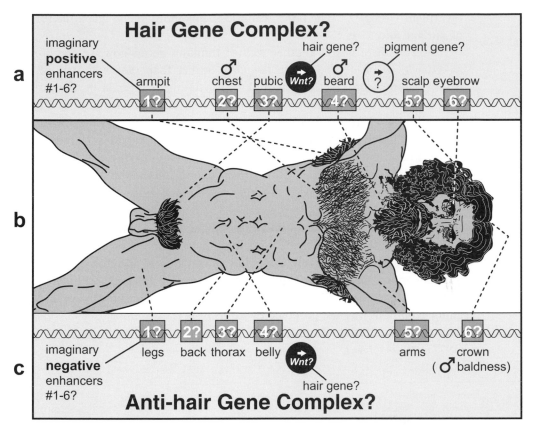

Fig. 6.6. Hypothetical Hair Headquarters (HHQ) for *Homo sapiens.* The hairy regions of humans (**b**) can be viewed from two different perspectives, in the same way that a figure–ground illusion can be parsed in alternative ways. Assuming that one of our 19 Wnt genes serves as the hair-promoting master gene (see Fig. 6.3), then it might be activated in hairy areas (**a**) or inhibited in hairless ones (**c**). In the latter case, our *cis*-enhancers would not be activators for hairy areas (scalp, armpits, groin, etc.) analogous to the AS-C, but rather would act as inhibitors to suppress hair formation in complementary parts of the body (torso, legs, forehead, etc.). Regardless of whether our HHQ operates in a positive or negative mode, its circuitry might resemble that of the AS-C (Fig. 6.2). If so, then it might be expected to have ~6 *cis*-enhancers (shaded rectangles) whose order along the chromosome is scrambled (dotted lines) relative to the regions they control. The HHQ might also contain a pigment gene (**a**) to assign hair color, though the pigment gene in the AS-C (*yellow*) does not appear to share enhancers with *ac* and *sc* [419,451,1163]. It would not make sense for a pigment gene to reside in an Anti-hair Gene Complex (**c**), since hairless areas do not vary much in pigmentation [631,1004]. Why our species has this odd landscape of hairy tufts is unclear, as is the mystery of what our extinct hominin relatives might have looked like [964], since hair does not fossilize. Adapted from Leonardo da Vinci's Vitruvian Man (ca 1492).

control, and (3) the hair-LOF and hair-GOF syndromes that have been described so far. Theoretically the HHQ could include positively acting *cis*-enhancers like those in the AS-C, but it might instead consist of *negatively* acting enhancers that erase hair from different parts of our surface – effectively functioning as an

"Anti-hair Complex." The latter scenario seems more likely, given the furriness of our ape ancestors.

However, there is still another way of thinking about how we acquired our nakedness, which depends upon *temporal*, rather than *spatial* control of hairiness. "Heterochrony" is the general term for timing changes in the development of a descendant relative to an ancestral species [708], and "neoteny" is a type of heterochrony where the descendant exhibits the same traits as the ancestor but with a much delayed schedule [467]. Humans are demonstrably neotenous with regard to our fellow apes [1141,1173]. The most obvious example is the flatness of our face, which resembles that of a baby chimp, minus the muzzle that characterizes the adult (Fig. 6.7). Another example is our wisdom teeth, which erupt late.

Our hair pattern fits nicely into this scheme [557]. To wit, it turns out that newborn chimps, gorillas, and orangutans have hair on their scalps but little elsewhere (Fig. 6.8), precisely like adult humans. It is therefore plausible that hominins postponed fur formation in most parts of the body so much so that modern humans never acquire a full covering during our entire lives. This intriguing hypothesis was first proposed by Louis Bolk (1926), embellished by Gavin de Beer (1958) and Ashley Montagu (1962), and popularized by Stephen Jay Gould (1977) [135,278,467,905].

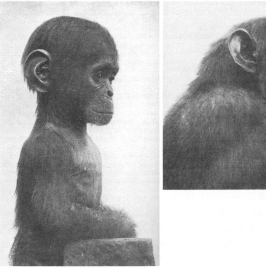

Fig. 6.7. Facial features of a baby chimp relative to an adult chimp. The striking resemblance of the baby's flat face to that of an adult human was one of the many clues that led Louis Bolk to propose "fetalization" as the primary driving force of hominin anatomical evolution [135,1173]. Another corroborating clue is the baby's relatively hairless chest (cf. Fig. 6.8). This argument animated Gould's 1977 monograph *Ontogeny and Phylogeny* [467], wherein he buttressed Bolk's hypothesis. From [933].

Fig. 6.8. Newborn great apes. All of these baby apes display profuse hair on the scalp but relatively sparse covering elsewhere – a pattern that differs greatly from their fur-covered adult stage but is eerily typical of humans at all stages of our lives. *Upper left:* Three-month-old chimp Vindi (born 17 February 2007) with her mother Jodi (photographed 21 May 2007), courtesy of Maureen O'Leary, Tulsa Zoo. *Bottom left:* Infant gorilla Goma (photographed in 1959), courtesy of Photo Archive, Basel Zoo; *right:* Baby orangutan Teliti with her mother Puteri at the Perth Zoo (photographed in 2009), courtesy of Samantha Finlay-Norman. From [558]; used with permission. (A black and white version of this figure will appear in some formats. For the color version, please refer to the plate section.)

Most important for our purposes, Bolk's "fetalization" hypothesis [135] has genetic implications that are worth considering in some detail, since they place the AS-C analogy in a decidedly different context:

1. Area codes may be activated in a specific sequence that is conserved among primates. If different area codes educe hair in a definite temporal order, then humans may not actually possess any unique area codes (*cis*-enhancers at a *Wnt* site?) of our own.
2. The gradual progression of hair acquisition during men's lifetimes could just be a slow-motion version of the proto-hominin "movie": scalp hair at birth, then beard, armpit, and pubic hair at puberty, then chest hair, back hair, and nose hair toward middle age (along with some loss of scalp hair), and finally bushy eyebrows – a trait so common that Aristotle himself was impelled to remark that "in old age [the eyebrows] often become so bushy as to require cutting" (*Parts of Animals*, Book 2, Pt. 15, p. 658, col. b, lines 19–20) [71].
3. The terminal area codes of the primate sequence may have been delayed so much that we no longer activate them at all, and this dormancy may have

persisted for millions of years. If so, then the respective enhancers (for, say, hair on our neck) may have decayed so much by the accrual of mutations that they can no longer be atavistically reawakened.

4. Hair graying comports with this scenario [1113]. Gorillas use silverback coloration during their peak reproductive years as a sign of male virility, and proto-hominins may have used gray scalp (and back?) hair in a similar way. Our graying occurs much later in our life cycle [1122], though our eyebrows (and certain other areas) remain dark much longer [886].

5. Male-pattern balding may have once helped attract mates (like the silverback trait) in our hominin ancestors, but then suffered a delay (past our prime) due to the systemic slowdown of other hair-related features, so that it is now mostly a useless vestige.

Thus, we may have become the only naked apes via heterochronic mutations that slowed down the utilization of our area codes without ever changing either their regional identities or their overall temporal sequence. In that case, we would have relied on a prior set of temporal – not spatial – enhancers at the primate HHQ, wherever it is. Unfortunately, we understand even less about how genes are regulated in time than we do about how they are regulated in space [362,509,1177], so it may be some time before we decipher how evolution tweaked the hominin clockwork so as to make humans the most anomalous simians on the planet [331,1366].

GP-12: The Notch pathway enforces binary decisions

The *cis*-enhancers of the AS-C do not pinpoint macrochaete sites at single-cell resolution. Rather, they merely denote rough areas around those sites. *How rough?* The "proneural clusters" (PNCs) that they evoke can contain as many as 20–30 cells [254]. Within each such equivalence group a single sensory organ precursor (SOP) cell must somehow be chosen [551,1069,1162], and nearby cells must be stopped from becoming SOPs [1304]. Otherwise, macrochaetes would sprout in tufts, as they actually do when the Notch pathway is disabled, as explained below.

Both of these fine-tuning steps (SOP selection and competitor suppression) were first deduced by Curt Stern from an astute investigation he carried out in about 1954 [1201]. His reasoning was based on one odd finding: a macrochaete can sometimes arise a short distance from its preferred site when that site is prevented from making a bristle by the presence there of *achaete* mutant tissue (Fig. 6.9).

Few of us would have paid much attention to one tiny misplaced fly bristle, but Stern was no ordinary researcher. Like the fabled Sherlock Holmes, he was attuned to eccentric exceptions, and he could follow clues like a bloodhound on a scent trail. He must have been smirking when he chose "Two or three bristles" as the title for his paper about this work, because it implied that his essay had nothing noteworthy to report [1202]. On the contrary, his article used the PDC displacement

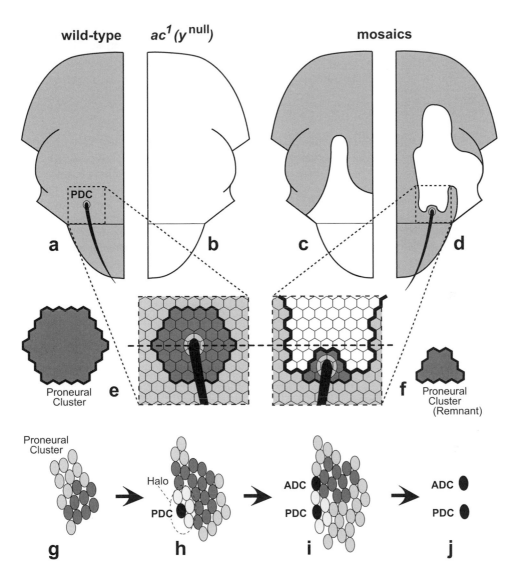

Fig. 6.9. The 1954 experiment that unveiled proneural clusters. a–f. Curt Stern's investigation of *achaete¹* (*ac¹*) "mosaics" [1201]. **a.** Half-thorax of a wild-type *D. melanogaster*, showing one of the 20 macrochaetes that exist on this side (see Fig. 6.2): the posterior dorsocentral bristle (PDC). **b.** Half-thorax of an *ac¹* fly, which lacks a PDC. Besides *ac¹*, the X chromosome also carried a null mutation (*y¹*) in the *yellow* gene. That allele causes bristles and cuticle to be yellow (depicted as white) instead of brown (gray in **a**), which allows the contours of any *y¹* territory to be traced in a wild-type background. The chromosome also carried a LOF mutation (*sn¹*) in the *singed* gene. That allele causes bristles to be gnarled. **c.** Half-thorax from a fly that began life as a *y¹ ac¹ sn¹/+++* (XX) female heterozygote. At the first mitosis of the zygote, one daughter nucleus lost its +++ X chromosome to become male (XO) and to express all three recessive mutations. At that point the fly became a mosaic (two genotypes in the same body), as well as a gynandromorph (half-male, half-female). In this case the XO tissue happened to cover the entire PDC region, suppressing the PDC via *ac¹*. **d.** Half-thorax from a similar mosaic, where XO tissue barely overlapped the PDC site.

to derive a monumental new theory of pattern formation called the "Prepattern" hypothesis [1203]. That theory later merged with Wolpert's "Positional Information" hypothesis to become the chief paradigm of developmental biology [899,1290,1403]. Hence, his modesty, feigned or not, was unwarranted. Stern envisioned the PNC as a miniature embryonic field wherein an incipient SOP blocks its neighbors from adopting its fate (boldface added):

> We have seen that no bristle may differentiate in a critical region if the area is occupied by achaete tissue. However, in the nearby non-achaete tissue a phenomenon often occurs which at first is surprising. In this nearby tissue, covering a region which would never form a bristle in non-mosaics, a bristle may form. … Yet **this astonishing result fits perfectly well into existing concepts of the embryologist. He has discovered the existence of pre-patterns which he calls embryonic fields.** These are areas in which a specific differentiation may occur anywhere. Actually, under normal circumstances, the differentiation takes place in only a limited part of the whole field, at a peak, figuratively speaking. **Once differentiation has set in at the peak, no other differentiation occurs within the larger field.** If, however, differentiation at the peak is suppressed, then a lower region of the field may differentiate. This differentiation itself will exclude other differentiation within the remainder of the field. Thus an embryonic field is a dynamic, flexible phenomenon. … If the influence of the achaete gene at the normal point prevents differentiation of a bristle, other parts of the field may assume the properties of peaks and differentiate bristles. [1202]

How does the nascent SOP stifle its rivals? Stern didn't know. The answer only became clear from an experiment published in 1978 by David Shellenbarger and Dawson Mohler, who studied a temperature-sensitive *Notch* allele, *Notch*^{ts1} (N^{ts1}). They used N^{ts1} to turn the Notch protein OFF at will by exposing mutant

---

←

Fig. 6.9. (*cont.*)

Surprisingly, a wild-type bristle arose at the edge of the yellow area. **e, f.** Illustrations based on Stern's interpretation of bristle displacement. Small hexagons are cells. Dashed line marks the normal PDC site. **e.** Stern reasoned that the ability to become a PDC must extend over a wider area than the single-cell site where the PDC normally arises. That area of bristle potential was later termed a "proneural cluster" (dark shading) [254]. **f.** As for the ectopic PDC (**d**), Stern deduced (1) that the mutant tissue must have blocked most cells in the cluster from becoming an SOP, leaving only a tiny cohort of "competent" XX cells, one of which became the SOP, and (2) that an SOP must have the ability to inhibit its neighbors from becoming SOPs. The only riddle that Stern did not figure out was why the cell in the center of the cluster (or its remnant) becomes the SOP. **g–j.** Confirmation of Stern's 1954 logic by histological evidence in 1989 [1079]. Drawing of a real cluster on a right half-thorax at ~25, 20, 10, and 0 hours before pupariation, as revealed by antibodies to Scute [254,1079,1171]. Ovals are cell nuclei. The DC cluster (1) has an irregular shape, (2) grows, and (3) yields two macrochaete SOPs along its edge. **g.** Some cells express more Scute (dark shading) than others. **h.** The SOP for the PDC (black) arises in a high-Scute area and acquires a halo of low-Scute cells. **i.** As the cluster grows, the high-Scute area shifts and the anterior dorsocentral bristle (ADC; no halo yet) arises. What causes this shift is unknown. **j.** In this cluster, non-SOP cells stop expressing Scute before SOPs, which do so later (before dividing). Growth widens the ADC-PDC gap. As for why both bristles form on the medial (dorsal) side of the cluster, the reason appears to be that a diffusible bristle-promoting signal (Dpp) enters from that side [1024,1296]. From [555].

flies to heat pulses at different times during development. When they administered pulses at the end of the third larval instar, tufts of bristles sprouted at macrochaete sites, and similar temperature-sensitive periods (TSPs) were found for microchaetes [1139]. Evidently, SOPs use the Notch receptor to suppress competitors during those times. Later studies confirmed the role of the Notch pathway in mediating such "shouting contests" across the animal kingdom in a process called "lateral inhibition" [1161,1168].

Here we encounter an apparent paradox. As mentioned above, the PNC contains 20–30 cells [254], but the Notch pathway should require contact between the signaling and receiving cell, so how can a single SOP inhibit more than the ~6 cells next to it [223,612,887]? A useful clue emerged in 2003 with the finding that SOPs extend filopodia studded with Delta over several cell diameters [281]. Might they be using those tendrils to subdue distant PNC cells [910]? That idea could explain *microchaete* spacing, but it was disproven for *macrochaete* SOPs in 2015 when (1) their radii of inhibition were shown to be a single cell diameter, and (2) only a small subset of the 20–30 PNC cells were shown to be capable of becoming an SOP [1304].

GP-12 tangent: The bristle

Strangely, the N^{ts1} TSP for extra bristles that Shellenbarger and Mohler found was followed almost immediately by a TSP for *missing* bristles (Fig. 6.10). The latter TSP made no sense, because Notch should have finished its job once a mature SOP emerges victorious. *What's going on here?* It turns out that the Notch pathway plays a third major role in development aside from delineating boundaries (see GP-8) and mediating lateral inhibition (GP-12) [146]. This third chore, which is related to lateral inhibition insofar as it also involves binary choices, is to force cells to adopt alternative fates within a fixed cell lineage – here, the offspring of the SOP.

All of the daughter cells in the SOP pedigree vie with their sisters, but these contests differ from the lateral inhibition scenario because, instead of being mildly biased by the amount of Ac or Sc, the outcome is totally predetermined by a supervisor called Numb [1416]. Numb is a protein that localizes to one specific side of each mother cell (cf. PCP proteins; Chapter 4) [94,297,455]. Thereafter, it segregates into only one of the two daughters at each mitosis [1063], where it prevents Notch from relaying its signal [427,1136].

Figure 6.11 peers into the genetic machinery of the SOP lineage previously diagrammed in Figure 1.3e. Each cell records whether its Notch receptor was activated in the previous mitotic cycle, presumably by letting Notch go to its nucleus and turn ON a memory gene of some kind, and the ON or OFF states from successive cycles are then combined into an overall binary code that dictates each cell's fate. For example, the socket code would be "11," where the repeated 1 indicates that Notch was activated in both of the preceding mitoses, whereas the shaft code would be "10," meaning that Notch was activated in the penultimate mitosis (bit on the left) but suppressed in the most recent mitosis (bit on the right).

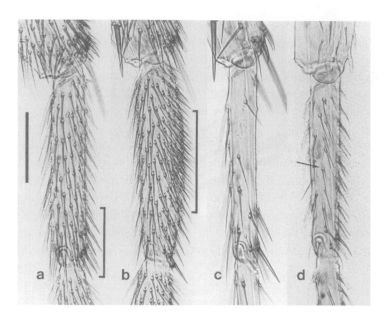

Fig. 6.10. Bristle abnormalities in temperature-sensitive mutants. Each photo shows the anterior side of the right second-leg basitarsus from a Notch pathway *D. melanogaster* mutant exposed to a pulse of high temperature (29 °C) during the pupal stage. **a**, **b**. Basitarsi from N^{ts1} males exposed to a 6-hour pulse of 29 °C starting at 2 or 7 hours after pupariation (hAP), respectively. Brackets denote areas of highest bristle density, which make sense based on the maturation sequence of bristle sites [550]. Total numbers of bristles per segment are 150 (**a**) and 316 (**b**), which constitute a doubling or quadrupling of the usual number (~75; see Fig. 1.4a) [551]. Note that the extra bristles are distributed fairly evenly within affected areas, rather than forming discrete tufts, as they do at macrochaete sites [888]. **c**. Basitarsus from a N^{ts1} male exposed to a pulse starting at 13 hAP. Most bristles are missing, except the most dorsal and distal ones. **d**. Basitarsus from a *shibire*^{ts1} male exposed to a pulse starting at 15 hAP. Most of the bristles (e.g., the one indicated by the arrow) have two shafts but no socket or bract – a defect expected if the socket cell, which is needed for bract induction (see Fig. 1.3), were to transform into a shaft cell (see Fig. 6.11). The same phenotype is seen in N^{ts1} males pulsed at the same time, but at lower frequency. Scale bar (**a**) = 100 μm. From [550]; used by permission of *Springer Nature/RightsLink*.

We can now see why disabling Notch during the last mitosis (via N^{ts1}) changes the socket (11) into a shaft (10), thereby causing double shafts (Fig. 6.10d).

As discussed in Chapter 1, Sydney Brenner contended that there are two cardinal modes of pattern formation: intercellular communication – the American Plan – and invariant pedigrees – the British Plan [1060]. SOP differentiation blurs this distinction by blending the Notch pathway – a canonical communication channel – with a strict cell lineage [1031,1051]. However, the SOP algorithm is more British than American because Notch is in a default ON state that is deaf to external input [669]: fates are dictated entirely by the presence or absence of internal Numb.

The Notch pathway is unusual relative to the other pathways that animal cells use to converse (e.g., Hedgehog, Wnt, TGFβ, Fgf, and EGFR) [795]. As stated before, it requires direct cell contact instead of using diffusible ligands [128]. Second, Notch serves as both a surface receptor and a transcriptional co-activator: when it binds a ligand, it leaves the membrane and goes to the nucleus, where it regulates target genes [369,888]. But its main utility is to force cells to adopt alternative (binary) states by using feedback loops to amplify slight initial differences [36,128,1161].

Until recently, the role of endocytosis in Notch signaling was unclear [133]. The first hint of the involvement of this engulfment device came in 1978 when N^{ts1} was shown to produce the same suite of bristle phenotypes in the same order as a temperature-sensitive mutation in the gene *shibire*, which encodes dynamin – an agent of endocytosis [1029,1139]. In 2000 this riddle was solved with the revelation of a bizarre, unsuspected process: the signaling cell must swallow the outer part of the Notch receptor (along with its own Delta ligand) in order for the inner part of the activated Notch receptor to effectively convey its signal to the nucleus of the listening cell [574,997].

One of the deepest questions we can ask about development is: How do genes specify anatomy? That is why the AS-C and its Notch-pathway partner are so useful as windows into the "mind" of the genome. The AS-C's *cis*-enhancers encode macrochaetes as a one-to-one map, but each site is only perceived as a blurry image. Every PNC must then be whittled down to a single SOP [888]. Thus, the genome achieves pinpoint precision for those sites by a process of successive approximation, where the final step uses Notch-mediated lateral inhibition as a fine-tuning "app" [887].

Microchaetes, however, are a different story. Unlike macrochaetes, they vary in position from fly to fly, and they even differ on the two sides (left and right) of a single fly. Evidently, the genome cares less about their arrangements, so to speak, than it does about the geometric constellations of macrochaetes [552]. The only features of microchaetes that seem to matter are their average densities and their axial alignments: most of them are evenly spaced within regular rows.

Microchaetes do not arise singly in proneural *spots* like macrochaetes but instead arise communally in proneural *stripes* [555]. Does the AS-C dictate where those stripes should form? Apparently not. An exhaustive search in 2017 for microchaete enhancers in the AS-C failed to uncover any at all, at least for the thorax [241] (their Fig. S1). Faced with that "dry hole," the researchers who conducted this search proposed an interesting model in which the proneural stripes of the thorax *self-organize* by using the Notch pathway in a dynamic manner involving epidermal cells mutually inhibiting one another [574,1131].

That model does not seem to apply to the legs, however, where proneural stripes are carved from a cylinder of *ac*-expressing cells in several discrete steps. Those steps are illustrated in Figure 6.12.

The process begins with the overall coordinate system of the leg disc, whose dorsal–ventral axis is governed by gradients of the morphogens Dpp and Wg (see Fig. 3.10). Those gradients elicit expression of the *ac* inhibitor *hairy* (*h*) in four equally spaced stripes around the leg circumference (Fig. 6.12b) [657]. The gradients

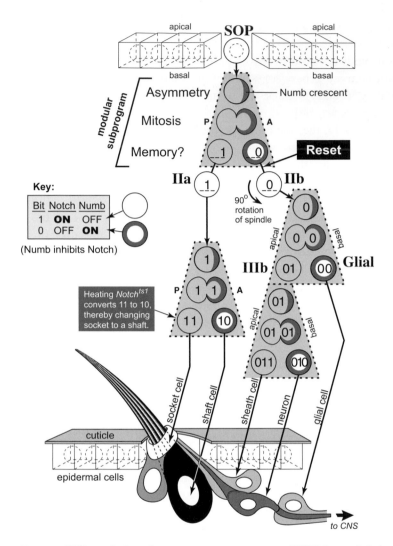

Fig. 6.11. Differentiation of a sensory organ precursor (SOP) into a bristle. At the top is a cutaway of the monolayer epidermis, with the SOP depicted as a sphere. The apical (cuticle-secreting) surface is above, the basal surface below. Anterior (A) is to the right, and posterior (P) to the left, as befits the thoracic or abdominal skin, though these directions would be proximal and distal, respectively, for the legs. The SOP arises from an ordinary epidermal cell within a proneural spot or stripe after having subdued its competitors via "lateral inhibition" (see text) [63]. It then undergoes four mitoses (gray trapezoids), each of which entails: (1) asymmetric allocation of Numb protein to a crescent at one pole (anterior or basal) prescribed by the PCP pathway [94,297,455] (cf. Fig. 4.2), (2) segregation of Numb into one of the two daughter cells, and (3) Numb-mediated stifling of Notch (1 → 0) in that daughter cell [427,1136]. "1" means activation of the Notch receptor (N*), transport of its inner part to the nucleus, and transcription of N* target genes, while "0" means suppression of N by Numb. The left-to-right order of 1s and 0s denotes each cell's memory of prior states preserved as a binary code. Cellular generations are marked by roman numerals (I, II, and III), with "a" and "b" signifying sisters. The IIa and IIIb cells use N* to switch *numb* back ON [1051], while the IIb cell must purge ("reset") its Numb state before it can reboot the

also apparently evoke the Notch activator *Delta* (*Dl*) at the same locations but in wider stripes. Then *h* inhibits *Dl* wherever they overlap so that the four wide *Dl* stripes split into eight narrower stripes that foreshadow the eventual bristle rows. (This echelon of the hierarchy resembles the pair-rule stage of the fly embryo, where a cohort of seven stripes later yields 14 parasegments [214].)

At this point one might expect the Dl ligand of the Dl-expressing cells to bind and stimulate the Notch receptor (N*) on all cells between the Dl stripes, but that only happens in five of the eight inter-stripes (Fig. 6.12c). For some unknown reason Notch fails to get activated in any of the *h* stripes except the ventral one [657]. At this stage all cells start to express *ac*, but they are blocked in the eight inter-stripes where either *h* or N* (or both) inhibits them (Fig. 6.12d).

Evidently, the AS-C long ago abdicated any meaningful role in delineating proneural stripes on the legs, having left this chore to the leg's coordinate system, which creates them using a complementary template of *h* and N* inhibitory stripes. This kind of subtractive ("negative template") logic is used in other aspects of imaginal disc development as well [241,555].

SOPs appear to percolate at fairly random locations inside each of the leg's proneural stripes (Fig. 6.12e), with fine-tuning cell movements (Fig. 6.12f) coming into play after an optimal density is reached [555]. The Notch pathway must be involved in limiting that density, because disabling it can quadruple the bristle number (Fig. 6.10) [550,551]. However, it may not be operating as it does inside a macrochaete PNC – i.e., by the AS-C/Dl-N feedback loop of the lateral inhibition process [37]. Rather, it may be functioning in more of a *mutual*-inhibition mode [241], where equivalent (bistable) cells engage in jousting matches that amplify any stochastic differences into alternative (SOP vs. non-SOP) outcomes [128]. That process has been likened to the patterning model proposed by Alan Turing, which will be discussed in Chapters 9 and 10 (boldface added):

> According to the currently favored model of SOP determination, the biochemical network that links proneural gene self-activation to Delta-mediated lateral inhibition is **topologically analogous to the reaction–diffusion chemical network envisaged by Alan Turing**, but for such a network to have long-range pattern-generating properties some sort of long-range inhibition is required. In this perspective, our finding that long-range lateral inhibition is mediated by Delta-promoted filopodia suggests that complex spatial patterns of sensory bristles might emerge from the dynamic behavior of the lateral inhibition genetic circuit. [281]

←───

Fig. 6.11. (*cont.*)

algorithm. The binary code (key) allows us to figure out the phenotypes obtained when N^{ts1} flies are exposed to heat pulses. Thus, a pulse during the first mitosis (not shown) would force both daughters (IIa and IIb) to adopt a "0" in their first binary register, causing all of their descendants to become glial, sheath, or neuron cells, leading to the "missing bristle" anomaly (Fig. 6.10c), whereas a pulse during the IIa mitosis would convert the socket cell (11) into a shaft cell (10), leading to the "double-shaft, no socket" anomaly (Fig. 6.10d). Adapted from [555]; though the key here is reversed (i.e., "1" means Notch-ON and Numb-OFF, not the other way around).

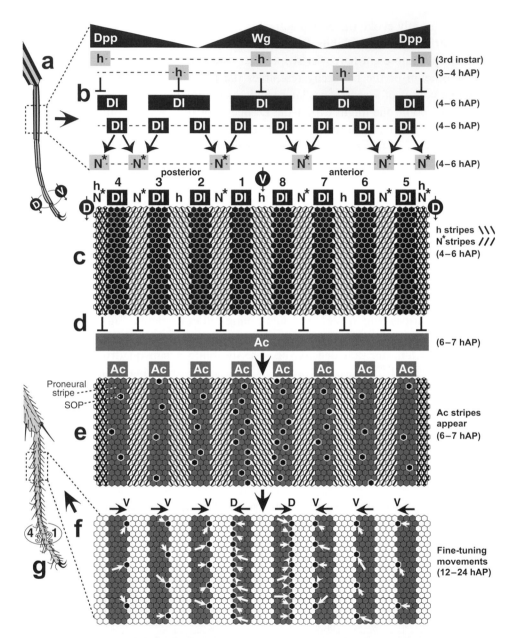

Fig. 6.12. Origin of microchaete SOPs from proneural stripes on the legs. a. Tarsus of a left *D. melanogaster* second leg, with three of the four zones of *hairy* (*h*) gene expression (black stripes) visible from the posterior side, and dorsal (D) and ventral (V) midlines indicated on the circle at the bottom. (Pupal legs are actually stockier than depicted here.) **b–f**. Panoramic surface of the basitarsus (dashed rectangle in **a**) that has been filleted along its D midline and laid flat, with cells drawn as hexagons. Times are given in hours after pupariation (hAP) along the right edge. **b**. Gradients (black triangles) of Dpp and Wg (cf. Fig. 3.10) dictate the positions of all *h* and *Delta* (*Dl*) stripes directly. Two *h* stripes arise at the D and V midlines in the early 3rd instar; the other two *h* stripes arise soon after metamorphosis starts (3–4 hAP). Four wide

Like the previous two chapters, this third chapter on fly development has explored the mechanisms of pattern formation – one of the three key processes that are used to construct embryos, the other two being growth and shaping. The three chapters of Part III, taking dogs as examples, will instead focus on growth and the devices that regulate it.

←───

Fig. 6.12. (*cont.*)

stripes of *Dl* expression (4–6 hAP) get carved into eight smaller ones by inhibition from *h*. Delta activates the Notch receptor (N*) on flanking cells. Arrows indicate activation; T-bars denote inhibition. Curiously, Dl fails to activate N inside *h* stripes, except at the ventral midline. As for why Dl fails to activate N inside its own (Dl) stripes, such *cis* Dl-N binding does not allow the N protein to be cleaved so that its inner part can go to the nucleus [574]. **c.** N*- and *h*-expressing stripes are hatched (key at right). Dl stripes are black. Bristle rows are numbered 1–8. **d.** Achaete (Ac) begins to be expressed around the whole segment but gets suppressed by N* and *h* stripes. **e.** The outcome is a kind of photonegative reversal, where eight proneural Ac stripes (shaded) emerge in the spaces between the preceding N* and *h* stripes. Each Ac stripe spans ~3–4 cells, just like each of the N* and *h* stripes, which raises the question of how Dl was able to activate N over a distance of more than one cell diameter. As discussed previously, such a feat is possible if Dl-expressing cells extend filopodia to span the requisite distance (laterally in this case) [281]. Within the eight Ac stripes, SOPs (black hexagons) emerge more or less randomly. The number of final SOPs per stripe decreases with distance from the V midline (cf. Fig. 5.3). **f.** Movements of SOPs in D (stripes 1 and 8) or V (stripes 2–7) directions, as inferred from cell-lineage analyses (see Fig. 12.6) [549]. These movements (white arrows) refine the alignment and spacing of bristles within each row [550,551,566]. **g.** Adult tarsus, showing rows 1–4. Revised from [555] based on data from [657,980,1159].

Part III

Dogs

7 The Shar-Pei

The domestic dog (*Canis lupus familiaris*) includes over 450 different breeds, most of which were created by artificial selection during the last several centuries [694]. They vary in size, shape, color, and behavior [995]. Most people are so familiar with the gamut of breeds that it is hard to see any of them as peculiar any more [601]. Nevertheless, some are arguably stranger than others. The Shar-Pei is a case in point. Shar-Pei puppies have rolling folds of skin all over their body, as if they had accidentally donned a cashmere coat several sizes too large and were gamely trying to make the best of it. Curiously, the skin becomes smoother as they age, except on the face (Fig. 7.1).

Shar-Pei dogs originated in China, where they have been prized for their hunting and guarding abilities for hundreds of years. A few were exported to the United States in the 1970s, and that stock was subsequently subjected to selection for the wrinkled-skin phenotype [978].

Remarkably, a similar syndrome exists in humans [1044], and there too the epidermis becomes smoother as the individuals "grow into their skin" (Fig. 7.2) [1426]. In both dogs and humans the wrinkling trait is due to excess amounts of a mucoid polysaccharide called hyaluronic acid (HA), which pools in the dermis [11,1044]. The puddles of HA loosen the connective tissue [1302,1427], but cutaneous flaccidity alone, which is so characteristic of old people [808], shouldn't be able to

Fig. 7.1. The Shar-Pei breed. At the puppy stage (left) Shar-Pei dogs (meatmouth subtype) have rumpled skin [978], but it smoothens with age so that the adult (right) looks more normal except for its face. Shar-Pei means "sandy skin," denoting a rough texture [922]. American Hairless Terriers have a milder wrinkling [994]. Photos (scale unknown) are used with permission from *iStock* Getty Images.

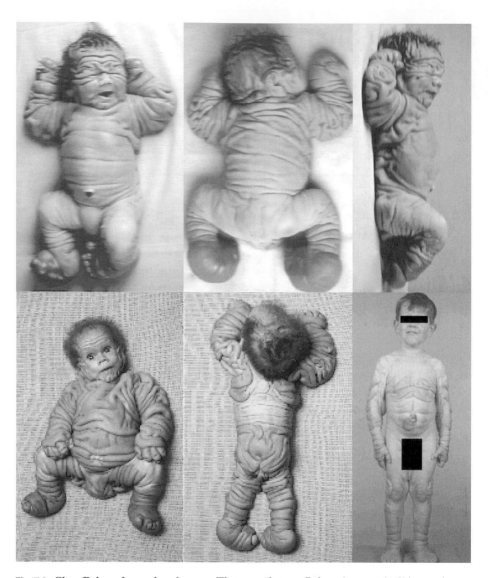

Fig. 7.2. Shar-Pei syndrome in a human. The same boy at 7 days (top row), 5½ months (bottom: left, middle) and 3 years (bottom right) [1044]. A similar "Michelin Tire Baby" syndrome has an entirely different etiology (viz., lipomatous nevus) [401,1086,1087], but a milder condition called scleromyxedema is likewise associated with increased mucin deposition and reddish-brown skin coloration [674,1081]. Photos courtesy of Professor Martin Delatycki, Victorian Clinical Genetic Services, Victoria, Australia. (A black and white version of this figure will appear in some formats. For the color version, please refer to the plate section.)

cause such severe folding without overgrowth occurring too [1407]. We do not yet know why the skin grows so much, nor why its growth rate slows after the first few weeks of life. The surplus HA has been traced to a 16 kb duplication of a regulatory DNA region upstream of the *HA Synthase2* gene [311,978].

Fig. 7.3. Shar-Pei syndrome in a cat? This hairless Sphynx cat is named Xherdan. In contrast to other cats of this breed, he has deep, labyrinthine creases in his forehead (as seen in some mutant pigs [609]) and excess, rumpled skin all over his body. Unlike a Shar-Pei dog, which he resembles, Xherdan has not grown less wrinkly as he has aged. These photos, shot by the Caters News Agency and used with permission, were taken in 2020, when he was nearly 7 years old [1197]. The etiology of his phenotype is unknown. (A black and white version of this figure will appear in some formats. For the color version, please refer to the plate section.)

In 2013 a male cat named Xherdan was born in Switzerland with the same kinds of undulating folds that epitomize Shar-Pei dogs. He belongs to the hairless Sphynx breed, which normally manifests some subtle wrinkling [393], as do some hairless dog breeds [994], but Xherdan is obviously an extreme case (Fig. 7.3). According to his owner (Sandra Filippi, personal communication), neither his parents nor his siblings are known to have this trait. He has been neutered, and there are no plans to biopsy his skin or sample his DNA, so we may never know whether his peculiar phenotype is due to an excess of HA synthase enzyme. Meanwhile, Xherdan has become an Instagram sensation, due in part to what people perceive as an angry expression on his face [1197], though that reaction may be an accidental consequence of how our species uses a furrowed brow to convey consternation [268].

GP-13: Patterns can emerge from physical forces

The corrugations on Xherdan's forehead are vaguely reminiscent of the convolutions on the surface of the human brain. In both instances, the ridges and valleys – called gyri and sulci in the case of the brain – have roughly uniform widths, despite meandering chaotically in a labyrinthine configuration (see Fig. 9.7 for a possible explanation) [537]. Moreover, in both cases, the pattern is asymmetric about the

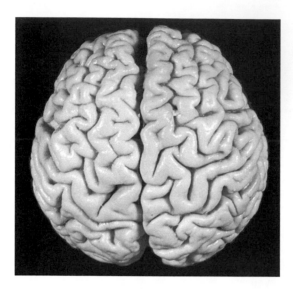

Fig. 7.4. Albert Einstein's brain. Note the cursory similarity of the labyrinthine pattern of crests and creases – termed gyri and sulci – to the ridges and furrows on the forehead of the Sphynx cat in Figure 7.3. The human brain raises its surface-area-to-volume (SA/V) ratio by pleating its surface [1260]. Thus, the gray matter near the surface (white matter is below) increases its information-processing capacity (i.e., thinking ability) relative to lissencephalic (smooth) brains of species like the mouse, whose SA/V ratio is much lower [1217,1438]. People whose brains are lissencephalic due to mutations are disabled to various extents [1080]. Einstein's brain size was normal, and this top view offers no obvious clues to his genius, but side views (not shown) reveal unusual gyral asymmetries [355,876,1236]. This image was provided by the Defense Health Agency Archives, National Museum of Health and Medicine, Silver Springs, Maryland.

midline. The extent of asymmetry for the brain in Figure 7.4, which happens to be that of Albert Einstein, is fairly typical of human brains in general.

Given this left/right disparity and the erratic twists and turns of the grooves in each hemisphere, it would be reasonable to presume that the surface is simply buckling randomly as it expands into the confines of a cramped cranial vault [942,1413]. Indeed, this conjecture was the prevailing explanation for cortical morphogenesis for many decades [361,618]. It was proposed by Wilfrid Le Gros Clark in 1945 [766,1217] and popularized by George Gray in a 1948 *Scientific American* essay about brain anatomy that was aptly entitled "The great ravelled knot" [481]:

> Beginning as an insignificant segment of the embryonic brain, this gray mantle eventually grows so large that it must fold in on itself in wrinkles to accommodate its expanding surface to the walls of the skull. [481]

If this overcrowding hypothesis were true, then the buckling should fail to occur if external pressures are removed [1382]. On the contrary, buckling can proceed in the absence of any restraint [361], as in the case of the normal girl with half a brain (see Chapter 2) [920,1299]. Moreover, lifelike folds arise in (1) mitotically growing organoids that simulate the nuclear motions and cytoskeletal forces of developing brain tissue [676,746] and in (2) bilaminar polymer gels that mimic the properties of gray versus white matter [164,1259,1260]. In both cases the folding occurs in sinuous arcs typical of the cerebrum, rather than in parallel ripples typical of the cerebellum [764,773].

In 2018 an even more convincing proof of autonomous folding was adduced. Brain tissue from aborted fetuses was cultured in vitro, and the tissue pleated itself into gyral motifs at the normal time on its own (sans skull) [818]. The investigators found that they could enhance or prevent the folding process by marinating the developing tissue in a cocktail of extracellular-matrix molecules [817]. Amazingly, the most potent ingredient within that cocktail turned out to be HA – the very same hyaluronic acid that causes Shar-Pei syndrome in dogs and humans [1382]:

1. LOF analysis: inactivation of HA (by increased degradation or reduced synthesis) leads to a decrease in gyri.
2. GOF analysis: treatment of fetal brain tissue with excess HA accelerates and accentuates the formation of gyri.

The ability of surplus HA to speed up the formation of native gyri is startling. Instead of the bulges forming at their normal pace of several weeks, exogenously administered HA can cause the neocortex to buckle within 24 *hours*, implying that the cortical plate is "spring-loaded" with a "hair trigger" that is released by HA. This Jack-in-the-box phenomenon is reminiscent of how quickly the trypsin enzyme can unleash the eversion of the imaginal leg disc in *Drosophila* (see Chapter 1). How HA facilitates buckling is not entirely clear, but it seems to be acting by affecting local stiffness within the cortical plate epithelium [818].

Not all of the grooves in the neocortex are as disorderly as they appear when viewed from above. When viewed from the *side*, the brain exhibits two major grooves in each hemisphere that are quite constant: (1) the central fissure, which runs vertically and separates the frontal lobe from the parietal lobe, as well as dividing the motor cortex from the sensory cortex, and (2) the Sylvian fissure, which runs diagonally and separates the frontal lobe from the temporal lobe [1141]. These invariant furrows recall the invariant shapes of the fly imaginal discs (see GP-2), each of whose folds emerges by differential growth [1301], and that growth in turn is controlled by coordinate systems of gene expression (cf. gut loops [944], heart loops [683], and tooth cusps [1105]).

The constancy of the major fissures implies that they are carved genetically [282], whereas the apparent fickleness of the other folds suggests that the sulci are etched by physical forces instead. Which deduction is correct? Both may be true, because the two primary grooves develop earlier (~27 weeks after conception) than the remaining secondary (~31 weeks) and tertiary (~37 weeks) ones [395,1217]. Indeed, the variability of any given groove's trajectory increases with its inception time [618,732].

The larger issue here is to what extent patterns in general are specified by the genome versus being forged by physical forces beyond the genome's control [545,812].

This debate about causation – genetics versus physics – has tended to polarize embryologists into opposing camps [921,1105] ever since D'Arcy Thompson's 1917 manifesto *On Growth and Form* [824,1280] showed that some aspects of anatomy can be explained by physical influences alone [510,625].

A familiar example of how physics affects development (albeit postnatally) is the ability of trabecular bones to remodel themselves in response to weight-bearing by fortifying their osteoid matrix along stress lines [26,334]. In the case of the crinkled cortex, as just discussed, both genetics and mechanics seem to be involved in morphogenesis at different stages [1382], and the same is true for organs that are shaped by feedback loops between epithelial strain and cellular signaling (e.g., intestinal villi, lung branches, and heart valves) [378,516,1002].

One anomaly where we can be fairly sure that epithelial strain is to blame is a fly mutant that was named *shar-pei* after the dog breed which it resembles (Fig. 7.5) [672,1267]. The fly's body is enlarged, as is its surface area, and its head

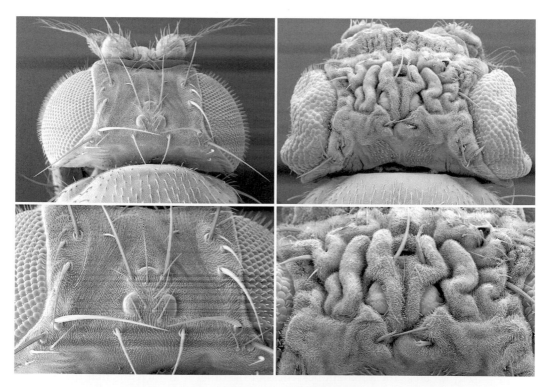

Fig. 7.5. A fly mutant that resembles the Shar-Pei dog. Scanning electron micrographs (low magnification above and high magnification below) of the top of the head of a wild-type *D. melanogaster* (left) and a predominantly *shar-pei* fly (right) [672]. Note the maze of cuticular ripples between the mutant's deformed compound eyes. (The furry texture is due to tiny trichoid hairs.) Flies also have three *simple* eyes – the dome-shaped bumps that form an equilateral triangle in the center of the head of the wild-type fly. These "ocelli" [731] persist in the mutant, though they are tucked between the folds. The *shar-pei* gene, also called *salvador*, belongs to the Hippo growth-control pathway, which is highly conserved in evolution [504]. Horizontal scan lines in some images are electrical artifacts. Photos are courtesy of Georg Halder.

cuticle is roiled into a maze of asymmetric ripples much like the face of a Shar-Pei dog (or Xherdan's forehead). The mutation responsible for this phenotype is located in a gene that regulates both mitosis and apoptosis. Hence, the mutant tissue not only fails to stop growing when it should, but it also fails to prune the excess cells it makes because its cell-death pathway is broken.

The apparent frivolity of the macaroni-like squiggles in the human brain (Fig. 7.4) may suggest the involvement of stochastic factors, but there is no way of knowing how much of the bilateral asymmetry is epigenetic by just examining individual brains [725]. In general, random asymmetries are called "fluctuating" [318] or "antisymmetric" [986], while genetic ones are termed "directional" [987]. Notable directional asymmetries exist that are associated with functional biases (Fig. 7.6).

Our own brain shows more directional asymmetry than that of our closest relative, the chimpanzee (*Pan troglodytes*) [1141] (boldface added):

> Broca's area in the inferior frontal gyrus, an important region for speech production, is usually larger in the left hemisphere, which is also the functionally dominant side. Furthermore, the central sulcus ... is deeper and longer with increased cortical folding in the left hemisphere of right-handers (i.e., the contralateral side responsible for the right body side). Similarly, the hand motor region within the central sulcus is more dorsally located in the left hemisphere of right-handers. **These patterns of "directional asymmetry" are shared among most individuals and considered to be under strong genetic control**, although a recent study did not find significant heritability in directional asymmetry. [946]

The surest way to gauge the extent of genetic involvement in any kind of patterning process is to compare monozygotic twins, and this approach has indeed been applied to the question of brain gyrification. It turns out that identical twins are significantly more alike in their brain anatomy than fraternal twins – proving some degree of genetic control – but they nevertheless differ substantially in the meanderings of their gyri [325,619,637] – indicating an appreciable level of "noise" in the system [1046].

GP-13 tangent: Fingerprints

Some of the "noisiest" patterns are literally at our fingertips. The pads on the undersides of our fingers, toes, palms, and soles display parallel ridges and grooves known as dermatoglyphs, which means skin carvings [256,1166]. Their configurations vary so much from person to person that they have been routinely used in crime investigations to identify individuals [168].

Why are they there? They evolved in our primate ancestors, who used them for gripping branches in arboreal habitats [256], and they arose independently in koalas – a tree-living marsupial that uses them in this same way (see Fig. 9.2) [257,573]. Anyone who has ever watched a monkey swing through a jungle canopy realizes why a good grip can mean the difference between life and death. The friction that they afford is enhanced by the presence of sweat glands (water helps

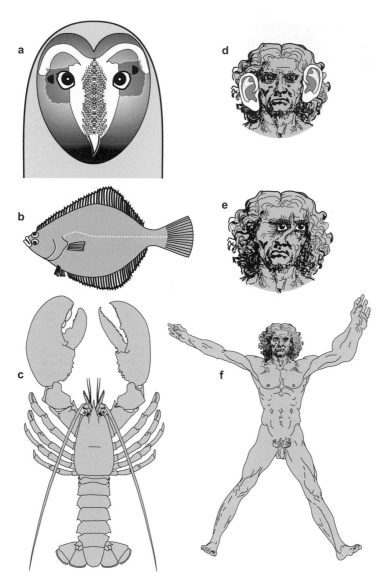

Fig. 7.6. Directional asymmetry versus antisymmetry. In directional asymmetry the biased feature consistently forms on the same side of the body (**a** and **b**), whereas in antisymmetry the trait randomly occurs on the left or right at equal frequency (**c**) [986]. **a**. Ears of the nocturnal barn owl (*Tyto alba*) [901], whose height disparity enables sounds to be localized precisely. **b**. Eyes of the flounder (*Platichthys stellatus*) [532], one of which migrates across the midline to allow binocular vision while the fish is lying flat. **c**. Claws of the lobster (*Homarus americanus*) [476], whose division of labor lets one (in this case, the left) act as a powerful "crusher" (note the molar-like teeth) and the other (right) act as a "cutter." **d**–**f**. Imaginary human phenotypes drawn to mimic the asymmetries of the animals on the left. Adapted from [557].

suction) and the absence of sebaceous glands (oil would cause slippage) and hair [1353]. Ridges tend to run perpendicular to the digit axis, which makes sense, analogous to how treads run across the axis of tire motion. These corduroy-like corrugations are also ideally suited for detecting textures, and the digit pads, not surprisingly, are densely innervated [1117].

How do they arise? Ridges appear in the dermis by week 10 of gestation and reach their mature conformation in the overlying epidermis by week 24 [52,734]. A separate "app" in the genome may be dedicated to sculpting them, since mutations can suppress them without disturbing any other aspects of anatomy [965,1314]. One of the genes that is involved (*SMARCAD1*) encodes a transcriptional regulator [200,314]. That gene is located on human chromosome 4, and its mutations are inherited as autosomal dominants [965], but we do not yet know what role the skin-specific isoform of this protein plays in the dermal sculpting process [877].

What causes the designs? No one knows. Many theories have been proposed [403,734], some of which are based on buckling forces [735]. What we *do* know is that the gadgetry which etches these grooves must be uncoupled from the processes that build *gross* anatomy. Our finger bones have an internal plane of symmetry (see Fig. 3.4a), but the radial and ulnar loops on our fingertips, which comprise 67% of all motifs, do not (Fig. 7.7c, d), nor do the double loops (Fig. 7.7e) or spiral (vs. concentric) whorls. Moreover, there is no consistent trend from finger to finger, and our left hand does not seem to care what our right hand is doing, or vice versa.

The ability of our skin to defy the symmetry of our skeleton is not confined to our digits. We also see it in the whorls of hair on the back of our scalp [780,904], which exhibit a blatant disregard for our body's midline [703,704,1107]. And just below the skin, the wayward paths taken by our blood vessels differ on the two sides of the body [1430]. Have a look at your wrists to see for yourself.

Dermatoglyphs and hair whorls have two things in common: they both arise in the dermis, and their component cells are polarized by the PCP pathway (see GP-7) [204,365,611,1172]. Under in vitro conditions these bipolar cells automatically align like magnets [353] into swirling streams that resemble whorls and loops (Fig. 7.7g) [343,482]. Such cells adhere to one another and communicate via cadherins [593,613] and integrins [283,825] on their front and rear [743], all the while responding to push–pull force vectors [354] via their actomyosin cytoskeletons [272,541,990]. Thus, dermatoglyphic designs may mostly be the emergent properties of fibroblast collectives [734,1083,1394] – not the fixed outputs of genetic blueprints.

This sort of "herd behavior" routinely occurs in embryos in various germ layers at various times [73,87,827,1023]. In all of these cases, the physical forces are generated locally, so that the patterns mainly self-assemble from the bottom up, rather than being passively constructed from the top down, which is the norm for morphogen gradients [1394]. Moreover, once cells have aligned themselves into chains, those linear arrays can be maintained during subsequent growth by the force-dependent orientation of mitotic spindles [1375], which helps explain why

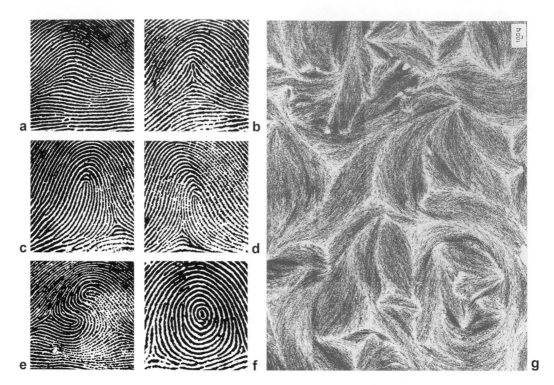

Fig. 7.7. Common types of fingerprint motifs and fibroblast patterns. Fingerprint designs resemble the altitude contour lines on a topographic map [1002,1166]. **a**. Plain arch. **b**. Tented arch. **c**. Radial loop, pointing toward the thumb. **d**. Ulnar loop, pointing toward the little finger. Strangely, ulnar loops are 10 times more common than radial loops, based on a survey of 24,518 Norwegians [734]. **e**. Double loop. **f**. Whorl, which in this case is considered plain because it lacks complicating features (e.g., a lateral-pocket loop) at the core of its concentric circles. **g**. Lawn of confluent fibroblasts that were cultured in vitro. The cells organize themselves into swirls (like Van Gogh's *Starry Night*) that resemble fingerprint loops and whorls [482]. Each fibroblast is ~100 μm long (scale at upper right). NB: Congenital absence of fingerprints (adermatoglyphia) is rare: only four families are known to display this anomaly [965]. Arches, loops, and whorls are also found on the finger pads of koalas (see Fig. 9.2) [257,573]. From [31] (**a–f**) and [343] (**g**); used by permission of *Springer Nature/ RightsLink*.

fingerprints remain constant once they are established in the 24-week-old fetus, despite later enlargement by an order of magnitude [52,734].

How much of any given fingerprint is dictated genetically? Patterns can't be 100% preordained because they differ between identical twins [1265]. In 1934 researchers were afforded a unique opportunity to address this question in some detail with the birth of five identical sisters – the famed Dionne quintuplets [106]. Table 7.1 lists the pattern types for all 10 fingers of each girl, categorized according to the standard scheme in Figure 7.7 [828].

These data indicate that dermal cells are relatively free to aggregate into arches, loops, or whorls, regardless of any bias imposed on them by the genome.

Table 7.1. Finger pattern types for the Dionne quintuplets.

Sister	L-5	L-4	L-3	L-2	L-1	R-1	R-2	R-3	R-4	R-5
Emilie	UL	UL	**TA**	UL	UL	**W***	UL	UL	**RL**	UL
Yvonne	UL	UL	UL	UL	**W***	**W***	**TA**	UL	UL	UL
Cecile	UL	UL	UL	UL	UL	UL	UL	UL	**RL**	UL
Marie	UL	UL	UL	UL	UL	UL	UL	UL	UL	UL
Annette	UL	UL	UL	UL	UL	UL	**UL^{TA}**	UL	**RL^{W}**	UL

Fingers are listed for the left (L) and right (R) hands of each sister along each row: 5, pinky; 4, ring finger; 3, middle finger; 2, forefinger; 1, thumb. Most of the patterns (see Fig. 7.8) are ulnar loops (UL). Exceptions (boldface) are: RL, radial loop; RL^{W}, radial loop tending toward a whorl; TA, tented arch; UL^{TA}, ulnar loop tending toward a tented arch; and W*, complex whorl that contains either a double-twin loop or a lateral-pocket loop. From [828].

To continue the herd analogy, the multiplying fibroblasts can evidently be "stampeded" by incidental ephemera to head in virtually any direction. Why the stochastic stimuli that "spook" them cause them to adopt one of the canonical pattern types instead of other possible ones remains to be determined. Figure 7.8 shows the fingers for which there were discordances among the quintuplets.

GP-14: Uneven growth rates can foster shape changes

Our tracing of the etiology of the Shar-Pei anomaly that launched this chapter led to an enzyme that makes too much of an extracellular-matrix component (HA). It is easy to see why flooding the dermis with that component might cause the skin to become loose, but, as mentioned before, it is harder to grasp why there is so much extra skin produced. Perhaps the excess HA prevents the skin from receiving the signals that would normally couple its growth to that of the underlying tissues.

It would only take a single extra mitosis throughout the epidermis to double its surface area relative to the structures below it, and two extra mitoses would quadruple it. Indeed, the similar phenotype of the *shar-pei* fly (Fig. 7.5) appears to be due to just such a defect in the control of cell division [672,1267]. In principle, disparities in relative growth rates could be causing not only buckling phenotypes in particular but dysfunctional proportions in general.

On the other hand, D'Arcy Thompson realized that growth disparities could be *useful* in enabling species to change their shape during evolution [465,571]. That insight was implicit in the title of his magnum opus *On Growth and Form* [5], and it was the guiding principle behind the graphical gimmick that he used to illustrate his "theory of transformations" [38]. That theory asserted that the differences between related species can often be reduced to global deformations, which in turn can be depicted by grid distortions (Fig. 7.9) [824].

In the most iconic illustration of this way of thinking, Thompson "magically" turned a porcupine fish (Fig. 7.9a) into a sunfish (Fig. 7.9b) by warping a rectangular grid into a hyperbolic one [286] – essentially the inverse of a Mercator map

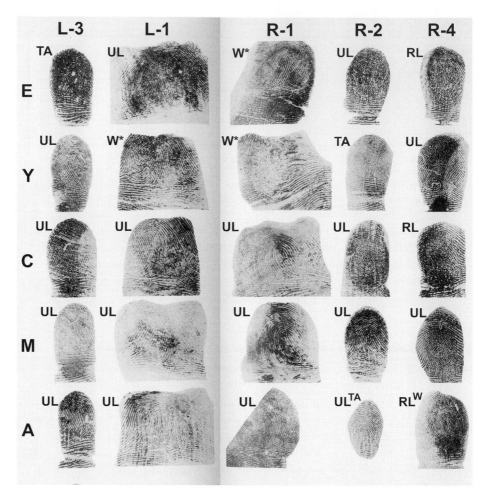

Fig. 7.8. Discordant fingerprints of the Dionne quintuplets. Abbreviations for names: E, Emilie; Y, Yvonne; C, Cecile; M, Marie; A, Annette. Abbreviations for fingers and pattern types are given in Table 7.1. Fingerprints are only depicted for those fingers whose patterns differ among the sisters. The remaining fingers all exhibited ulnar loops (UL; see Table 7.1). NB: The prints for L-1 and R-1 are wider because they were rolled from thumbs. Condensed from [828]; used by permission of the University of Toronto Press.

projection. The warping was achieved by imposing a "growth gradient" onto the grid such that the rate of growth was made to increase gradually from anterior to posterior. This trick was able to account for the sunfish's key traits, including its "rounded body, exaggerated dorsal and ventral fins, and truncated tail" [1280].

The explicit motive for such diagrams was to show, as William Bateson had attempted to do 23 years earlier [80], that Darwin's idea of incremental change as the driver for evolution was too narrow, and that organisms could alter their anatomy radically in one fell swoop [824]. Thompson expressed his intent succinctly (boldface added):

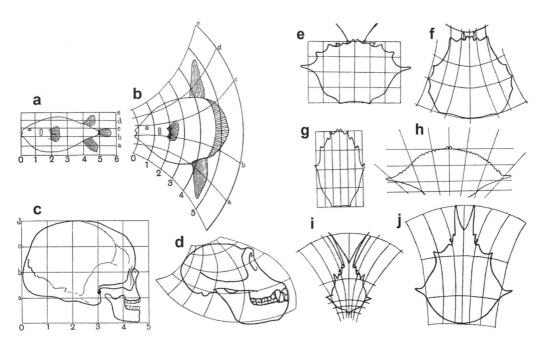

Fig. 7.9. D'Arcy Thompson's theory of transformations. Examples of how the anatomy of one species can be converted into that of a related species by a simple geometric deformation. **a, b**. Fish genera *Diodon* (**a**) and *Orthragoriscus* [= *Mola*] (**b**). **c, d**. Skulls of *Homo sapiens* (**c**) and chimpanzee (*Pan troglodytes*) (**d**). His portrayal of humans as "ancestral" to chimps was infelicitous [38], and this foible has been rectified in Figure 7.11. **e–j**. Carapaces of various crab genera: *Geryon* (**e**), *Paralomis* (**f**), *Corystes* (**g**), *Lupa* (**h**), *Scyramathia* (**i**), and *Chorinus* (**j**). From [1281]; reprinted with permission.

> This is no argument against the theory of evolutionary descent. It merely states that formal resemblance, which we depend on as our trusty guide to the affinities of animals within certain bounds or grades of kinship and propinquity, ceases in certain other cases to serve us, because under certain circumstances it ceases to exist. Our geometrical analogies weigh heavily against Darwin's conception of endless small continuous variations; **they help to show that discontinuous variations are a natural thing, that "mutations" – or sudden changes, greater or less – are bound to have taken place**, and new "types" to have arisen, now and then. [1281]

This rebuttal to Darwin never had much impact, though, for several reasons. First, the postulated conversions of species A into species B were based on the assumption that the former evolved into the latter, but most of those relationships were speculative [38]. Second, Darwin may not have envisioned growth gradients per se, but he was not as dogmatically in favor of mosaic evolution (i.e., each body part being capable of independent modification) as Thompson implied. Indeed, Darwin made room in his theory for a holistic gestalt wherein a change in one body part necessarily leads to changes elsewhere, just as pulling one thread of a spider's web affects the whole web. Modern geneticists refer to this phenomenon as "pleiotropy." Darwin called it the law of "correlation of growth" (boldface added):

There are many laws regulating variation, some few of which can be dimly seen, and will be hereafter briefly mentioned. I will here only allude to what may be called **correlation of growth**. Any change in the embryo or larva will almost certainly entail changes in the mature animal. **In monstrosities, the correlations between quite distinct parts are very curious**; and many instances are given in Isidore Geoffroy St. Hilaire's great work on this subject. Breeders believe that long limbs are almost always accompanied by an elongated head. Some instances of correlation are quite whimsical: thus cats with blue eyes are invariably deaf; colour and constitutional peculiarities go together, of which many remarkable cases could be given amongst animals and plants. ... Hairless dogs have imperfect teeth; long-haired and coarse-haired animals are apt to have, as is asserted, long or many horns; pigeons with feathered feet have skin between their outer toes; pigeons with short beaks have small feet, and those with long beaks large feet. Hence, **if man goes on selecting, and thus augmenting, any peculiarity, he will almost certainly unconsciously modify other parts of the structure, owing to the mysterious laws of the correlation of growth.** [267]

Most importantly, the growth gradients that Thompson used to buttress his theory lacked any empirical support [5,151]. A modicum of support would come 15 years later with Julian Huxley's *Problems of Relative Growth* [464,614], which documented gradient trends in crustacean limbs, plus copious cases of fixed organ/body growth ratios in various animal taxa. Even so, those trends and ratios were only descriptive [38], and their molecular and cellular bases remain elusive to this day [5,958,1151].

In 1936 Huxley and Teissier coined the term "allometry" to denote the change in proportions that results when body parts grow at different fixed rates [405,615,814]. One way to envision this phenomenon is to imagine the body as a set of balloons being inflated at different speeds [679]. Under such conditions the shape of the body will continue to change until growth stops [16,959]. Allometric rules, in combination with the scaling of body size [607], have led to a garish menagerie of grotesque traits across the animal kingdom [345,756,1150]. Among those odd traits are lobster claws, grasshopper legs, bat wings, beetle horns, and deer antlers (Fig. 7.10; cf. Fig. 7.6) [234,466,805,1077,1279].

GP-14 tangent: The human brain

The human brain also qualifies as an exaggerated trait, and researchers have long wondered to what extent this hypertrophy can be attributed to allometric trends within our primate order and more generally within our mammal class [213,367,908,1175]. Indeed, there is no greater holy grail in all of biology than the riddle of how humans acquired our intelligence [1088,1142], so it is worth at least a brief tangent to inquire how we got our big brain with its anomalous powers of thought and language [1088].

In Figure 7.11 D'Arcy Thompson's grid transformation of a human skull into a chimp skull (Fig. 7.9c, d) has been redrawn in reverse – with the chimp skull as the antecedent – to simulate something closer to the actual path of evolution, and the respective brains have been included so as to address the question at hand.

Many of the higher cognitive functions that we consider distinctly human are executed by our prefrontal cortex (PFC), located just behind our forehead [1185].

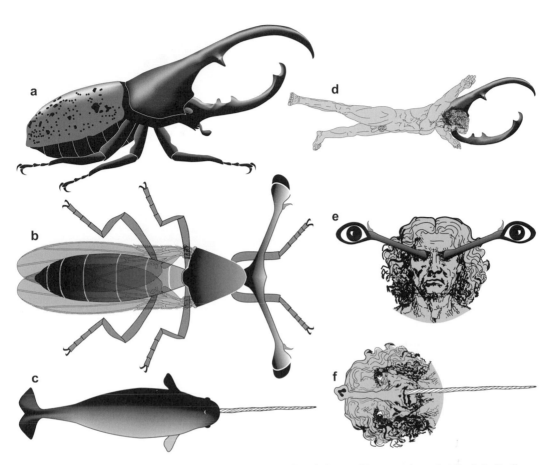

Fig. 7.10. Exaggerated traits in animals. a. Beetle horns (*Dynastes hercules*) [345]. **b**. Stalk eyes (*Cyrtodiopsis whitei*) [1354]. **c**. Narwhal tusk (*Monodon monoceros*) [967]. Each of these sexually dimorphic structures is used in male–male contests [346,1355] and appears to have evolved by positive allometry under the pressures of sexual selection [347,756,1152]. The tusk is also an instance of directional asymmetry, since it normally develops only on the left side (cf. Fig. 7.6) [967]. **d**–**f**. Imaginary human phenotypes drawn to mimic the anatomies of the animals on the left. **d**. The upper pincer of the Hercules beetle (**a**) comes from the dorsal thorax, while the lower one is a nasal horn, so they are here depicted as growing from corresponding sites of Vitruvian Man. Facial horns have evolved in vertebrates (e.g., rhinoceros) [936], and so have dorsal defensive spines (e.g., fish and reptiles) [345], so beetle-like pincers could perhaps evolve in a mammal. **e**. Stalk eyes have evolved in vertebrates (e.g., hammerhead shark), but not mammals [345]. **f**. The narwhal tusk is an overgrown canine tooth from the upper left side of the jaw [967], and its helical grooves twist sinistrally [697]. Those features are simulated here, as are the asymmetric blowhole (dorsal nostrils) and widely separated (lateral) eyes [248,579]. Adapted from [557,560].

This brain region is involved in abstract reasoning, social cognition, symbol manipulation, language processing, working memory, decision making, and fore-sight [316,317,1174,1441]. Based on its overarching importance we might expect the human PFC to be unusually large ... and it is.

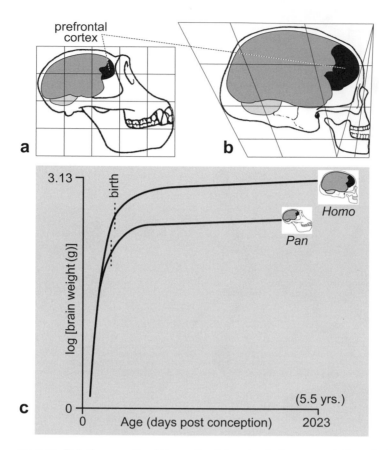

Fig. 7.11. Relative growth of the prefrontal cortex in human evolution. Outlines of chimp (**a**) and human skulls (**b**) are redrawn from *On Growth and Form* [1281], with permission, with brain schematics inserted from [557]. Shading denotes prefrontal cortex (black), remainder of neocortex (dark gray), and cerebellum (light gray). The starter grid was superimposed on the chimp skull (vs. the human skull; Fig. 7.9) as a proxy for our ancestor. Intersections of grid lines in **a** were re-plotted at corresponding points in **b**, and new (non-orthogonal) grid lines were drawn to connect those points. Our brain enlarged while our muzzle shrank (cf. Fig. 6.7). These opposite trends defy the gestalt that made Thompson's diagrams so esthetically appealing. The prefrontal cortex expanded at a faster pace than the rest of the brain. **c.** Human and chimp brain size (log plot in grams) versus time after conception (in days). Dashed lines denote birth times (240 vs. 270 days for chimp vs. human). An adult human brain weighs ~1350 g, while a chimp brain weighs ~400 g, indicating a ~3-fold difference in volume [530,576], whereas the prefrontal cortex expands by ~4.5-fold [316]. Redrawn from [1066].

Our PFC is about 4.5 times larger than that of a chimpanzee, but our brain as a whole is only about 3 times larger [530,576], so the PFC comes to occupy 1.2 times as much of our brain volume [316]. This increase in bulk is consistent with the allometric trend of the great apes, but it differs from the corresponding trends in more primitive primates and other mammals [1174,1334]. Hence, the relative growth of the PFC must have somehow gotten bolstered in our anthropoid lineage ca 15–19 million years ago. Sean Rice, who plotted the comparative trajectories that have been redrawn here (Fig. 7.11c), deduced a comparable event affecting the brain as a whole based on those data in 2002 (boldface added):

> It is generally argued that postnatal growth of the human brain is simply an extension of the fetal pattern of growth beyond our premature (relative to other primates) birth date, either with a simple extension or through sequential hypermorphosis. The data considered here suggest that this is not the case. ... Putting all of these results together on a phylogeny suggests that, sometime between the common ancestor of Old World monkeys and apes, around 25 million years ago, and that of chimpanzees and humans (around 5 million years ago), **a novel phase of brain growth appeared in the hominoid line. This growth phase begins at around the time of birth** and continues for approximately nine months in chimpanzees and one year in humans. [1066]

We may not know what environmental pressures led to this novel growth phase in the distant past, nor how the alteration was implemented genetically. However, the *recent* past since we diverged from chimp-like ancestors is now becoming clear. The extreme enlargement of our brain in general and our PFC in particular could have simply been a byproduct of an increase in body size, given the allometric ratios that were already hard-wired into our genome [908].

The evolutionary tweaking of the timing or rate of developmental events is a hallmark of heterochrony – a phenomenon that was discussed in Chapter 6 (see Fig. 6.7) [708]. It is ironic that a disharmony in growth like the one that disfigured the Shar-Pei may have given us our big brain, and it is sobering to think that our ability to think may have been a lucky accident [407,474].

8   The Bully Whippet

Whippets are racing dogs related to Greyhounds [694]. The fastest Whippets, which can run up to 56 km/h (35 mph), turn out to be heterozygous for a LOF mutation in the *myostatin* gene, which encodes a protein in the TGFβ family of growth-promoting factors. Unlike other members of that family (e.g., Dpp; see Chapter 3), myostatin *inhibits* growth (hence the "statin" suffix), and it does so mainly in muscle tissue (hence the "myo" prefix). As a consequence, skeletal muscles over-grow when the gene is disabled. Dogs that are homozygous (–/–) have so much muscle that they cannot run as fast as heterozygotes. They display a "double-muscling" (DM) phenotype and comprise a sub-breed called "Bully" Whippets (Fig. 8.1). The experiments that deciphered this etiology were published in 2007 by Dana Mosher *et al.* [917].

Fig. 8.1. Bully Whippet (right) and normal Whippet (left). Wendy (the larger dog) is a Bully Whippet owned by Ingrid Hansen in Victoria, British Columbia. She is homozygous for a LOF mutation in the *myostatin* gene, whereas ordinary Whippets like Foxy (the smaller dog) are either heterozygous or lack the mutant allele [917]. Heterozygosity confers speed, while homozygosity does not. The mutation is a 2-base-pair deletion that truncates the myostatin protein. No such mutation exists in Greyhounds – another racing breed to which Whippets are related [694]. Photo by Stuart Isett (©2007); used with permission. (A black and white version of this figure will appear in some formats. For the color version, please refer to the plate section.)

The *myostatin* gene itself had been discovered in mice 10 years earlier by Alexandra McPherron *et al.* [872]. When they mutated the gene artificially, they saw a dramatic increase in skeletal muscle mass throughout the body (Fig. 8.2a–h). They traced the overall increase to (1) prenatal proliferation of muscle cell precursors and (2) postnatal enlargement of the differentiated muscle fibers – i.e., a combination of hyperplasia and hypertrophy [9]. These "mighty mice" [1378] are, for all intents and purposes, rodent versions of human body builders, and the resemblance is enhanced by a concomitant reduction in their overall body fat [871].

In 2004 a case report was published in the *New England Journal of Medicine* that was eerily reminiscent of the "mighty mice" manufactured by McPherron *et al.* [1129]. The paper, by Markus Schuelke *et al.*, reads like a detective story. It begins blandly enough but takes a dramatic turn when the obstetricians realize that the baby that they have delivered is no ordinary mortal. In the excerpt that follows "myoclonus" denotes muscle spasms, which in this neonate subsided after 2 months:

> A healthy woman who was a former professional athlete gave birth to a son after a normal pregnancy. The identity of the child's father was not revealed. The child's birth weight was in the 75th percentile. Stimulus-induced myoclonus developed several hours after birth, and the infant was admitted to the neonatal ward for assessment. He appeared extraordinarily muscular, with protruding muscles in his thighs and upper arms. [1129]

The doctors could not help but notice that the boy was more muscular than any baby they had ever seen (Fig. 8.2i, j). They confirmed this hypertrophy by ultrasound when he was 6 days old, and they continued to monitor him every 6 months thereafter. By age 4½ he could hold two 3 kg (~7 lb) dumbbells with his arms extended. Hints of uncommon strength were also evident in other members of his family, including his mother and uncle, and his grandfather was a construction worker who unloaded 70 kg (~150 lb) concrete curbstones by hand.

Having read the article by McPherron *et al.*, they suspected that the boy might have a *myostatin* defect, so they sequenced this gene from him and his mother. They found nothing amiss in the protein-coding parts of the gene but did notice a G→A substitution at an mRNA splice site in both alleles of the boy and one allele of his mother. No such change was seen in 200 alleles that they tested from controls. Sure enough, when they studied expression in transfected cells, the allele with the substitution caused 69% of the *myostatin* mRNA to be misspliced and non-functional. Based on these results, they concluded that the boy was homozygous for a LOF allele of the *myostatin* gene, whereas his mother was heterozygous – exactly like Whippets, where homozygotes have the DM phenotype, but heterozygotes, albeit stronger than non-mutant individuals, do not [917].

The DM trait was first recognized in cattle more than a century ago [191]. The Belgian Blue breed represents the epitome of this phenotype (Fig. 8.3) [491,517,870]. Following the discovery of *myostatin*, an association between the gene's naturally

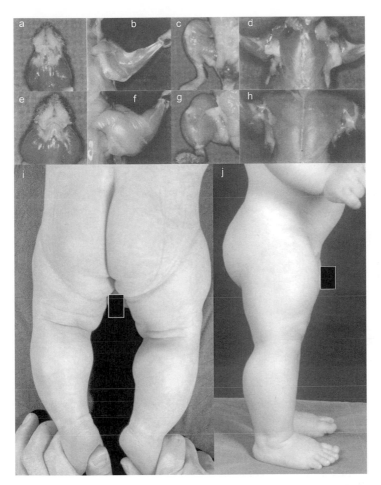

Fig. 8.2. "Mighty mouse" and "Superboy." In each case (mouse and human), homozygosity for a null allele of the *myostatin* gene doubles the size of skeletal muscles throughout the body. Muscle mass can actually be *quadrupled* by overexpressing the myostatin-binding protein follistatin in conjunction with disabling myostatin (not shown) [770,771]. **a–h.** Face, arm, leg, and pectoral muscles of a skinned wild-type mouse (**a–d**) and a homozygous null mutant for *myostatin* (**e–h**). **i, j.** Buttocks and legs of a boy at 6 days (**i**) and 7 months (**j**) of age [1056]. Note his bulky thighs and calves. His quadriceps were 7 standard deviations above the mean in cross-sectional area relative to controls, while the thickness of his subcutaneous fat was 3 standard deviations lower. His bones were normal. This mouse–human juxtaposition illustrates why mice have been such a powerful model system for studying clinical syndromes in general: they resemble us anatomically, they are easily manipulated genetically, and they develop relatively quickly [155]. Photos **a–h** are from [872], used by permission of *Springer Nature/RightsLink*. Photos **i** and **j**, provided by Markus Schuelke, are adapted from [1129] and used by permission of the Massachusetts Medical Society. (A black and white version of this figure will appear in some formats. For the color version, please refer to the plate section.)

Fig. 8.3. Belgian Blue bull and Italian Heavy draft horse. The excessive muscularity of both animals has been traced to particular LOF alleles of the *myostatin* gene [265,1378]. Photos used by permission of *Wikimedia/Creative Commons* and *Elsevier/RightsLink*, respectively.

occurring alleles and excess muscle mass has been documented not only in dogs, cows, and humans, but also in sheep [492], goats [9], pigs [1210], rabbits [1417], chickens [9], ducks [9], and pigeons [330]. The most extensive studies, however, have understandably been conducted on racehorses, where any competitive advantage is fiercely exploited [140].

The best predictor of a Thoroughbred racehorse's optimal track distance (sprint, mile, etc.) turns out to be the allelic status of its *myostatin* gene [580] and the amount of functional myostatin produced [1082]. Correlations between particular *myostatin* alleles and body type have been demonstrated for non-racing horse breeds as well [211], including the Icelandic horse [374] and the Quarter horse [881,1014]. One of the most interesting examples is the Italian Heavy Draft horse, which looks like an equine version of a Bully Whippet (Fig. 8.3). Stocky ("brachymorphic") horses of this kind have higher frequencies of certain *myostatin* alleles in general [264], and the muscularity of this breed in particular has been traced to specific single-nucleotide polymorphisms [265]. It is hard not to feel jealous of draft horses and other DM animals who can apparently keep their physique without ever having to exercise, though more muscle mass does not necessarily mean greater strength [1327].

Why should the actin–myosin machinery of our muscle cells need regular renewal by exercise? Muscle maintenance depends on myostatin, as does muscle development: at high levels of myostatin, extant muscles waste away, while at low levels they grow and strengthen [586,985]. Strangely, aerobic exercise alone does not suppress myostatin; it must be combined with resistance training in order to accomplish that, so as to increase muscle mass. Clinicians are obviously interested in anti-myostatin interventions as a way of treating muscle-wasting diseases [687] as well as aging-related muscle deterioration [740], and progress is being made on both fronts [352,799].

GP-15: The Hippo pathway controls organ growth

The notion that organs can limit their growth by secreting inhibitory factors was proposed by William Bullough in 1962 [167]. He called such paracrine agents "chalones" [341]. They were supposed to be emitted in proportion to tissue mass, rather than to cell number [277,769], because tetraploid organs grow to the same size as diploid ones, despite having only half as many cells [356,819]. Myostatin certainly qualifies as a chalone, but it may be the exception, rather than the rule [1013]. Conclusive evidence for the involvement of tissue-specific chalones other than in skeletal and cardiac muscle is negligible [819].

Other tissues may not be relying on their own unique rheostats for the simple reason that they are all using a common generic mechanism. For example, at least six signaling pathways help determine the sizes of imaginal discs in flies, regardless of the type of disc [521,945]. The most critical among them is the highly conserved Hippo pathway [462], which was discovered in flies [1307] and later documented in mice [504,1359]. In fact, the fruit fly's *shar-pei* gene, which was discussed in the previous chapter (see Fig. 7.5), is a charter member of that pathway [671].

As is true for *shar-pei*, the LOF phenotype for most of the other Hippo pathway components involves tissue overgrowth [522] – implying that they normally serve as growth inhibitors like myostatin. Indeed, dysfunctions in Hippo signaling cause a variety of ghastly cancers in humans [1026]. However, one key difference that distinguishes the Hippo pathway from the myostatin cascade is that it tends to restrain organ size by sensing *mechanical* forces, rather than *chemical* signals [1429,1434]. Those forces take the form of compression or stretching, and they are sensed by intercellular junctions where critical kinases of the Hippo apparatus are situated [462,499].

GP-15 tangent: The mammalian liver

One of the most remarkable organs from the standpoint of size regulation is the mammalian liver [61,1001]. When two-thirds of a mouse liver is excised, the remaining third instigates mitosis within hours [434,1269] and replaces the missing mass within about a week [880]. When the liver of a small dog is transplanted into a big dog it grows to suit its larger host, and when the reverse is done, the donor liver shrinks to fit its smaller host [373]. Similar size adjustments are observed for human transplants [685]. No other organ possesses such scaling or regenerative abilities in any adult mammal [61].

> Both the embryonic and adult liver rapidly recover mass following cellular loss, either through cell growth (hypertrophy), proliferation, or both ... A remarkable feature of this process is that the regenerated liver neither exceeds nor falls short of its original size, a sign of a regulatory feedback mechanism that has been termed the "hepatostat." [878]

A bewildering array of signaling pathways is involved in liver regeneration [61,1315], but the Hippo pathway is as pivotal here as it is in imaginal discs [839]. Indeed, artificially boosting the expression of its components in organs that do not normally regenerate can magically give them this ability [919], and that empowerment could lead to clinical treatments for injuries, diseases, and aging. The control circuitry for the liver has only been deciphered quite recently [878]. It involves an intricate feedback loop between bile acids secreted by the liver that are sensed by the intestine, which in turn emits the growth factor FGF15 into the bloodstream to stimulate the mitosis of liver cells via the Hippo pathway [644].

> Metabolites produced by the liver (i.e., bile acids) are sensed by another organ (i.e., the intestine), causing the regulated release of a circulating growth factor (i.e., FGF15) to influence hepatocyte proliferation and metabolic output through the Hippo kinases Mst1 and Mst2. [878]

GP-16: The insulin pathway nurtures sexual frills

Adequate nutrition is a prerequisite for cell proliferation, so it should come as no surprise that growth depends not only on the Hippo "spark plugs" that keep the mitotic engine running but also on the metabolic energy that fuels the cell cycle [461]. Insulin is widely known as the hormone that regulates glucose metabolism [955], but its cousins, the insulin-like growth factors (IGFs) and insulin-like peptides (ILPs), are less well known.

Unlike insulin, the IGFs, which operate in vertebrates [339], and the ILPs, which operate in insects [154,811], are more concerned with growth than with metabolism per se, though they use the same downstream effectors as insulin [335]. In each case the ligand binds a membrane-bound receptor, which triggers the activation of an enzyme known as phosphoinositide 3-kinase (PI3K). The importance of this pathway is illustrated by two especially striking phenotypes: (1) LOF flies are half the size of normal flies [131,772], and (2) GOF humans suffer from a horribly disfiguring syndrome of tumorous outgrowths that luckily appear to be curable by a simple drug treatment [1326].

Doug Emlen et al. [347] have postulated that the insulin signaling pathway mediates the evolution and development of exaggerated traits that are (1) sexually dimorphic [1150] and (2) positively allometric [1151,1355] (see Fig. 7.10). Those traits consume a lot of energy in their construction and hence come at a high cost in terms of nutritional resources [452]. In short, they are a luxury. Only the healthiest animals can afford to make them, and their size is therefore a useful indicator of an individual's fitness [347]. The fittest competitors have the most sex appeal and the best chance of securing a mate who is smitten by their prowess. Mostly what we are talking about here is strutting males and swooning females, rather than the other way around [1286,1355].

GP-16 tangent: Deer antlers

Among all of the gaudy ornaments that are flaunted by vertebrate males, buck antlers may be the most quintessential, given their use as weapons, with peacock fans coming in a close second. Another aspect of their uniqueness is the fact that they are shed and regrown every year in preparation for the rutting season [74], with their growth rate rivaling that of an aggressively malignant tumor: ~2 cm/day [793]. No other bone in any other vertebrate grows this fast [1032].

This breakneck speed and the neck-wrenching weight of the final rack are dependent on the buck's vitality, and that vitality turns out to be a function of its nutritional state as proclaimed by IGF. In 1985 the amount of IGF1 in the blood was shown to be correlated with antler elongation rate [1239]. In 1994 IGF1 was found to provoke proliferation of antler cells in vitro [792,1033]. In 2007 IGF1 was detected in chondrocytes (cartilage-making cells) and osteoblasts (bone-making cells) along the entire shaft of the growing antler, most strongly in the tip and upper part [495]. And in 2017 exogenously added IGF1 was reported to affect the mitotic rate of antler chondrocytes via the transcriptional activator Runx1 [1418].

In summary, deer antlers support the hypothesis of Emlen *et al.* regarding the role of the insulin pathway in exaggerated (primarily male) traits across the animal kingdom [347]. However, the value of this model system extends far beyond its utility as a vehicle for probing the mechanisms of sexual dimorphisms. The mesenchymal stem cells that renew the antlers so quickly every year have attracted the attention of clinical researchers, who see them as a possible way to enable humans to regenerate amputated limbs in a relatively short period of time [350,1349].

The antler story is therefore a useful parable. It proves that playful curiosity is not always an idle pastime. Our childlike instinct to explore strange phenomena can sometimes lead us down rabbit holes to Wonderlands of practical applications in terms of translational medicine. The Bully Whippet was another case in point. That dog breed led us to investigate the basis for excessive tissue growth stemming from signaling pathways that malfunction as a result of simple mutations. A better grasp of those mutations and their cascading effects could someday lead to cures for certain types of cancer. If so, then the epithet "man's best friend" could take on a whole new meaning.

9 The Great Pyrenees

Great Pyrenees are big dogs, standing almost 90 cm (3 ft) high at the shoulder and weighing more than 45 kg (100 lb) when full grown. They were bred for centuries in the mountainous border area between France and Spain to assist shepherds in the guarding and herding of their flocks. Their calm demeanor makes them good pets. One of the most singular traits of this breed is the double dewclaws (extra toes) on their hind feet (Fig. 9.1) [224].

Fig. 9.1. Great Pyrenees; double dewclaws on hind feet. Extra toes like these are common in large dog breeds, but rare in small ones except the Norwegian Lundehund, which has them on its forepaws as well [993]. The trait was initially attributed to a deletion of repeated DNA coding sequences [371] but was later traced to a *cis*-enhancer of the *Shh* gene [993] (see text) – a discrepancy that has not yet been fully resolved. Photos are from *Wikimedia/Creative Commons* (above) and Jane Harvey of *JaneDogs.com* (below).

In 1986 these anomalous digits figured prominently in an essay published in *Natural History* magazine [14]. The essay was entitled "Possible dogs," and it was written by Pere Alberch, who was then an associate professor at Harvard. He had recently visited his parents in Spain, where he was surprised to see six toes on each hind foot of their new Great Pyrenees pet – the ordinary four on the paw plus a pair of toes (dew claws) hovering above (boldface added):

> I had never seen a six-toed dog . . . and I assumed that my parents' pet was a mutant. Soon I was corrected by a friend, a dog breeder who pointed out that our dog was not a freak, but that, in fact, it exhibited the morphology required by the standards for the breed. . . . So I decided to find out a little more about how the number of toes, or digits, varies among dogs. I learned that in most breeds, and in wolves and jackals, there are four digits on each hind limb and five (one being a short vestigial toe, or dewclaw) on the front limbs. Some breeds, most commonly the larger ones, also have dewclaws on the hind feet. In addition, **some of the biggest breeds, such as Great Pyrenees, St. Bernards, and Newfoundlands, have a tendency to develop a sixth toe on their hind limbs**. . . . Conversely, medium- and small-sized breeds rarely, if ever, develop this double dewclaw. [14]

Double dewclaws on the larger breeds are a case of preaxial polydactyly [993], analogous to the two thumbs on the hand of the koala (Fig. 9.2) [288]. The foot of the koala, it so happens, is even stranger than its hand: it exhibits a fusion (syndactyly) of two adjacent toes. How and why the koala evolved these quirks is a mystery.

Alberch wondered whether there could be a causal link between body size and digit number. He had noticed just such a trend in the amphibian limbs which he had been studying in his scientific research: the smallest frogs have fewer digits. Taking all of the evidence into consideration, he surmised that an allometric constraint of some kind must affect vertebrate limbs: bigger bodies mean proportionally larger limb buds that can sprout more digits, while smaller bodies mean proportionally smaller limb buds that can lose digits. There is even a Biblical reference to such a trend in the story of David and Goliath (boldface added):

> Goliath, the Gittite, is the most well known giant in the Bible. He is described as "a champion out of the camp of the Philistines, whose height was six cubits and a span" (Samuel 17:4). . . . A literal interpretation of the verses suggests that his brother and three sons were also of giant stature. The name of Goliath's third son does not appear in the Bible, so we have named him Exadactylus as it was said that **"he had on every hand six fingers, and on every foot six toes"** (Samuel 21:20–21). Goliath's family tree is suggestive of a hereditary autosomal dominant pituitary gene, such as AIP. [319]

Alberch was postulating a constraint that would foster extra toes in bigger dogs – hence "*Possible* dogs" – while making them *im*possible in smaller ones [13]. Developmental constraints were just starting to emerge as a major theme in the nascent field of evolutionary developmental biology – "evo-devo" for short – and Alberch was one of its main proponents [12,15,858] and a founder of that field [306,1057]. His life was tragically short (1954–1998), but he left an enduring legacy [1048].

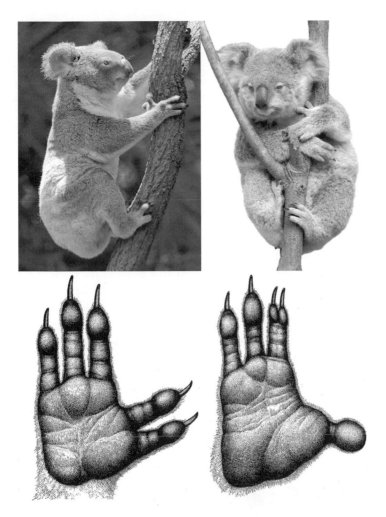

Fig. 9.2. Kooky digits of the koala. The hand of the koala (*Phascolarctos cinereus*) (lower left) has two thumbs, while its foot (lower right) has a stubby hallux and a peculiar fusion of its second and third toes. That fusion, known as syndactyly, is a rare anomaly in placental mammals, but for some unknown reason it became fixed as a feature not only in koalas but also in the distantly related macropodid family of marsupials (kangaroos, wallabies, wallaroos, pademelons, and quokkas) [206]. Although not depicted in these drawings, koalas have fingerprints that are strikingly similar to those of humans (see Fig. 7.7) [573]. Those ridge designs evolved independently of ours [257] and illustrate the general principle of evolutionary convergence [863]. Even so, one difference between koala and human hands stands out: they have claws instead of fingernails [1300]. Birds also have claws and a hallux. For the strange saga of how the bird got its hallux see [138]. Photos (above) are from *iStock* Getty Images. Drawings (below) are from [288]; reproduced with permission, ©1980 *Scientific American*, a division of *Nature America Inc.*, all rights reserved. (A black and white version of this figure will appear in some formats. For the color version, please refer to the plate section.)

GP-17: Anatomy can be influenced by body size

Why are some breeds so big and others so small? A genomic screen in 2007 found that body size in small dogs is mostly due to a single partial-LOF allele of the same *IGF1* gene that promotes the growth of deer antlers, as we saw in Chapter 8 [1238]. A follow-up study in 2012 showed that most small breeds also contain a LOF mutation in the IGF1 receptor [602].

The Great Pyrenees stands near one end of the dog size spectrum, with the Chihuahua at the other end. This range of size spanned by *Canis lupus familiaris* (~50-fold in weight) is greater than that of any other vertebrate species (Fig. 9.3) [1071], so it offers us a convenient window into anatomical scaling effects. Alberch highlighted one trait (digit number) that seems to be affected by body size. Are there others?

One simplistic question we should be able to address is whether intelligence varies with brain size [368,597] – an unmistakable trend in the evolution of hominids (cf. Fig. 7.11) [456,862], though such trends *between* species cannot meaningfully be compared with correlations *within* species [145]. For dog breeds, anecdotal accounts abound, but reliable scientific evidence has been hard to come by, partly because it is difficult to measure intelligence separately from obedience. Clouding the issue still further are the instincts (herding, fetching, guarding, etc.) that artificial selection has etched into certain breeds over the eons. Hence, no definitive answer exists yet.

Fig. 9.3. Size spectrum spanned by dog breeds. Dogs in each series are depicted at the same scale. Dogs vary in weight over a 50-fold range from the Chihuahua, which weighs about 2 kg (~4 lb), to the English Mastiff, which can be over 100 kg (~220 lb; not shown) [1124]. Photos from *iStock* Getty Images; used with permission. (A black and white version of this figure will appear in some formats. For the color version, please refer to the plate section.)

Fig. 9.4. The tabby and the tiger. Note the difference in pupil shape – i.e., slit versus round (cf. Fig. 11.1 and Fig. 12.1). Whether this difference is merely correlated with body size or is somehow caused by it is unclear. Photos from *iStock* Getty Images; used with permission. (A black and white version of this figure will appear in some formats. For the color version, please refer to the plate section.)

Domestic cats have nowhere near the size range of dogs, but there is a comparable range when we widen our lens and consider the felid family as a whole. It turns out that big and small felids do differ in at least one consistent, albeit subtle, trait: the shape of the pupil in their eye. The pupils of lions, tigers, and other large felids are round, whereas the pupils of house cats are vertical slits (Fig. 9.4), and the middle-size lynx has an intermediate pupil shape [838]. This size–shape correlation appears to be more accidental than causal, however. The slit pupil helps cats see color images more sharply because their lens is multifocal (i.e., having concentric zones of varying focal lengths embedded inside), but for some reason larger felids lack this ability [838].

GP-17 tangent: Fly bristle patterns

Despite what seemed like goldmines in the span of their body sizes, neither dogs (species level) [1071] nor cats (family level) [969] have furnished any support for a size–anatomy linkage, aside from the ingot found by Alberch in the Great Pyrenees. It is here that fruit flies (genus level) come to the rescue [18,32]. Roughly 4000 *Drosophila* species have been described worldwide, a quarter of which live on the Hawaiian islands [971]. The Hawaiian fly fauna is fascinating for many reasons. For one thing, its males rival birds of paradise in the exuberance of their courtship rituals and in the extravagance of the ornaments that they flaunt while dancing their passionate jitterbugs [670]. But what concerns us here is their size. While not actually as big as birds, those island flies are huge compared with their mainland cousins (Fig. 9.5) [189,1209].

Studies of leg bristle patterns across this size spectrum show that the number of longitudinal rows remains constant, regardless of leg circumference (cf. Fig. 5.3) – implying that this aspect of the pattern is wedded to the segmental coordinate system (see Fig. 3.10) [1400], like a tattoo that simply expands as you gain weight [548].

Fig. 9.5. Range of size in the *Drosophila* genus. The smaller fly in the upper panel is a *D. melanogaster* (body length ≈ 2 mm); the larger is a *D. planitibia* (body length ≈ 7.2 mm). Dashed lines in the lower panel delimit midleg basitarsi of *D. saltans* and *D. planitibia*. Despite the difference in basitarsal size (length ≈ 235 vs. 1617 μm; circumference ≈ 94 μm vs. 245 μm) both leg segments have eight longitudinal rows of bristles, but the number of bristles per row varies in proportion to segment length [548]. The proportionality is due to constancy of bristle interval within each row, which in turn is attributable to a likely involvement of Turing-like devices for bristle spacing (see GP-20). Scale bar for lower panel = 300 μm. Photos of flies are courtesy of *Wikimedia/Creative Commons* and Luc Leblanc; the photo below is from [548] and is used by permission of *Springer Nature/RightsLink*. (A black and white version of this figure will appear in some formats. For the color version, please refer to the plate section.)

However, the number of bristles per row is *not* constant. Instead, it varies linearly with segment length – implying that the genome is enforcing a fixed bristle *density* (= sensory acuity), regardless of the overall area, like a corduroy fabric that keeps its ridge spacing no matter how big the pants are that are cut from it [549,1209].

Frontispiece. Cyclopic goat.

Fig. 1.1. A frog whose eyes developed inside its mouth.

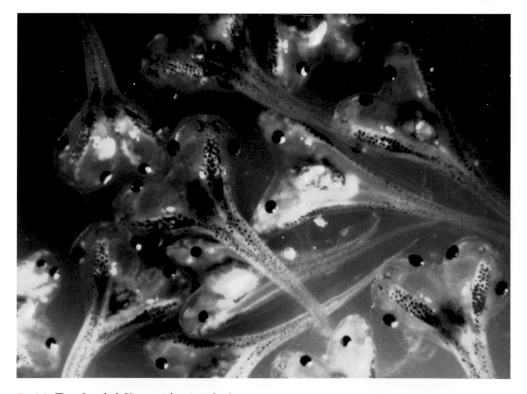

Fig. 2.1. Two-headed *Xenopus laevis* tadpoles.

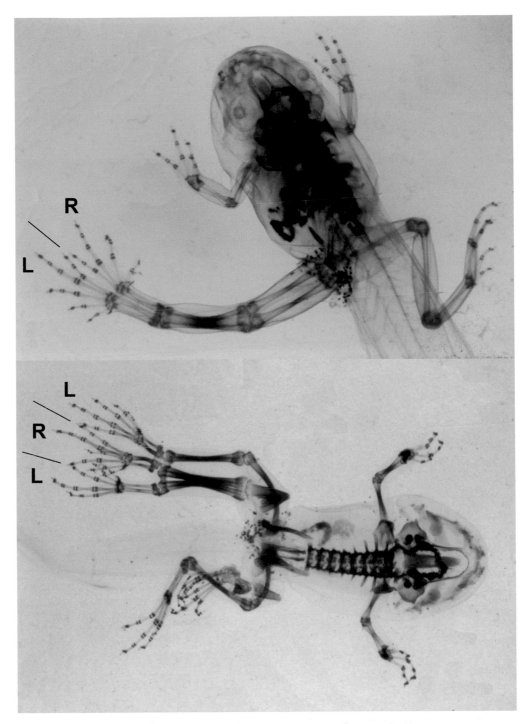

Fig. 3.2. Frogs with planes of symmetry between adjacent legs.

Fig. 6.1. Wild-type fly and $ac^3\ sc^{10-1}$ mutant fly.

Fig. 6.8. Newborn great apes.

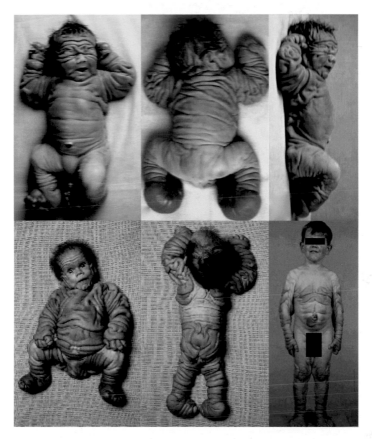

Fig. 7.2. Shar-Pei syndrome in a human.

Fig. 7.3. Shar-Pei syndrome in a cat?

Fig. 8.1. Bully Whippet (right) and normal Whippet (left).

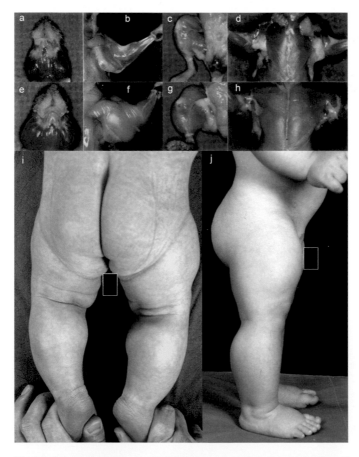

Fig. 8.2. Normal mouse (a–d), "Mighty mouse" (e–h), and "Superboy" (i–j).

Fig. 9.2. Kooky digits of the koala.

Fig. 9.3. Size spectrum spanned by dog breeds.

Fig. 9.4. The tabby and the tiger.

Fig. 9.5. Range of size in the *Drosophila* genus.

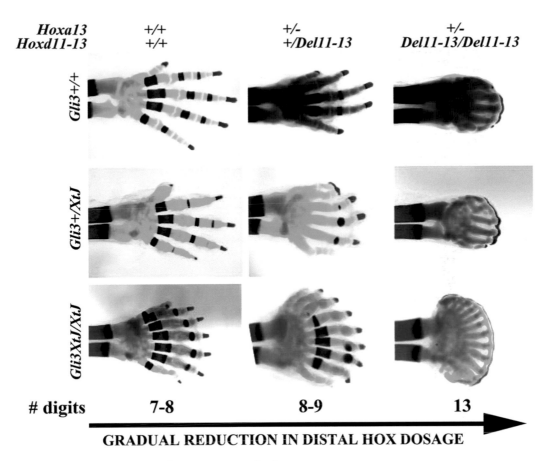

	Hoxa13 +/+ Hoxd11-13 +/+	+/- +/Del11-13	+/- Del11-13/Del11-13

Gli3+/+

Gli3+/XtJ

Gli3XtJ/XtJ

 # digits 7-8   8-9   13

→

GRADUAL REDUCTION IN DISTAL HOX DOSAGE

Fig. 9.8. Skeletal phenotypes of mutant neonatal mice.

Fig. 10.1. Tabby cats and cheetahs each come in two varieties.

Fig. 10.2. Spotted versus king cheetah.

Fig. 10.3. Mammal and insect pigment patterns differ in symmetry.

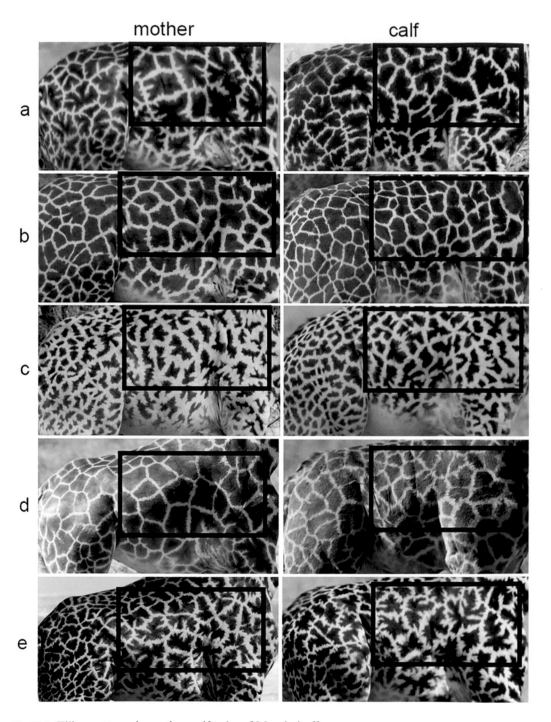

Fig. 10.4. Tiling patterns in mother–calf pairs of Masai giraffes.

Fig. 10.5. Spotted zebras.

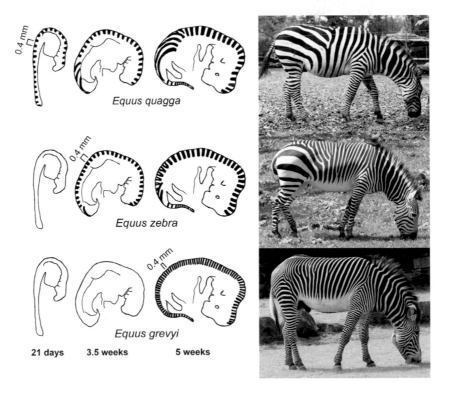

Equus quagga

Equus zebra

Equus grevyi

21 days   3.5 weeks   5 weeks

Fig. 10.6. How zebras get their stripes.

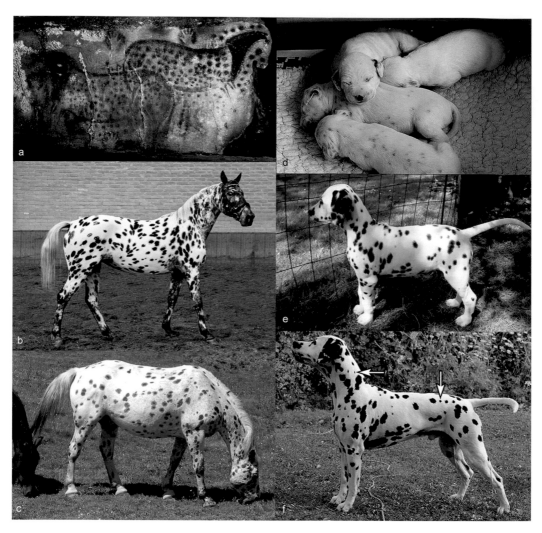

Fig. 10.8. Spotted horses and spotted dogs.

Fig. 12.1. Calico cat.

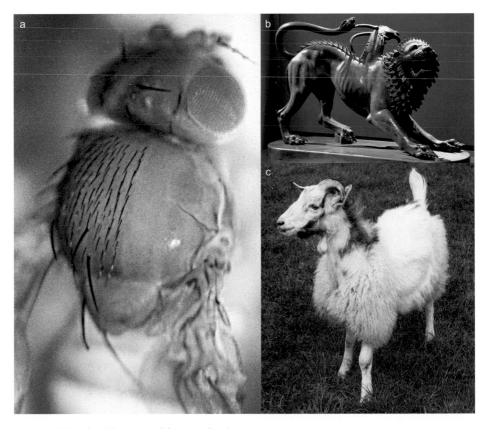

Fig. 12.2. Mosaics (a) versus chimeras (b, c).

Fig. 12.3. Gynandromorph butterflies and a mosaic lobster.

These inter-species trends match the size-dependency of bristle patterns within *D. melanogaster* itself as a function of nutrition [547], leading to a firmer understanding of how bristles are sited in general [554,562]. Here we can clearly see the synergy of combining developmental and evolutionary analyses, which is the hallmark of the evo-devo field [186].

GP-18: The Hedgehog pathway controls digit number

Polydactyly is not confined to dogs like the Great Pyrenees and Saint Bernard. It is also found in cats. The breeds where it is most prevalent are the Maine Coon and Pixie-Bob [513], and there is a strain known as Hemingway cats, named for Ernest Hemingway [749]. The trait is generally inherited as an autosomal dominant, but its manifestation varies from cat to cat, and from foot to foot. In his 1894 book [80], Bateson analyzed polydactyl cats extensively [749] and defined four "conditions" of the hind foot from the point of view of a right foot, based on the asymmetric orientations of successive claws (R = right-handed digit; L = left-handed digit; A = ambiguous digit type):

> Condition 1 [normal situation]: In the normal hind foot of the Cat there are four fully formed toes, commonly regarded as II, III, IV and V, each having three phalanges. In the place where the hallux would be there is a small cylindrical bone articulating at the side of the internal cuneiform. As usually seen, all the four digits are formed on a similar plan, each having its claw retracted to the external or fibular side of the second phalanx, the four digits of a right foot being all right digits and those of left feet being all left digits. The rudimentary hallux has of course no claw. [External to internal]: V(R), IV(R), III(R), II(R), I(rudiment).
>> Condition 2: [External to internal]: V(R), IV(R), III(R), II(Indifferent), I(L).
>> Condition 3: [External to internal]: V(R), IV(R), III(R), II(L), A(L), A(Hallux-like).
>> Condition 4: [External to internal]: V(R), IV(R), III(R), II(L), A(L), A(L). [80]

Among the abnormal types, condition 4 was most common according to Bateson, and he illustrated this anomaly using a foot marked "II" in Figure 9.6. That foot manifests an obvious plane of mirror symmetry, like the human arm in Figure 3.3c, which is also from his book.

Polydactyly in cats has been traced to mutations in a *cis*-regulatory enhancer of the *Shh* gene – the ZPA regulatory sequence (ZPS) – in the Hedgehog (Hh) signaling pathway [778,1244]. In such mutants, the extra digits arise via a second (anterior) site of expression of the Shh morphogen in the limb bud opposite the native (posterior) site (cf. Fig. 3.7). These two sources yield double-posterior, mirror-image feet (Fig. 9.6) [1282,1437]. The full spectrum of extra-toed cat phenotypes, the underlying genetics, and the alternative explanations for the deviant etiologies are cogently summarized in a 2014 review by Axel Lange *et al.* [750].

The double-dewclaw trait of dogs (Fig. 9.1) was also mapped to the ZPS region – but, surprisingly, the mutation in that case lies outside the ZPS itself, and it fails to cause any ectopic Shh within the limb buds [993]. Hence, the etiology of that oddity remains unclear in Great Pyrenees and other large breeds.

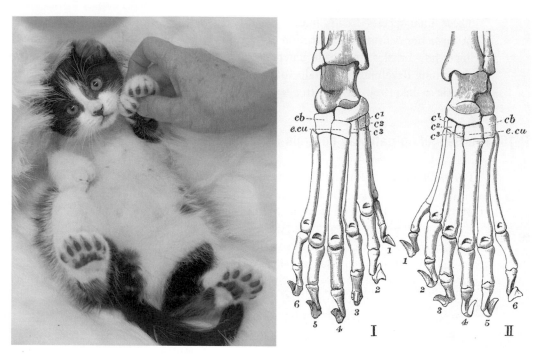

Fig. 9.6. Polydactyl cat and foot skeletons. Drawings are hind feet of a specimen in the Oxford University Museum (Fig. 87 of Bateson's 1894 book [80]): II, the left foot, has a plane of mirror symmetry such that toes 4–6 are left-handed, while toes 1–3 are right-handed; I, the right foot, is similar, though its digit 3 is ambiguous. Photo by Roham Sheikholeslami; provided courtesy of *Creative Commons*.

GP-18 tangent: Mice with thirteen toes per foot

The rarity of polydactyly in dogs, cats, and other vertebrates raises the question of how development manages to be so precise as to make a human hand, for example, with exactly five fingers – no more and no less – in almost every baby [391,1252]. In 2012 a team led by Rushikesh Sheth figured out the answer [1143]. Their explanation is based on a theory of pattern formation proposed 60 years earlier by Alan Turing [1306], and that theory will be recounted briefly before describing Sheth's solution of the five-finger problem.

Turing (1912–1954) was a mathematical genius. He is famous for inventing the modern digital computer (albeit not single-handedly) and for helping the Allies defeat Hitler by cracking the Nazi's Enigma code [235]. The latter story was portrayed in the biopic *The Imitation Game* [590]. Shortly before he died, Turing published a paper titled "The chemical basis of morphogenesis" [58,1306]. It proposed a new way of thinking about how cells make patterns. He called the chemicals "morphogens" and postulated that they react with one another and diffuse throughout the tissue being patterned. Hence, his theory and its various derivatives have come to be known as reaction–diffusion (RD) models [153,721,875].

Turing's RD equations invoke two morphogen molecules [151,197], which were later codified as a slowly diffusing Activator (**A**) that stimulates its own production and a rapidly diffusing Inhibitor (**I**) that represses **A** (Fig. 9.7) [874]. Initially, **A** and **I** are distributed uniformly across the territory, but this situation is unstable, and statistical fluctuations eventually cause **A** to exceed **I** at random sites [700]. Wherever such blips occur, **A** can feed on itself, like a fire that flares on an island above sea level. As the amount of **A** rises, the amount of **I** does too because another

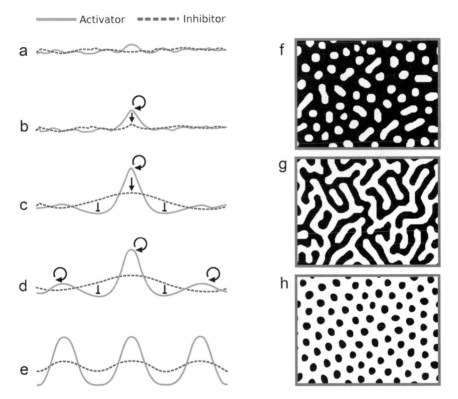

Fig. 9.7. Turing's reaction–diffusion (RD) model. a–e. Stages of pattern formation. Two types of morphogen molecules are involved: (1) an activator **A** that provokes its own production by a feedback loop (circular arrow) and (2) an inhibitor **I** that is stimulated by **A** (straight arrow) but suppresses **A** (T-bar). Levels of **A** and **I** are drawn as solid or dashed lines, respectively. **a**. Initially the levels of **A** and **I** are wobbly, so a blip of **A** will rise above baseline somewhere just by chance. **b**. This blip of **A** escapes inhibition by **I** long enough to feed back on itself, which causes it to rise further. **c**. As the blip grows, **A** not only makes more **A** but also more **I**, whereas **I** diffuses away quickly to form a shallow hump instead of a tall peak. The hump stifles further blips of **A** within a certain range (the RD wavelength). **d**. Beyond this range new blips of **A** can arise and repeat the process. **e**. Eventually the profile stabilizes with peaks of **A** at regular intervals along the axis, though generally not as regular as depicted here. This standing wave can then induce elements such as fingers. **f–h**. Examples of periodic patterns that can emerge in 2D sheets of cells instead of along a single axis [1156]. Labyrinthine patterns (**g**) might explain fingerprints (see Fig. 7.7) [403]. From a 2017 primer by Nogare and Chitnis [962]; used with permission.

aspect of the model is that **A** promotes **I**. The peaks of **A** stay sharp, but **I** can only form rounded hills because it spreads so quickly. The hills of **I** prevent more blips of **A** from emerging within a certain diffusion range (the RD wavelength), which dictates the interval between the final peaks. The model has always had a mystical aura about it because it creates heterogeneity from homogeneity, with periodic patterns emerging suddenly from a placid background like a surfacing submarine.

In summary, small deviations from homogeneity get amplified into a periodic array of singularities [569,843]. Peaks of **A** can then go on to induce structures such as fingers [516]. When the model is extended to two dimensions it can yield spots or stripes [723,926,927,928] or labyrinthine motifs (Fig. 9.7f–h) [113,1155]. Three-dimensional shapes can even emerge if the **A**-type cells grow faster than non-**A** cells [1105], and "morphodynamic" processes of that kind could be responsible for patterns such as fingerprint ridges [403], tooth cusps [643], or palatal rugae [332], but the quintessential proof of the Turing mechanism's involvement in

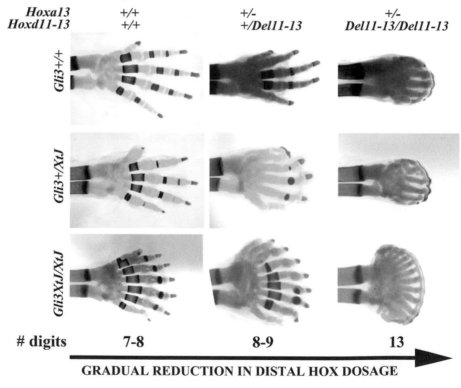

Fig. 9.8. Skeletal phenotypes of mutant neonatal mice. A wild-type foot with five toes is at the upper left. The most extreme case with 13 toes (*Gli3*-null *Hoxa13*-null *Hoxd11–13*-null) is at the lower right. Digit number increases with decreasing dosage of both *Gli3* (top to bottom) and distal Hox genes (left to right). Digit tips in the right column are connected by a thin band of cartilage and bone. From [1143]; used with permission from *Science/RightsLink*. (A black and white version of this figure will appear in some formats. For the color version, please refer to the plate section.)

developmental patterning is the mouse foot, where the morphogens appear to be BMP and Wnt [1047,1440].

Until 2002 the conventional wisdom was that the zone of polarizing activity (ZPA) at the rear edge of the limb bud specifies digits via a gradient of the Shh

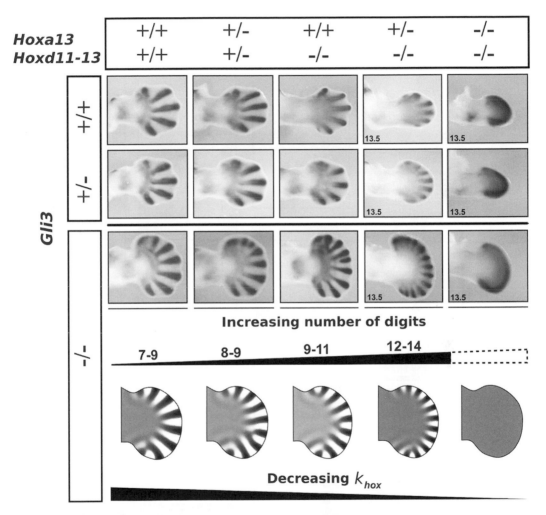

Fig. 9.9. Simulation of mutant mouse phenotypes. The top three panels show cartilage condensations (dark rays) in the handplates of mice carrying the same mutations as in Figure 9.8. Digit number increases along the same axes as in Figure 9.8, but digits fail to form when Hox function is totally gone (right column). The bottom panel depicts a Turing model of digit formation (k_{hox} = Hox dosage) that faithfully mimics (1) mutant trends in digit number, (2) shortening and thinning of digits with Hox dose from left to right, (3) dissolution of digits in Hox-null embryos, and (4) occasional digit bifurcations. Digit spacing is controlled by RD wavelength [1015]. Modifications to this model have recently been recommended based on the fact that digit condensations actually begin as spots that elongate into stripes, rather than as stripes ab initio [585]. From [1143]; used with permission from *Science/RightsLink*.

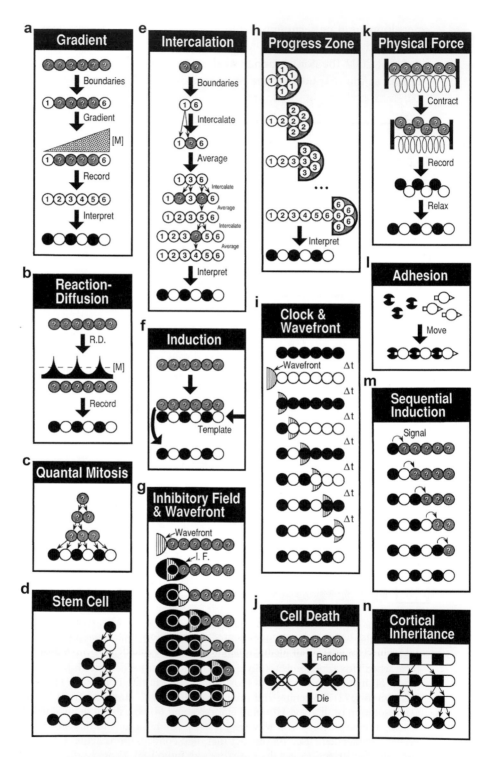

Fig. 9.10. Models for periodic patterning. Each mechanism is portrayed as leading to the assignment of alternating black and white cells in a line of six cells. Gray circles with an

morphogen that is secreted by the ZPA (cf. Fig. 3.7p) [621,1437]. In that year this paradigm was challenged when digits were found to develop even when *Shh* and its key downstream effector in the Hedgehog (Hh) pathway – *Gli3* [1348] – are disabled [806,1273]. In the *Shh*-null *Gli3*-null double mutant the number of digits increases, and they all look alike. Hence, the default state of the mouse foot was shown to be a fan of radiating spokes resembling the fin of a fish. Evidently, the Hh pathway serves to constrain digit number and impose differences as a function of the distance of each digit from the ZPA reference point [520]. Thus, the essential axiom of the paradigm survived intact: positional information (PI) does establish a coordinate system wherein cells can adopt different digital states [585].

←——

Fig. 9.10. (*cont.*)

inscribed question mark are naïve and uncommitted cells. Inscribed numbers represent positional values. **a**. Lewis Wolpert's source–sink Gradient model of positional information [1400]. Cells at termini (1 and 6) become a source (right) or sink (left) for the production or consumption of a morphogen (M). The steady-state gradient (triangle) allows cells to record [M] as positional values (2–5) that get interpreted as black (odd) or white (even) states. **b**. Alan Turing's Reaction–Diffusion model (see text) [1306]. Peaks of activator morphogen (M) arise at regular intervals, and any cell whose [M] exceeds a threshold (dashed line) becomes black; else white. **c**. Howard Holtzer's Quantal Mitosis model [599]. Cells proliferate until a final mitosis wherein each cell creates two different daughters: black (left) and white (right). **d**. Stem Cell model, wherein a stem cell (right) oscillates between black and white states as it creates daughter cells that keep their state. **e**. French *et al.*'s Intercalation model [377]. Cells adopt terminal states (1 or 6). Discontinuities greater than 1 cause cells to divide, and naïve daughters adopt a state that is the average of their neighbors until a scalar series (1–6) arises and is interpreted as in **a**. **f**. Induction model: e.g., notochordal induction of the nervous system in vertebrates. A template layer induces states in an overlying naïve layer. **g**. Donald Ede's Inhibitory Field and Wavefront model for hexagonal patterning [333], depicted here in one dimension. A wavefront (hatched semicircle) progresses from left to right, endowing cells with the ability to become black and to set up an inhibitory field (black oval), which prevents any cell therein from becoming black. **h**. Summerbell *et al.*'s Progress Zone model for the chick wing [1231]. Cells in a progress zone (half-circle) advance their state from 1 to 2, etc., based on a clock, whose time freezes when a cell leaves the zone, yielding a scalar array (1–6) that interprets values as in **a**. **i**. Cooke and Zeeman's Clock and Wavefront model [56,233]. Cells oscillate between black and white states until a wavefront (hatched hemicircle) reaches them, whence they keep that state. **j**. Cell Death model, based on pruning of neurons in the central nervous system [979]. Cells adopt states randomly, but then cells that find themselves next to an identical neighbor commit suicide asynchronously. **k**. Physical Force model: e.g., buckling of neocortex into gyri and sulci (see Chapter 7) [942]. **l**. Adhesion model, based on the ability of cells to sort out by type [1198]. **m**. Sequential Induction model: e.g., slime mold waves [903]. Black induces white, and vice versa. **n**. Cortical Inheritance model, based on cell fate determinants [312,834]. Alternating black or white determinants in a progenitor cell segregate into separate daughters. This menagerie of theoretical mechanisms is abstracted from *Models for Embryonic Periodicity* [553], which should be consulted for background, details, and further references. For other compendia see [478,1131,1383]. Hybrid versions of these models abound as well (e.g., PI–RD combinations; see text).

In 2012 a Turing RD amendment was added to this Wolpertian PI scenario (cf. GP-6) when experiments by Sheth *et al.* uncovered an unexpected role for Hox genes in digit patterning [1143]. Three genes at one end of the *Hoxd* complex (see Fig. 5.7b) – *Hoxd11–13* – along with the distalmost gene in the *Hoxa* complex – *Hoxa13* – were found to be critical for reducing digit number from an even more copious fish-like state of 13 digits per foot down to the (still excessive) *Shh*-null *Gli3*-null level of 7–8 digits per foot (Fig. 9.8).

Such severe polydactyly had never before been seen in any tetrapod foot, and it led Sheth *et al.* to investigate the involvement of a Turing mechanism. They tinkered with several variables of Turing's RD equations and were able to devise a simulation that mimicked virtually all aspects of the mutant digit trends. That simulation presumes that the distal Hox genes modulate the RD wavelength as a function of distance along the proximal–distal limb axis (Fig. 9.9) such that digit intervals widen as the digits approach their tips. The similarity of the maximal digit phenotype to the fins of sharks and other primitive fish implies that the same generative rules of the Turing algorithm are conserved in fins and limbs alike. Evidently, the tetrapod limb acquired its five fingers when (1) an Shh gradient was added along the anterior–posterior axis and (2) a Hox gradient was added along the proximal–distal axis [217,816,1015]. The Shh gradient reduced digit number somewhat (from ~13 to 7) and assigned identities, while the Hox gradient reduced the number still further (from ~7 to 5) and liberated the digit tips from one another.

A key axiom in the Sheth *et al.* simulation is the tuning of RD wavelength by agents outside the Turing mechanism itself. A tuning device of some sort had been envisioned to regulate the number of wavelength peaks by Jonathan Cooke in 1975. Cooke questioned the ability of Wolpert's recently proposed PI theory [1400] to explain repeated elements like fingers on its own (boldface added):

> The p.i. concept admittedly throws the burden of control, in production of normal pattern, onto the intracellular interpretation mechanism. The theory must then face the **problems posed by patterns with highly controlled species-typical numbers of repeating, similar elements.** . . . [I think that] **p.i. may in some systems necessarily regulate the dimensions (including the "wavelength" of repeating elements) of prepatterns.** . . .
>
> In straight p.i. terms, such patterns require a discrete, regularly distributed subset of values for the p.i. state variables to be similarly interpreted, while the subset of intermediate values is differently interpreted or "ignored." In terms of molecular mechanisms for switching circuitry, this makes numbers of elements up to, say, five perhaps plausible, but for multiple regular units it is patently inadequate. **Even for the five fingers in the hand rudiment, for instance, it seems more likely that other processes among spatial state variables cause the similar elements to arise as singularities in some sort of prepattern, while a p.i. gradient then causes their anterior/posterior character (thumb → little finger) as five *different* gradient values.** [232]

It turns out not only that Cooke was right about the blending of PI with RD – an idea that has been explored extensively in recent papers [478,483,537,1008] – but that he was also prescient in a theory that he proposed in 1976 for how the somite precursors of vertebrae are created from head to tail in the paraxial mesoderm of vertebrate embryos [233]. Both of the key components of his Clock and Wavefront model were eventually verified 30 years later at the molecular and cellular levels [56]. Alberch, whose musings launched this chapter, is better known than Cooke [307,1057], but his work has likewise been largely overlooked.

Scholars are obliged to credit their predecessors, but those who work in developmental biology may be forgiven for not wading into literature that has been flooded with a deluge of theoretical models for 50 years or so. Models for periodic patterning in particular have multiplied like rabbits, so there may be some value in trying to classify them for ease of reference and comparison. Figure 9.10 attempts to do that for readers who may feel bewildered or overwhelmed at this point.

Part IV

Cats

10 The Blotched Tabby

Breeds of the domestic cat (*Felis catus*) often have a tiger-like striping pattern, and the mackerel tabby is a prime example (Fig. 10.1) [340]. However, some "classic" tabbies exhibit irregular blotches instead of parallel stripes. In 2012 the genetic basis for this difference was revealed. Classic tabbies turn out to be homozygous for a recessive mutation in a gene called *Taqpep* (short for *Transmembrane amino-peptidase Q*) that affects skin pigmentation [664]. Such a finding, insightful though it may be, would not have aroused much interest had it been published alone, but the *Science* paper by Kaelin *et al.* that reported this result furthermore showed that the same type of mutation also explains a rare type of cheetah from southern Africa called the king cheetah [632,664]. Unlike the common spotted cheetah, king cheetahs manifest the same kinds of blotches as classic tabbies plus several distinct stripes down their back (Fig. 10.2). Those longitudinal stripes are reminiscent of the coat

Fig. 10.1. Tabby cats and cheetahs each come in two varieties. Tabby cats (left) can be either mackerel (above) or blotched (below), while cheetahs (right) can be either spotted (above) or blotched (below). Blotched cheetahs are also called king cheetahs. The blotched phenotype in both cases turns out to be caused by similar mutations in the same (*Taqpep*) gene [664]. Cheetahs and other spotted felines tend to have striped tails for reasons having to do with the quirky behavior of Turing models on narrow cones and cylinders [66,929]. Cat photos were taken by Helmi Flick; all were kindly supplied by Greg Barsh and are used with permission from *Science*/*Rightslink*. (A black and white version of this figure will appear in some formats. For the color version, please refer to the plate section.)

Fig. 10.2. Spotted versus king cheetah. A common cheetah (*Acinonyx jubatus*) (left) is compared with a king cheetah (right). In the king cheetah the spots have fused into blotches, and there are parallel stripes (in this case, three full ones and two partial ones) along the back. Surprisingly, these stark differences are due to a single mutation. Note that the normal cheetah has smaller dots interspersed between its regular spots, which may be indicative of a second round of spot formation that stopped before it could make full-size spots [560]. Photo courtesy of Greg Barsh. (A black and white version of this figure will appear in some formats. For the color version, please refer to the plate section.)

patterns in chipmunks and related rodents [648,836], and all of these dorsal designs may share a similar etiology [664].

How could a simple mutation create a blotched pattern from two entirely different starting points: a *striped* tabby on the one hand and a *spotted* cheetah on the other? Here is where Turing's reaction–diffusion (RD) model, which was introduced in Chapter 9, proves its mettle. According to that model, spots and stripes are not really as different as our eyes would lead us to believe, nor do they differ in any significant way from the quasi-labyrinthine designs that we perceive as "blotchy" [340]. All three types of pattern (spot, stripe, and blotch) can easily emerge from the witch's brew of activator (**A**) and inhibitor (**I**) molecules in Turing's chemical cauldron (see Fig. 9.7). Indeed, his model can shift from one type of outcome to another with only minor tweaks in the coefficient parameters [22], territory topography [929], or tissue orientation [66,584].

The Taqpep protein itself cannot be acting as either of the diffusible agents (**A** or **I**), because it is firmly anchored to cell membranes, but Kaelin *et al.* found evidence that a separate factor could be playing such a role. They discovered that the *Edn3* gene is expressed more intensely in the black versus yellow skin territories (2× higher in tabbies and 5× higher in cheetahs). This gene encodes a paracrine hormone that could be playing an RD role of some sort. However, the endothelin pathway, to which *Edn3* belongs, is also known to increase the number and density of the dermal melanocytes that are responsible for skin pigmentation [1318,1319], so the topography of dark and light areas in feline skin could actually be relying more on *cell* behavior than on the kinds of *molecular* antics envisioned by Turing's model [212,349,1131].

GP-19: Mammal coat patterns use Turing-like devices

In the seven decades since Turing proposed his archetypal hypothesis, a host of derivative models have been concocted [235], all of which retain his tenets of local activation (LA) and long-range inhibition (LI) [748,928]. This "LALI" family of mechanisms [962] was championed by Alfred Gierer and Hans Meinhardt [429,874,875] and by James Murray [927], who specifically adapted them to mammal coat patterns [925,926]. The descendant versions range from chemical to cellular to physical incarnations [537], as well as a gaggle of hybrid spinoffs [962], and it can be somewhat difficult to distinguish them empirically [583,584].

Regardless of how they are formulated, all of these Turing-like devices differ from Wolpert-like devices insofar as they envision patterns as emerging "bottom-up" from actions on a local level [1052], rather than responding to "top-down" signals that enforce global coordinate systems [143,537,1324]. They share a core set of assumptions about how patterns develop. For example, they predict that the two sides of the body should look different because the LALI mechanism is susceptible to random perturbations – i.e., "noise" [66,700,1046]. Mammal coat patterns are indeed asymmetric (Fig. 10.3). In contrast, cuticular patterns of insects (e.g., butterflies) are symmetric, presumably because they use morphogen gradients (see GP-6) [924].

Giraffes take the property of asymmetry to its logical extreme because their tiling patterns spread over their back without even a hint of respecting the midline (Fig. 10.3). The exact form of the tiling differs among the ~9 subspecies [101,159], with reticulated giraffes having neat polygons and Masai giraffes having radiating splotches inside the polygonal compartments. The Masai markings look like lobate leaves, with the amount of incision around the foliate perimeters varying from one individual to the next.

The breadth of lobate shape diversity within the Masai subspecies is shown in Figure 10.4, which also illustrates another Turing-like property: heritability of patterns is quite low because the LALI process operates largely beyond the control of the genome. Indeed, if giraffes were cloned like dogs, then we would not expect the identical offspring to be more similar than mother–calf or sib–sib pairs [695]. This same argument was made earlier for the labyrinthine designs of human fingerprints (see Fig. 7.8).

The amount of ruffling around the edges of the splotches in this sample (Fig. 10.4) varies from a minimum for the mother in panel *d* to a maximum for the calf in panel *e*. Indeed, the mother in panel *d* looks more like a reticulated giraffe than a Masai one, despite its verified Masai identity. In surveying these motifs it is easy to see how the lobed splotches could transition into neat polygons if each were to expand in all directions until it abutted its neighbors on all sides [560].

According to this notion, skin patterning would occur in two distinct phases. In phase 1 a LALI process would create a cheetah-like array of spots in the embryo.

Fig. 10.3. Mammal and insect pigment patterns differ in symmetry. The tiling patterns of
giraffes, the striped patterns of zebras and tigers, and the spot patterns of jaguars (upper
row) are left–right asymmetric (cf. Fig. 10.2), whereas analogous color patterns in
butterflies exhibit bilateral symmetry (lower row). The dichotomy is attributable to a
reliance of these respective taxa on Turing-like versus Wolpert-like patterning
mechanisms (see text). The giraffe belongs to the subspecies *Giraffa camelopardalis
reticulata* [159,1158], which has polygons instead of splotches (cf. Fig. 10.4). Those
polygons disregard the body midline – implying that the skin is patterned as a unit, with
no partition between left and right sides as in other mammals. Butterflies (left to right):
Rhaphicera dumicola (ventral), *Callicore atacama* (ventral), *Marpesia eleuchea* (dorsal),
Timelaea albescens (ventral). From [560], reprinted with permission. (A black and white
version of this figure will appear in some formats. For the color version, please refer to
the plate section.)

In phase 2 each spot would start to spread in all directions at different rates along
its margin like a coffee stain in a linen shirt, giving it an irregular lobate
appearance. Spreading would arbitrarily cease at different stages in different
subspecies, causing the blotches to be "caught in the act" to varying extents in
each giraffe subspecies. This same sort of two-step cascade could also help to
explain the rosette arrays of leopards and jaguars [722,809]. In those cases, phase
2 would entail splitting of the original spot into 3–5 remnants during the spreading process [58,1372].

How each of these phases might have been programmed into the genomes of different mammalian lineages over the eons is unclear [212], but what *is* clear is that one cell type holds the key to unlocking this unsolved mystery. The melanoblast arises from the migratory neural crest (cf. Fig. 1.2) [961] and matures to become the pigment-producing melanocyte [752,755,1321]. Ergo, if we want to know how the leopard got its spots, the zebra its stripes, and the giraffe its polygons, then we will have to figure out the games that melanoblasts play [722] in terms of oriented mitoses, directed movements, limited suicides, targeted signaling, and other "if/ then" instructions that they obey based on subtle local cues (see GP-5) [1025,1115].

GP-19 tangent: The spotted zebra

Zebras offer puzzles that are at least as beguiling as those of a giraffe's pelage. How do their melanoblasts make stripes with such even spacing and sharp edges? Some of the best clues we have in trying to solve these puzzles come from very rare individuals whose stripes are replaced with spots. Only two cases of fully spotted zebras have ever been reported, and both involve the plains zebra (*Equus quagga*). One was documented in 1967 [66], the other in 2019 (Fig. 10.5). The patterns of those two zebras are remarkably similar, with the spots aligning in rows which follow the normal contours of the stripes that are characteristic for this species.

One perennial riddle about zebras is whether they are black horses with white stripes or white horses with black stripes [471]. These two spotted zebras appear to settle the issue in favor of the former conjecture, because . . .

> It is only possible to understand the pattern if the white stripes had failed to form properly and that therefore the "default" color is black. The role of the striping mechanism is thus to inhibit natural pigment formation rather than to stimulate it. Further evidence for this view comes from the now-extinct *quagga* [today considered a subspecies of plains zebra]: this animal is striped on the head and shoulders alone, the unpatterned rump is dark. [66]

The ability of the zebra's white stripes to dissolve into white spots reinforces the conclusion we were forced to adopt in contemplating the reciprocal case of the king cheetah whose spots fuse into stripes along its back (Fig. 10.2) – namely, that stripes and spots are not really as different from the perspective of the patterning machinery (a Turing-like device?) as they may appear to us [340,688] or to natural predators [181,981]. We don't yet know how zebras become spotted, nor whether the gene(s) involved are related to the disabled gene (*Taqpep*) that makes cheetahs striped.

GP-20: Turing-like devices have a fixed wavelength

Over the years theoreticians have tried to figure out why zebras are striped in the way they are [480,642]. Some of their observations and speculations are at least worth mentioning.

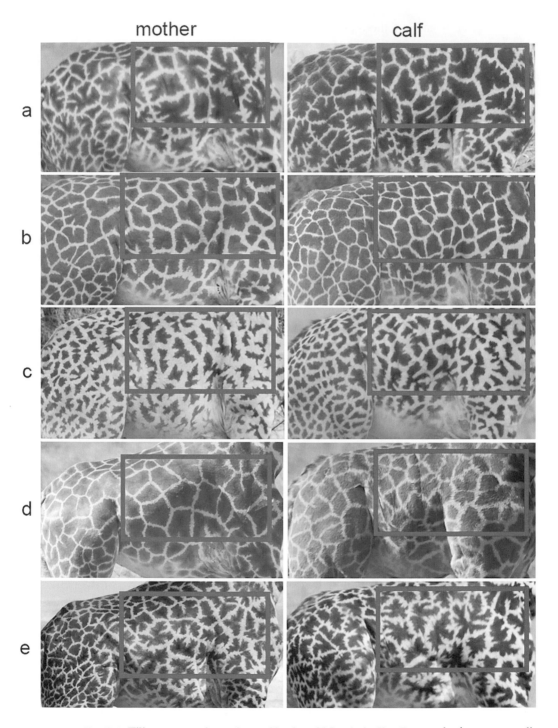

Fig. 10.4. Tiling patterns in mother–calf pairs of Masai giraffes. Rectangles in corresponding parts of each panel were used to estimate heritability of spot shapes. The regular polygonal tiling of the mother in panel **d** is typical of reticulated giraffes [159], but this individual was

1. Zebras encounter a problem where the vertical stripes encircling the body meet the horizontal stripes encircling the legs [925]. All three zebra species solve this problem in the same way at the shoulder: one body stripe typically bifurcates to allow intrusion of the transverse stripes coming up from the leg, yielding an inverted-Y "triradial" zone that resembles the chevrons on a sergeant's sleeve. One species – *Equus grevyi* – uses the same strategy for its hindleg, but the other two let the transverse stripes from the hindleg continue into the thigh and rump, leading to one or more *non*-inverted-Y triradii near the middle of the flank [66].

2. The three species of zebra differ noticeably in the spacing of their stripes. Stripes are furthest apart in *E. quagga*, less so in *E. zebra*, and closest together in *E. grevyi* (Fig. 10.6). In 1977 Jonathan Bard interpreted this trend in terms of another fundamental property of Turing-like devices – viz., a fixity of wavelength due to constant widths of LALI inhibitory fields. He overlaid gratings of constant interval (0.4 mm) upon drawings of horse embryos at different stages to estimate when skin cells decide whether to later become black or white (Fig. 10.6). Once those decisions are made, the number of stripes would stay constant, while their intervals would widen for the remainder of the embryonic growth period.

3. Some races of *E. quagga* tend to have gray "shadow" stripes between the black stripes on the haunches [66,1030], suggesting a second round of Turing-like stripe formation that aborted prematurely [65,560]. Fred Nijhout put it this way: "It looks like the patterning system is attempting to fit additional stripes in where the separation between existing stripes becomes large" [954].

If Bard is right that all zebras use the same RD wavelength to make stripes (point #2 above), then the differences in stripe spacing among the three species should be due to different *times* when stripes get imprinted in the embryonic skin (Fig. 10.6). It is therefore no surprise that when species with the largest and smallest spacings (*E. quagga* and *E. grevyi*, respectively) are crossed, their hybrid offspring resemble members of *E. zebra*, whose intervals are intermediate, implying that the imprinting time can slide along that continuum [238].

If Nijhout is right that a second phase of striping might be able to start before the window of opportunity for imprinting closes completely (point #3 above), then the *duration* of the patterning period must be taken into account for Turing-like devices as well. The extreme situation would be where the duration is infinite, and new stripes – or spots – keep popping up between previous ones during

←————————————————————————————————

Fig. 10.4. (*cont.*)

definitely a Masai giraffe (*Giraffa camelopardalis tippelskirchii*) (Derek Lee, personal communication). The spots on several of the giraffes (especially panel **e**) have fractal shapes that could be created by branching algorithms [660]. For variations in tiger striping see [821]. Contrast-enhanced from [768]; used with permission of *Creative Commons*. (A black and white version of this figure will appear in some formats. For the color version, please refer to the plate section.)

Fig. 10.5. Spotted zebras. The polka-dotted foal trotting behind its mother is named Tira after Anthony Tira, the Masai guide who first noticed him. The photo was taken ca 13 September 2019 by Rahul Sachdev in Kenya's Masai Mara game reserve, which is famous for its wildebeest migrations. The foal was about 4–5 months old at the time. The inset shows a mature spotted zebra seen in Botswana in 1967. Both are plains zebras (*Equus quagga*) (cf. Fig. 10.6). Note the similarities of their spot patterns. For other anomalous zebras see [21]. Large photo is courtesy of Monika Braun, founder of Matira Bush Safari, Ltd., Kenya. Inset, published in South Africa's *Star* newspaper on 30 December 1967, is courtesy of Jonathan Bard [66]. (A black and white version of this figure will appear in some formats. For the color version, please refer to the plate section.)

growth. That sort of perpetual percolation was noticed by Vincent Wigglesworth long ago in insects that shed their cuticle periodically as they grow [1390,1391]. When he saved the shed cuticle from one instar (juvenile stage) and compared it to the next instar, he found that new bristles always arise in the largest gaps of the previous pattern (Fig. 10.7).

This sort of intercalation is consistent with a Turing-like LALI mechanism, assuming (1) each bristle has an inhibitory field and (2) each field has a fixed diameter. In that case, the first epidermal cells to have a chance to become bristles would be those midway between neighboring bristles, because inhibitory fields will move apart and cease to overlap as the skin grows from one instar to the next. Hence, the density of bristles will remain constant. This same sort of argument was previously made for fly bristles, which rely on Achaete and Scute for their activators and the Notch pathway for their inhibitor (see Chapter 6). Indeed, LALI logic also explains why the number of bristles in each longitudinal row varies with leg segment length during evolution (see Fig. 9.5): bristles are obeying an invariant Turing wavelength, even if they're not utilizing a chemical RD recipe sensu stricto.

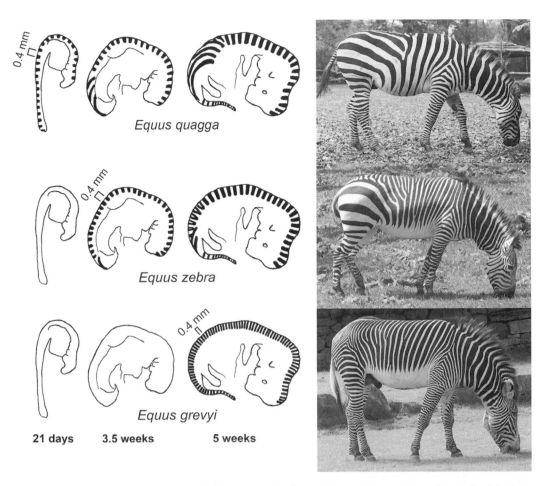

Fig. 10.6. How zebras get their stripes. The three zebra species are shown [470,1030] with their inter-stripe intervals decreasing from top to bottom. At the left are drawings of horse embryos at various stages. Jonathan Bard proposed that stripes are determined by a Turing-like device whose wavelength is ~0.4 mm no matter when it comes into play [65]. Based on species differences, Bard estimated that cells commit to black or white fates at 21 days in plains zebra (*Equus quagga*), 3.5 weeks in mountain zebra (*E. zebra*), and 5 weeks in Grévy's zebra (*E. grevyi*) (cf. the effects of incubation temperature on striping patterns in alligators [930]). One of the many riddles posed by zebra coat patterns in general is how the edges of the stripes are able to become so sharp and straight, despite the inherent noisiness of the LALI devices that are supposed to paint them [1046]. Nevertheless, a gradient mechanism would be even worse at trying to draw dozens of stripes at once without blurring them horribly [634]. Drawings were furnished by Jonathan Bard. Photos (top to bottom) by jgorinjac, GeekAaron, and Johnwobert; used with permission of *Foter.com*. (A black and white version of this figure will appear in some formats. For the color version, please refer to the plate section.)

Between their obvious spots, cheetahs have subtle dots that may correspond to the incipient bristles of insects [560]. If so, then a second round of spot formation would start before the patterning window slams shut and stops the dots from becoming spots (see Fig. 10.2). Curiously, a process of this sort has been

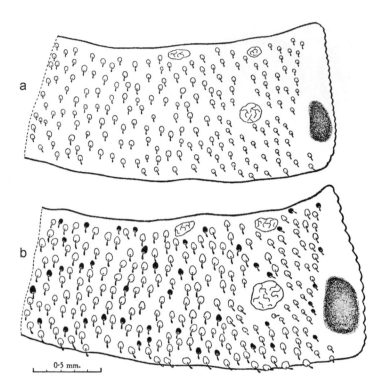

Fig. 10.7. New bristles arise via a LALI process during insect growth. Cuticle of the right half of the third abdominal tergite (dorsal part of the abdomen) of the hemipteran bug *Rhodnius prolixus*. The same individual is shown at successive stages of its development: fourth instar (**a**) and fifth instar (**b**). Old bristles persist in the epidermis from one instar to the next (stubby shafts protruding from white sockets), but new ones (black sockets) arise de novo as the insect grows. Note that new bristles develop in the largest gaps of the previous pattern. Such a trend is consistent with the LALI model of patterning because the inhibitory fields around bristles should separate as the skin expands, thereby allowing new bristles to sprout in the biggest gaps. From [1390]; used with permission from *Company of Biologists/Copyright Clearance Center*.

documented in spotted dogs. Investigators monitored spot locations in individual Dalmatians over several years and found that new spots do arise as the dog grows (Fig. 10.8) [320]. This discovery may not be relevant to the LALI school of models, however, because Dalmatian spots, unlike cheetah spots, are not evenly distributed. On the contrary, they seem to be almost random [320].

Whatever the (non-LALI?) mechanism may be for dappling a Dalmatian, it probably acts similarly in spotted horses, which also display irregular spot patterns (Fig. 10.8) [95,538,1193]. Indeed, spotted horses look a lot more like Dalmatians than like their own cousins – the rare spotted zebras (Fig. 10.5). Cave paintings of dappled horses from ~25,000 years ago suggest that they were prevalent in the Paleolithic [1038], but DNA analyses of Siberian fossils show that spotted breeds waned in the Middle Ages for reasons that remain hard to discern [1412].

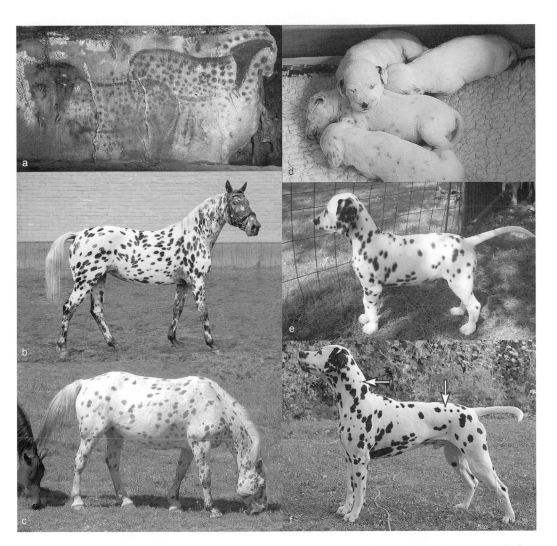

Fig. 10.8. Spotted horses and spotted dogs. a–c. Spotted horses. **a.** Ice-Age cave painting (~25,000 years old) of two dappled horses from the Pech Merle cave in southern France [1038]. The upper right corner of the panel wall is shaped like a horse's head, enhancing the verisimilitude of this rendition in which each horse is ~1.6 m (5¼ ft) tall [23]. However, the spots could be merely symbolic [23]. **b.** Knabstrupper breed [1362]. **c.** Appaloosa breed [1041], which suffers from night blindness [1362]. **d–f.** Dalmatians [320]. **d.** Litter of puppies 2 weeks after birth. Spots are just beginning to appear. **e.** Puppy at 10 weeks of age. **f.** The same dog at 6 years. Arrows indicate new spots. Photos were provided courtesy of Monika Reißmann (**b**) or used with permission of *Open Access/ Creative Commons* (**a**), *Wikimedia/Creative Commons* (**c**), *Springer Nature/RightsLink* (**d–f**), with high-resolution images of **d–f** furnished by László Pecze. (A black and white version of this figure will appear in some formats. For the color version, please refer to the plate section.)

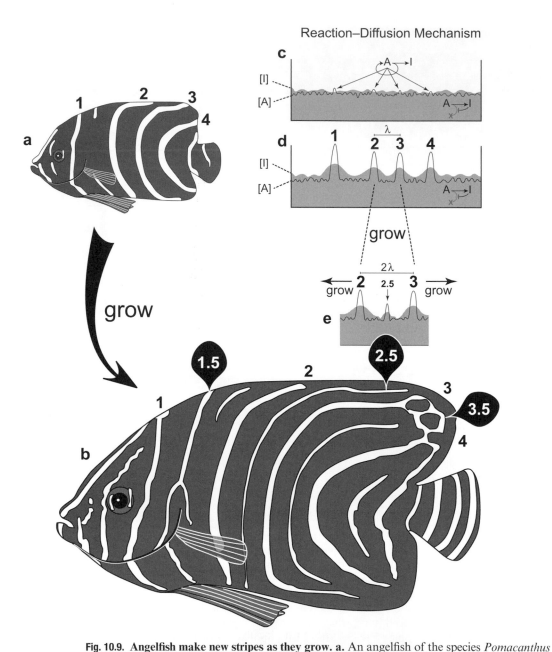

Fig. 10.9. Angelfish make new stripes as they grow. a. An angelfish of the species *Pomacanthus semicirculatus* at ~6 months of age (body length ≈ 4 cm). Major stripes are numbered (1–4). **b.** The same individual at ~12 months (body length ≈ 8 cm). New stripes (1.5, 2.5, and 3.5) have intercalated midway between prior stripes (1, 2, 3, and 4) so as to maintain a constant interval. New stripes are initially thinner than old ones but gradually broaden.
c–e. Simulation of striping (*x* = body axis) based on Turing's theory (see Fig. 9.7) [874].
c. A slowly diffusing activator (**A**; thin line) elicits a rapidly diffusing inhibitor (**I**; gray fill) and catalyzes itself (loop), but **I** suppresses **A** (T-bar) to thwart its feedback (**X**). Levels of

GP-20 tangent: The angelfish

One of the clearest examples of a pattern conforming to Turing's rules is the striped skin of the angelfish *Pomacanthus semicirculatus* [720] (Fig. 10.9). New stripes arise midway between old ones, analogous to insect bristles (Fig. 10.7), so as to maintain their initial periodicity. Whether **A** and **I** chemicals are actually diffusing throughout the fish's skin to ensure a constant inter-stripe wavelength remains to be determined, as do their identities, assuming they exist. As discussed above, though, other variations on this theme could be involved instead that use cells rather than molecules to calibrate distances and ensure regularity.

The angelfish study was conducted by Shigeru Kondo and Rihito Asai at Kyoto University. They published their results in 1995 in an article featured on the cover of *Nature* with the headline "Turing patterns come to life" [720]. In that same paper they also applied the Turing model to another species in the same genus whose stripes run horizontally instead of vertically. They showed that stripes can split via branchpoints that slide across the skin like a jacket zipper. Overall, the essay was a milestone in this field. Kondo has subsequently analyzed the stripe pattern of zebrafish from a LALI perspective [1358], and he has written seminal reviews on the entire field of animal pattern formation that should be digested by anyone interested in pursuing research in this arena [721,722].

←

Fig. 10.9. (*cont.*)

A and I (*y* axis) are comparable but vary locally. **d.** Fluctuations (waviness) allow blips of **A** to sprout as they escape inhibition, like brush fires above sea level. The **A** peaks stay high, but the **I** that they make spreads to form mounds. On average, peaks will be separated by a certain wavelength (λ) enforced by **I**. **e.** As the fish grows, peaks move apart so that new blips can emerge beyond the mounds of **I** around each **A** peak. Here a new peak (2.5) rises between previous ones (2 and 3). If the striping were controlled by a Wolpert-like positional information (PI) device instead of a Turing-like reaction–diffusion (RD) one, then the number of stripes should remain constant as the fish grows [179,634,983,1310]. Adapted from [560]. Fish in **a** and **b** were redrawn from [720].

11 The Siamese Cat

Siamese cats are born white but acquire a dark nose, ears, feet, and tail as they mature (Fig. 11.1) [617]. This "pointed" phenotype is due to a recessive LOF mutation in the *tyrosinase* (*tyr*) gene [823]. Comparable mutations in the *tyr* gene cause similar age-dependent phenotypes in Burmese and Tonkinese cats (different alleles from Siamese) [663,823], "Himalayan" mouse [90,741], Syrian hamster [484,1072],

Fig. 11.1. Siamese cat. Its dark markings are due to a mutation that partly disables a melanin-synthesizing enzyme (tyrosinase) – allowing it to function in cool parts of the body (extremities) but not warm ones. Null mutations cause an albino (pure white) phenotype regardless of temperature [620]. Partial-LOF alleles are relatively common in mammals (see text), so it is surprising that there are no "Siamese dogs" [663]. Though not obvious here, Siamese cats – and indeed albino mammals in general – have crossed eyes [105,1362,1419] because loss of eye pigment causes excess crossing of nerves at the optic chiasm [867,1034,1064] and a miswiring at higher levels of the visual system [43,594,661]. (Oddly, such miswiring is even seen in albino *heterozygotes* [779], and the auditory brainstem is affected as well [249].) The opposite defect was detected in Belgian Sheepdogs that lacked an optic chiasm entirely due to an unmapped mutation [1034,1395]. Despite the correlation of pigment with vision at the organism level, the story is different at the genome level, where separate *cis*-regulatory "area codes" are used to express tyrosinase in the eye and skin (cf. GP-11) [923]. Photo from *iStock*; used with permission.

Mongolian gerbil [1019], "albino" guinea pig [1409], American mink [98], and hamadryas baboon [716].

Also on that list is the Himalayan rabbit, whose genetics was first analyzed by Alfred Sturtevant [1223], about whom more will be said in the next chapter. Experiments with the rabbit helped clarify the etiology of all such "pointed" mammals [1203]. When fur was plucked from white areas flanking the spine (a few cm^2 per spot), it grew back white in all rabbits ($n = 50$) exposed to outside temperatures $> 11\ °C$, but grew back *black* in all rabbits ($n = 20$) exposed to outside temperatures $< 6\ °C$ [680].

In analogous studies with Himalayan mice the hair grew back black at $15\ °C$ and white at $30\ °C$ [691], and in "albino" guinea pigs it grew back black at $16\ °C$ and white at $32\ °C$ [1399]. Hence, the reason for the distinctive pigment pattern of Siamese cats and their lookalike counterparts is that the mutated tyrosinase works fine at low temperature, but not at high temperature [617]. There is even a family of humans with the same kind of mutation, which in their case deprives the enzyme of function above $35-37\ °C$ [696]. They mimic Siamese cats in the time course of their pigment maturation, albeit more subtly. The lower leg hair on the otherwise white patient in this excerpt was dark brown:

> The current proband and her two affected brothers completely lacked melanin pigment at birth and initially appeared to have classic tyrosinase-negative OCA [oculocutaneous albinism]. After puberty, however, they developed a unique pattern of pigmentation: relatively warm body parts (eyes, skin, hair of the scalp and axilla) remained unpigmented, but less warm parts (facial and pubic hair) developed slight pigmentation and relatively cool parts (arm and leg hair) became well pigmented. [428]

Tyrosinase is a membrane-spanning enzyme [663] that uses copper to catalyze production of melanin from the amino acid tyrosine (whence the name) [237,776,973]. The mutation that makes it vulnerable to warming is a substitution of a conserved glycine by arginine in the middle of the protein. That missense change might be weakening the enzyme's backbone so that it unfolds when agitated by Brownian collisions [823,1118], just as fried eggs turn white when their ovalbumin proteins denature [1303]. However, we don't yet know the exact cause of the temperature sensitivity.

> The substitution might affect the enzymatic efficiency, the trafficking, or the interaction with an inhibitory molecule. Any of these changes can occur directly or indirectly, through modification of the secondary structure of [the tyrosine protein]. [1118]

The mutation that makes tyrosinase heat-sensitive arose spontaneously, but geneticists realized decades ago that they could artificially create such mutations in any genes of interest and then use them to ascertain when those genes are utilized [1303]. This clever approach was first used in microorganisms [139,529] and bacteriophages [640] and later applied to higher eukaryotes, beginning with *Drosophila* in 1970 [1242]. The method proved to be a powerful way of pinpointing the times and places when otherwise lethal genes like *Notch* are deployed (see Fig. 6.10) [502], as well as dissecting pleiotropic syndromes that had seemed too complex to comprehend [1139,1365].

GP-21: Temperature acts as a toggle in some species

Some butterfly species develop alternative color patterns on their wings when their caterpillars metamorphose at different times of the year [956], suggesting that they might be using the outside temperature to steer development in one direction or the other [144], and in several species this suspicion has been verified directly (Fig. 11.2) [957]. The most noteworthy such confirmation was reported in 1875 by August Weismann (1834–1914), famous for his germ plasm theory [1196], who noticed that European map butterflies (*Araschnia levana*) emerge with different wing designs in spring versus summer. He tried cooling the pupae of the summer variety, and thereby induced them to manifest the spring design [430].

Richard Goldschmidt (1878–1958) cited the seasonal forms of butterflies as key evidence for the "reaction norm" concept in his magnum opus on evolutionary mechanisms, *The Material Basis of Evolution* [447,1204]. The basic idea, which was proposed in 1909 by Richard Woltereck [432], was that every genotype actually has a range of phenotypes that it can produce under different environmental conditions:

> In early Mendelian days Woltereck introduced the term "norm of reactivity" (*Reactionsnorm*) to describe one of the basic conceptions of genetics. The genotype cannot be described simply in terms of the phenotype, since the description must contain the whole range of reactivity of the phenotype under different external or internal conditions. [447]

Fig. 11.2. Seasonal differences in a butterfly species. Two males of the same species (*Precis octavia*) collected in South Africa at different times of the year [1037]. The summer form is on the left, the winter form on the right [404]. Both outside temperature and day length dictate which pattern develops [1037]. These "ecomorphs" are so distinctive that they were originally described as separate species [404]. Indeed, Goldschmidt cited intra-species disparities of this kind as evidence for his theory of "hopeful monsters" [304]. In his opinion, "macro" mutations like those leading to seasonal types could change anatomy so drastically that a new species could easily be formed by the interbreeding of the mutant outliers [447]. Darwin had argued that evolution normally relies on the incremental accumulation of "micro" mutations over many generations [267], but Goldschmidt, like Bateson before him [80], thought otherwise and argued instead for a *discontinuous* mode of progression [435,449,1017]. This "jerky" idea was not received warmly at the time [305,468], but it has experienced a revival lately [209,307,498,1275], and there is growing evidence that it may be valid in some isolated instances [310,582,767] such as turtle shells [1070]. Photos courtesy of Nipam Patel.

Certain mammals change color with the seasons also, which enables them to stay camouflaged [182]. This strategy is most critical in northern climes where snow exposes dark animals to predators. For example, the following brown-colored species turn white in the winter: arctic fox, collared lemming, long-tailed weasel, stoat, snowshoe hare, mountain hare, arctic hare, white-tailed jackrabbit, and Siberian hamster [889]. If they were using the same tyrosinase trick as the Siamese cat, however, they should all turn white in the *summer*. Moreover, they are probably relying more on the monitoring of day length than on the calibration of ambient temperature to trigger the necessary physiological changes [801,1439]:

> The physiology of an organism can have a large impact on color production, mediated by hormones. For example, melatonin and prolactin play a key role in regulating seasonal moulting in various animals: when days get shorter, specialized photosensitive ganglion cells in the eye retina convey a signal that ultimately reaches the pineal gland, which controls secretion of melatonin. Melatonin is produced at night at rates that are inversely proportional to day length. During the winter, as days get shorter, higher levels of melatonin inhibit the production of prolactin, leading to the production of white winter fur. [182]

GP-21 tangent: Turtle sex

Temperature may not be decisive in circannual mammalian coloration, but it *is* the direct trigger for the adoption of alternative anatomies in some animal taxa. Alligators are notorious for temperature-dependent (vs. chromosomal) sex determination [251], but turtles are just as odd in that regard [638]. The most thoroughly studied case is the red-eared slider turtle (*Trachemys scripta*). Weber *et al.* recently used a multitude of clever approaches to piece together all the steps in the pathway that assigns sexual identities in that species [1363]. The bifurcating cascade appears to start with the kind of ion channel (transient receptor potential or TRP protein) that lets our tongue detect the spiciness of a hot chili pepper [745], and it ends with the same ancient effector protein that governs sex determination in virtually every animal on earth (Dmrt) [728,1379].

GP-22: Other external factors can dictate embryo fate

Woltereck's concept of the reaction norm, mentioned above, is now more commonly known as "phenotypic plasticity" [387,744] or simply "polyphenism" [1036]. It denotes the ability of a single genotype to produce multiple alternative phenotypes, depending on the conditions prevailing during the organism's development. Aside from the examples already discussed (butterfly and mammal "ecomorphs" [430] and turtle sex [1363]), other salient instances include the following, all of which except the locust (#3) are known to be mediated by the insulin signaling pathway (cf. GP-16) [240,1036]:

1. Queen versus worker castes in honey bees as a function of whether the larva is fed royal jelly [528,540].

2. Soldier versus worker castes in the damp-wood termite as a function of social interactions [536].
3. Migratory versus sedentary locusts as a function of crowding – a transition of biblical proportions [1010].
4. Long- versus short-horned beetles as a function of the amount of nutrients available (cf. GP-16) [898].
5. Long- versus short-winged morphs of the brown planthopper as a function of environmental cues [1414].
6. Winged versus wingless morphs of the pea aphid as a function of crowding (maternal effect only) [479].

GP-22 tangent: Killer tadpoles

One of the most dramatic examples of a polyphenism is seen in the tadpole stage of spadefoot toads (*Scaphiopus* spp.) [430]. These North American toads live in an especially hostile environment – the Sonoran Desert. The adults hibernate underground until spring rains awaken them, and, after feverish mating, their tadpole offspring have little time to grow and metamorphose before their ephemeral pools dry up [1012,1020]. If the pond is big enough, there is little competition for food, and the tadpoles develop at a leisurely pace. However, in smaller puddles that evaporate quickly, a lethal rivalry ensues which induces some of the tadpoles to transform into cannibals [784]. Those muscular morphs use their powerful jaws to eat their smaller pondmates in a bloody parody of a Biblical morality play [372]. The only redeeming aspect of this anuran apocalypse is that the cannibals tend to refrain from eating their own siblings [1021,1022].

Figure 11.3 distills the essence of the polyphenism phenomenon into two types of schematic diagram. The first type is based on the "epigenetic landscape" analogy of Conrad Waddington (1905–1975), a revered patriarch of developmental genetics [67,505,753]. In his metaphor, development can be thought of as a rumpled tarp whose grooves represent the paths that a cell – here a ball – can take as it rolls down the slope to its final fate (bone, muscle, skin, etc.) [1435]. In the modified version shown here, the ball denotes the embryo instead, and the grooves lead to alternative anatomies. Internal changes (mutations) can deform the landscape and deflect the ball along other paths, but the point of these cartoons is that *external* cues can do so as well. Those cues can only dictate the destiny at certain times when the ball is nearing a fork in the landscape, and here is where the second type of (flow chart) diagram comes in handy. It depicts sensitive periods as binary decision boxes. Based on the cues that are inputted, the embryo will be shunted to one outcome or another.

Another phenomenon that is related to polyphenism is called "phenocopy," and it was the subject of Waddington's most famous experiment [252,1343]. The idea is that some external agent, applied at just the right time to an embryo, can cause the resulting newborn to display a certain mutant phenotype [448,1376]. Literally this term denotes the ability of that outside factor to copy a phenotype. In clinical

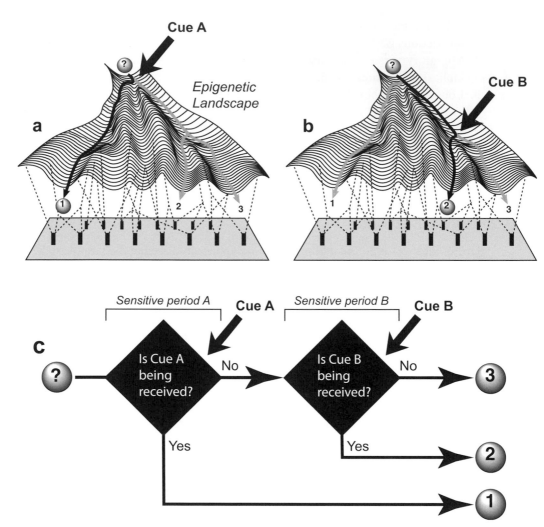

Fig. 11.3. Schematic depictions of polyphenism. a, b. Waddington's "epigenetic landscape" metaphor, where the possible developmental trajectories that an embryo (rolling ball) can follow on its journey from egg to newborn (top to bottom) are represented by grooves in a sloping surface. The deeper the groove, the more the trajectory is robustly buffered against perturbations (see GP-4) – either internal (genetic) or external (epigenetic) [307,329]. That buffering is commonly known as canalization [508]. Pegs represent genes, with dotted lines denoting their connections with one another and their "pull" on the contours of the corrugated landscape (tent canvas) above. If a gene gets disabled by a mutation, then the absence of its function can distort the landscape and deflect the ball into a different channel. However, the point of this sketch is to show that the path can also be affected by *external* influences ("cues"), such as ambient temperature, day length, nutritional supply, etc. **a.** Left to its own devices, the embryo should follow the deepest valley and end up adopting the rightmost fate (3), but in this case Cue A hits the ball (like a pool cue) just as it is entering Valley 3 so that it gets shunted to Valley 1 instead. **b.** The embryo has gone halfway down the slope unperturbed when it encounters Cue B just as it is approaching another fork. As a result it gets pushed into Valley 2 instead. Waddington's metaphor shows how a gentle

circles the best-known case is the drug thalidomide, whose awful side effect, when given to women in the first month of pregnancy, is to stunt the growth of her baby's arms and legs in the same way as the phocomelia mutation, so that the baby is born looking eerily like a Dachshund [192].

Waddington's research concerned fruit flies, and he was fascinated by the ability of ether to induce a bithorax phenotype (see Fig. 5.1) in wild-type fly embryos when they are exposed to its vapors during the first few hours after fertilization – the so-called bithorax phenocopy [178]. The purpose of Waddington's experiment, published in 1956, was to test a hypothesis he had proposed earlier:

> Some years ago it was suggested ... that if selection was practised for the readiness of a strain of organisms to respond to an environmental stimulus in a particular manner, genotypes might eventually be produced which would develop into the favoured phenotype even in the absence of the environmental stimulus. A character which had originally been an "acquired" one might then be said to have become genetically assimilated. [1343]

In fact, his experiment succeeded in confirming his "genetic assimilation" proposal. Starting with a stock of wild-type flies, Waddington treated embryos with ether vapor during the sensitive period and kept doing so, generation after generation, until by generation 8 he witnessed an untreated fly that had a mild bithorax phenotype (enlarged haltere). In generation 9 he found 10 such flies and began crossing them with one another to create a few "haltere effect" (HE) strains. By generation 29 some of the untreated flies had a nearly complete transformation, and the stock he established from those flies kept producing bithorax offspring at a frequency of 70–80% – thus proving that the initially "acquired" trait had been genetically "assimilated" into the population.

The theoretical implications of Waddington's interpretation for the field of evolution are beyond the scope of the present book, but suffice it to say that they have been substantial [115,252,306,1182]. His hypothesis may have had Lamarckian overtones, but it differs from that discredited school of thought [1376]. Regrettably, the physiological basis for the bithorax phenocopy remains a mystery [425].

←
Fig. 11.3. (*cont.*)
nudge from an outside force can make a huge difference in the individual's anatomy (cf. Fig. 11.2) [744,1311]. **c**. Logic diagram summarizing the cartoons in **a** and **b**. Diamond boxes are branch points where the embryo uses environmental inputs to decide whether to remain on its default path (3) or switch to a different path (1 or 2). It is only susceptible to external cues (A or B) at certain times (sensitive periods) [1036,1376]. We do not yet know how such algorithms are actually written in animal genomes. Drawings in **a** and **b** are adapted from [553,1342,1344]. The flow chart in **c** is modeled after [64,335] and based on [953,956].

12 The Calico Cat

A crazy quilt of colors with a cute little face. That's a calico cat in a nutshell, and such cats are almost always female (Fig. 12.1). Why? Female mammals carry two X chromosomes, but one gets inactivated early in development as a way of compensating for the extra dosage of X-linked gene products relative to those of males (XX vs. XY) [656,883]. The inactivation occurs randomly [822], and the X-linked O allele, which causes the orange color, is dominant over the wild-type o^+ [1119]. Hence, half of the embryonic cells in an O/o^+ heterozygote remain orange because they shut off the X with the o^+ allele, while the other half become black because they shut off the X with the O allele. If that were the whole story, then the cat would be a mix of orange and black alone – the so-called "tortoiseshell" design. However, calicos also carry a dominant allele (S) of the gene for white spotting that resides on an autosome, and the white splotches which are created willy-nilly by that allele suppress any pigment (orange or black) in its vicinity – the phenomenon of epistasis [663].

Fig. 12.1. Calico cat. The patchwork of black, orange, and white colors is typically only found in females because it arises via random inactivation of the X chromosome – a mechanism of dosage compensation in mammals (see text). The rare males with such a pattern arise by non-disjunction (XXY), chimerism, or somatic mosaicism [911,1009]. The stochastic nature of the process is most clearly shown by the fact that cloned cats differ in their patch patterns [1046]. Photo from *iStock*; used with permission. (A black and white version of this figure will appear in some formats. For the color version, please refer to the plate section.)

If all female mammals undergo X inactivation, then why don't human females look as kaleidoscopic as calico cats? The only reason is that, unlike *Felis catus*, we have no pigment genes on our X chromosome; otherwise they would.

GP-23: Cells sometimes "flip a coin" to chart their fate

Given the precision with which animals are constructed, the notion that anything happens at random catches our attention [655]. We have already discussed the fickleness of fingerprints (see GP-13), but that is occurring at the tissue level, as are the improvised paths of arteries and veins [1052]. *Genes* ought to be behaving themselves better inside hard-wired circuits, rather than deciding what to do based on flipping a coin, so to speak [884].

X inactivation may be purely random insofar as which X gets turned OFF, but safeguards exist to ensure equivalent dosage if extra Xs happen to intrude. Thus, babies that are born with XXX or XXY aneuploidy keep only one X active, and tetraploids, whose autosomes and Xs are doubled, keep *two* Xs active [931]. Hence, mammals must have a mechanism for counting their chromosomes [950] (cf. [216]). Moreover, marsupials always shut OFF the paternal X and spare the maternal X, and the same is true for certain cells in the mouse embryo [822], so mammals must also have ways of discriminating between the two Xs under certain circumstances. Why they normally opt for a coin flip seems rather odd.

GP-23 tangent: Other ways that cells gamble

There are many games of chance aside from those that involve binary outcomes, as anyone who has ever gambled at a casino knows well, and there are other instances in the animal world where the probabilities are tilted away from a 50:50 mean. For example, the R7 photoreceptor cells in the lower half of a fly's eye use 70:30 betting odds to select a particular opsin (Rh4 vs. Rh3) for their photon-capturing machinery [93,244,1374], and a similar bias governs human photoreceptors also [592]. Then there are all the "roulette wheels" that are used to select receptor alleles for sensory cells in our olfactory system [833] and to diversify the antibodies in our immune system [984]. If these examples, which I've explored before [561], are too arcane, then look no further than the life-or-death games we all played inside our mother's ovaries and oviducts, where one follicle each month was randomly picked to ovulate and then be fertilized by a victorious sperm. Luckily, you won both of those lotteries, or else you wouldn't be reading this now.

GP-24: Gynandromorphs are a weird kind of mosaic

Calicos are "mosaics." A mosaic is a mixture of multiple genotypes in the same body, which develops from a *single* zygote [515,1410]. In contrast, a "chimera" is a composite of multiple genotypes from *multiple* zygotes (Fig. 12.2) [639,1094].

Fig. 12.2. Mosaics versus chimeras. Mosaics and chimeras are both composite individuals containing more than one genotype in the same body, but they differ in that mosaics (**a**) come from a single zygote, while chimeras (**c**) come from the fusion of two or more initially separate embryos [849,1227]. **a.** Gyandromorph mosaic *D. melanogaster* whose left side is wild-type but whose right side lacks bristles because it is hemizygous for a deficiency at the AS-C locus (cf. Fig. 6.1) [562]. **b.** The Chimera of Arezzo, ca 400 BCE, found in Arezzo, an ancient Etruscan and Roman city in Tuscany; Museo Archeologico Nazionale, Florence. According to Homer's Iliad, it was "lion-fronted and snake behind, a goat in the middle" [85,1410]. **c.** An inter-species sheep–goat chimera [1228] created by fusing an 8-cell sheep embryo with three 8-cell goat embryos [358]. This "geep" was about a year old, with large patches of sheep wool cranially and caudally but goat hair elsewhere, plus goat-like horns that are twisted like sheep horns. Sheep and goats have been known to hybridize, but only rarely [1003]. Photo **a** by the author. Photo **b** by Following Hadrian on *Foter.com*; used with permission of *Creative Commons*. Photo **c** from the 16–22 February 1984 cover of *Nature* (Vol. 307, No. 5952) [358]; used with permission of *Springer Nature/RightsLink*. (A black and white version of this figure will appear in some formats. For the color version, please refer to the plate section.)

Mosaics arise via a genetic alteration such as non-disjunction, somatic mutation, or X inactivation, while chimeras come from a fusion of separate embryos [229]. In humans there are many debilitating syndromes that involve dysfunctional mosaicism [110,375,518,798,909]. For example, mosaic variegated aneuploidy (MVA) entails

(1) premature chromosome separation during mitosis, (2) abnormal numbers of chromosomes in the daughter cells, and (3) a propensity for those aneuploid cells to incite cancerous tumors in the first three years of life [1230]. It is caused by a LOF mutation in a mitotic-checkpoint gene called *BubR1* [673].

The fly in Figure 12.2a is partly male and partly female – a type of mosaic called a gynandromorph [914,1203]. It resulted from the loss of an X chromosome at the first mitosis of an XX heterozygous female zygote, yielding an XO nucleus that begat ~50% of the cells in the adult, the remainder deriving from its XX sister nucleus [549]. Flies use XX to specify female identity and XY to specify male identity, as in humans, but maleness is not dictated by the presence of the Y chromosome as it is in our species [896]. Rather, it is dictated by the ratio of X chromosomes to autosomes [216]. Hence, an XO nucleus is genetically male in flies, while it would be female in humans. The male tissue in this particular fly is easily discernible because it carries a recessive X-linked mutation that eliminates bristles [562]. Without such mutations gynandromorphs would be hard to detect in flies because there are so few sexually dimorphic aspects of adult anatomy. In contrast, gynandromorphs are much easier to identify in butterflies, because males and females tend to have different color patterns on their wings (Fig. 12.3).

Depending upon where the male/female boundary meanders inside the body, gynandromorphs can exhibit peculiar behaviors. Flies that look mostly male may try to court other males if they happen to have a female brain, or vice versa [1415], and the problem of cognitive dissonance is exacerbated if the two *halves* of their brain are of different genders. No matter what the brain decides to do, the situation at the other end of the body – the genitalia – can present further difficulties that thwart its desires [635].

Just as the phenomenon of temperature sensitivity (as embodied in the Siamese cat) has been harnessed by scientists as a tool for analyzing gene function (see Chapter 11), the phenomenon of somatic mosaicism (as embodied in the calico cat) has been adapted in a similar way (see Fig. 6.9) [507,1203,1290]. The ability to artificially juxtapose tissues of two different genotypes without the need for any transplantation has proven to be one of the most powerful tools in the tool chest of developmental geneticists – especially in *Drosophila*, where the cornucopia of technical approaches is unrivaled [418].

The male/female boundary of the butterflies in Figure 12.3 coincides with the body midline, just as it does in the fly in Figure 12.2a, and in general such congruence seems to be more common in both arthropods and vertebrates than would be expected by chance alone [47,904]. In flies the process of boundary elaboration has been studied in great detail (Fig. 12.4). The orientation of the first mitosis is random relative to the axes of the egg, but the wing discs come from opposite flanks, so the male/female border is likely to fall between them more often than not in a population of mosaic flies.

Because the wing and other imaginal discs arise at odd sites in the embryo's ectoderm, the male/female boundaries we observe in the *adult* skin do not reflect

Fig. 12.3. Gynandromorph butterflies and a mosaic lobster. Bilateral butterfly gynandromorphs from top to bottom (all are dorsal side up) are: *Anthocharis cardamines* (left side male; right side female), *Morpho didius* (left side female; right side male), *Heliconius melpomene* (left side male; right side female). Loss of a sex chromosome during an early embryonic mitosis accounts for most of such cases. Whether the mosaic lobster is a gynandromorph is not known. In all of these cases the mosaic boundary runs along the midline (see text), and it is remarkably straight in the lobster. Butterfly photos courtesy of Nipam Patel; lobster photo courtesy of Patrick Shepard, Maine Center for Coastal Fisheries. (A black and white version of this figure will appear in some formats. For the color version, please refer to the plate section.)

their original paths in the embryo [1424]. This discrepancy was clearly grasped by Janos Szabad, Trudi Schüpbach, and Eric Wieschaus (another Nobel laureate), who realized that it made more sense to trace those dividing lines in *larval* skin instead. That skin is just an expanded version of the embryo's outer surface [1245]. Figure 12.5 shows a panel of 48 such larval gynandromorphs, where the female tissue (XX) is black, and the male tissue (XO) is white. They offer a clear picture of how the male/female border wanders randomly within the epidermis.

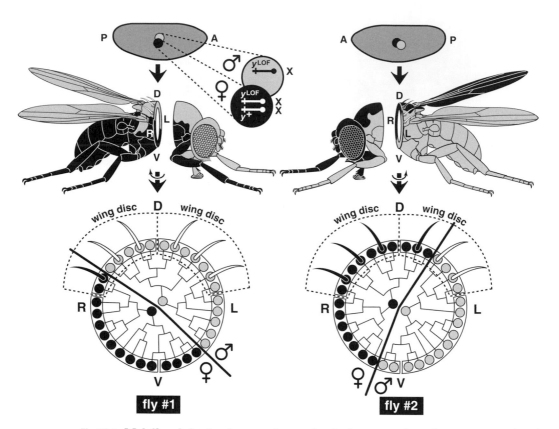

Fig. 12.4. Male/female borders in gynandromorphs. At the top are *D. melanogaster* eggs shortly after fertilization, drawn at the first mitosis of the zygote. The black nucleus has retained both X chromosomes, while the gray nucleus has lost one due to an instability that manifests itself mainly at this stage [998]. Because sex in flies is determined by the number of X chromosomes, the black nucleus stays female, while the gray one becomes male. The latter nucleus expresses a LOF mutation that turns bristles yellow (y^{LOF}) because the wild-type allele on the other X was lost [1386]. Hence, the epidermal disposition of male tissue can be easily mapped in adults based on the distribution of yellow bristles [398]. The bottom schematics are cross-sections through the thoraxes of the flies in the middle panel that grew from embryos at the top, with the "hour hands" representing differing orientations of the original mitosis. The inscribed "trees" are imaginary depictions of mitoses from the original two nuclei to the adult epidermis, with the derivatives of wing discs outlined by dashed lines. Genotypes are assessed based on bristle color in adults. A, anterior; P, posterior; D, dorsal; V, ventral; R, right; L, left. From [555].

GP-24 tangent: Mosaic mapping methodology

The potential utility of this randomness for studying development did not go unnoticed by Alfred Henry Sturtevant (1891–1970) [86], who was arguably the greatest student ever trained by Thomas Hunt Morgan (1866–1945) [1226], and he in turn trained Ed Lewis (1918–2004) of bithorax fame (see Chapter 5) [253,1397]. It is worth noting that all three generations of this noble lineage won Nobel prizes [1396].

Sturtevant is famous for realizing that random events of crossing over between homologous chromosomes afforded an opportunity to measure relative distances between genes, and he was the first to use frequencies of recombination to chart a gene map in any species [785]. After he found a mutation that made gynandromorphs and shared his observations with two colleagues (see below), he understood that he could use the same logic to construct fate maps of adult structures at the embryo stage [1224].

Sturtevant conducted a pilot study along these lines as a side project and tucked away his diagrams of 96 gynandromorphs after publishing a report in 1929. He subsequently collected an additional 379 such flies but never found the time to examine them and incorporate them into a larger sample with greater resolution. That task fell to Antonio García-Bellido and John Merriam, who picked up the project where Sturtevant had left it [421]. García-Bellido reminisced about the handover several decades later, using the shorthand term gynanders for gynandromorphs:

> At Caltech, they had invited me to give a series of seminars about what was going on in Europe at the time. I mentioned the data published by Müller on gynanders, which suggested that some elements were determined early, and that the fly was a kind of mosaic. Then Sturtevant offered me his collection of about 300 drawings of gynanders he had obtained in *D. simulans*. John Merriam happened to be there as a post-doc and he immediately entered the work on gynanders ... John also taught me a lot of practical mechanical genetics that helped me very much. [421]

He goes on to say that the key insight of using the specimens to build a fate map was actually his, rather than Sturtevant's, to whom he refers by his nickname:

> The idea that you could use the data to fatemap the entire embryo – this is mine. That was a very stimulating moment, because gynanders provided a way of mapping the black box at the beginning of development. Sturt followed the whole thing day by day – he was very excited! [421]

Finally, García-Bellido shares an epiphany about how mosaics in general (not just gynandromorphs) could be used to find out all sorts of things about how cells build anatomy – an insight that eventually blossomed into a cottage industry of fruitful experiments with his colleagues back in Spain. The italics are his own:

> I remember when I saw that in Caltech, before the talk I gave in Chicago, I said: *This is it!* This is the radiography of development – you can describe development at the level of cells, and other things start appearing because there were certain limits that the clones would not cross. Also the possibility of associating mutants with cell markers on the same chromosome to produce and analyze mosaics at the single cell level – the whole idea of clonal analysis and of studying the genetics of somatic cells was there. [421]

Since we are so near the end of this narrative, I would beg the reader's indulgence to briefly share an epiphany of my own along these same lines. When I was a graduate student I was fascinated with gynandromorphs, and I thought it might be fun to use them in larger numbers to construct an even higher-resolution fate map down to the level of individual microchaetes. I examined 1000 second legs of

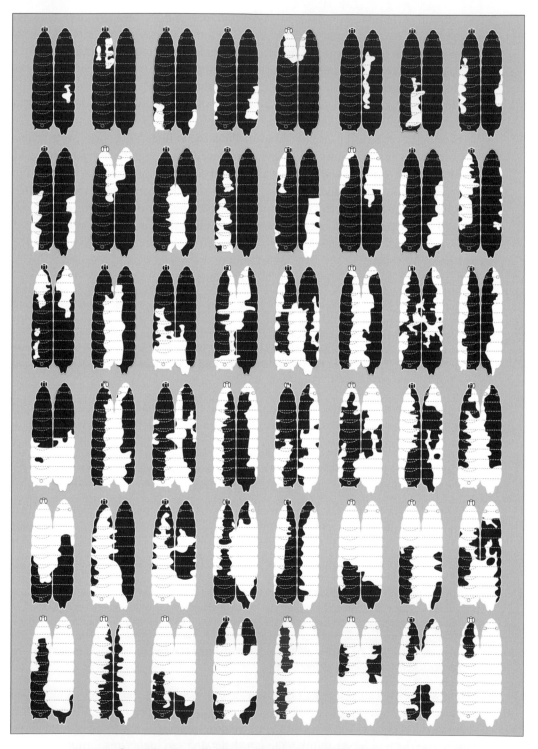

Fig. 12.5. Gynandromorph fly larvae. Distribution of XX (female, black) and XO (male, white) tissue in the skin of 48 *D. melanogaster* larvae, with the XX/XO tissue ratio decreasing from

Drosophila gynandromorphs and scored the genotypes (yellow vs. brown and gnarled vs. straight) of every single bristle (and bract) on their basitarsi [549].

During the process of data collection, I began to notice an unexpected trend. As the male/female boundaries made their way down the leg, they would often weave between bristles and bracts like a slalom skier avoiding trees, so that most of the bristles along a row would be on the female side of the line, and most of their bracts would be on the male side of the line, or vice versa (Fig. 12.6a–c). Why should the boundary be so wavy? At some point, it dawned on me. The bristles and bracts must be born some distance apart but then *move* laterally toward one another.

By collecting more data and plotting the map of boundary frequencies, I realized I could determine the relative directions. When I finally constructed the map (Fig. 12.6d), I could see that the bracts for two of the rows (1 and 8) were *dorsal* to their bristles (cf. Fig. 5.3), while the bracts for the remaining rows (2–7) were *ventral* to their bristles. Unbeknownst to me at the time, Peter Lawrence, Gary Struhl, and Gines Morata were coincidentally discovering the same phenomenon for row 1 using a different kind of mosaic technique [763]. Why different rows should be using different vectors of relative movement I did not know, nor has anyone yet figured out in the intervening four decades.

At that instant I experienced what all scientists crave – a glimpse into the inner workings of Nature that no one has ever seen before, notwithstanding the awareness that few people would ever hear about it, and fewer still would care. There is a sanctity in the intimacy of those apogees that is hard to describe. As I sat at my microscope relishing my good fortune, I hoped to have many such revelations in the future. Sadly, I must report that only about one comparable *Aha!* ever happened per decade since then in my lab. However, as Robert Frost once mused, "Happiness makes up in height for what it lacks in length."

An analogous deduction about bristle cell movements had been made 17 years earlier (in 1962) by Chiyo Tokunaga – an apprentice of Curt Stern's [1291] – regarding the sex comb on the male foreleg (see Fig. 5.3) [1289]. She noticed that when marked clones of cells traverse the sex comb, they bend at a right angle, implying that the sex comb must rotate to a similar extent during its development – a stunning inference at the time, which was later confirmed by observing that shift directly [567]. The diverse ways that different *Drosophila* species have handled their sex combs has offered a mother lode of enticing nuggets for the burgeoning field of evo-devo (see Fig. 5.5). That area of research has been spearheaded by Artyom

←

Fig. 12.5. (*cont.*)
upper left to lower right. Each larva is drawn as if slit along its right flank, pried open, and laid flat, with its ventral half on the left (denticle belts are outlined) and its dorsal half on the right. Dashed lines are segment borders. Although the XX/XO boundary is often irregular, XX and XO cells tend to occupy solid domains, except for cases like the fourth and fifth larvae in the second and sixth rows, which have more than one XX or XO patch. Some of the patterns resemble Paint horses [1362], though the etiology is entirely different. Clearly there is much cell mixing [1387]. Redrawn from [1245].

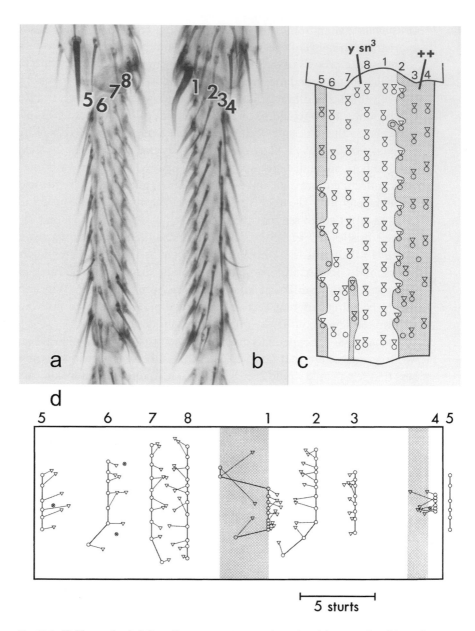

5 sturts

Fig. 12.6. Evidence for bristle cell movements. a–c. Anterior (**a**), posterior (**b**), and panoramic (**c**) views of one second-leg basitarsus among 1000 such leg segments of gynandromorphs that were analyzed in a study of cell lineage published by the author in 1979 [549]. Male (XO) tissue is marked with two different LOF mutations: *yellow* (*y*), which turns bristles (shown as circles) and bracts (triangles) yellow, and *singed³* (*sn³*), which makes bristles look gnarled but has no effect on bracts. Female (XX) tissue is wild-type (++) and is shaded in **c**. A stripe of male tissue runs through the segment, spanning about half the circumference from bristle row 6 to 1. Oddly, the stripe also includes bracts within bristle rows 5 and 2. The wobbly shapes of these margins imply that bristle cells move laterally relative to the cells that they later induce to become their bracts (see Fig. 1.3). Moreover, the directions of those

Kopp, whose papers on evo-devo in general [726] and sex combs in particular [727,728] offer clinics for novices in the art of scientific writing.

One of the greatest wizards of scientific writing ever to grace the field of developmental biology was Seymour Benzer (1921–2007) [27], whose career was lionized in the book *Time, Love, Memory* by Pulitzer Prize winner Jonathan Weiner [1365]. Together with Yoshiki Hotta, Benzer deftly used the gynandromorph-mapping method to localize the sites of action of a variety of behavior-affecting mutations [605,606]. That was but one of his many contributions to the nascent field of behavioral genetics [99,100].

In the heart of every would-be scholar is a virtual Valhalla of heroes, past and present, who sustain us along our journey. Benzer and Bateson and Stern are a few of mine, and others have been cited throughout this saga. The brilliance of their contributions endures as a pilot light when our own flame starts to flicker. The least we can do to thank them for their insights is to preserve their legacies for future generations, as Homer did for the godlike Greeks whose feats abide in our memory. If this book achieves that, then it will have been worth the effort.

Fig. 12.6. (*cont.*)
movements can be deduced: for rows 5 and 2 the bristle cells must move ventrally (toward the middle of the diagram in **c**). Peter Lawrence and his collaborators in Cambridge, England, independently found that the bristle cells of row 1 must move in the opposite (dorsal) direction, and they published their results in that same year (1979) [763]. **d**. High-resolution map of cell-lineage relationships among bristles and bracts on the second-leg basitarsus. The map was constructed by measuring the number of times that male/female boundaries crossed between neighboring (bristle–bristle or bristle–bract) structures in a sample of 1000 basitarsi. The position of each structure was then triangulated relative to its neighbors. Frequencies are given in "sturt" units, named for Alfred Sturtevant, where one sturt equals one separation by a male/female boundary per 100 basitarsi [606]. Line segments connect bristles to their own bracts and to neighboring bristles in the same row, which allows the directions of cell movements to be deduced (see Fig. 6.12). Shaded bands denote boundaries between the anterior and posterior cell-lineage compartments of the leg [763,1199], which cause a widening of the map at those locations. Circles inscribed with an X are bractless (chemosensory) bristles, which reside between the longitudinal rows. Photos and diagrams are by the author [549]; used with permission of *Springer Nature/RightsLink*.

Epilogue

Developmental biologists are a special breed. They are mainly driven by curiosity about how simplicity becomes complexity, as expressed in the title of Philip Ball's book *The Self-Made Tapestry* [57]. Like Lewis Carroll's Alice, they love exploring for its own sake, and they rejoice in being able to share their discoveries with like-minded souls. Attending a conference of one of their societies is like going to a religious revival meeting. Secular though they may be, they are as passionate as any preacher about spreading the news of Nature's beauty and glory and artistry. In short, embryologists are evangelists, and their enthusiasm has imbued this field with an ethos of energy, vitality, and purpose.

William Bateson once advised budding scientists to "treasure your exceptions" [219], and his magnum opus about animal anomalies brimmed with oddities of every sort imaginable. The present book was written in that same spirit, but with one key difference: after more than a century of further scientific experimentation, we are now in the position of being able to savor the sweeping significance of the treasures that those exceptions had held hidden for so long. Only when embryos are put in embarrassing situations – either naturally (by some accident) or artificially (by the hand of the experimenter) – are they forced to reveal their secrets [1202]. However, when they do so, the revelations are often whispered rather than shouted, and it takes an astute detective to notice the clues, piece them together, and follow the trail to solve the mystery.

Many real-life avatars of Sherlock Holmes have graced the history of this field, and some of their investigations have been revisited here. Collectively their exploits constitute a "how to" training manual for the next generation of would-be treasure hunters. Luckily, there is plenty of booty left to be found, and one goal of this book was to draw as many Xs on the treasure map as possible. And for readers who have no intention of ever becoming researchers the narrative may still have had some value as a spectator sport. Watching Olympians at the top of their game can be as thrilling as actually participating oneself. Regardless of whether one plays on the field or just sits in the stands, the riddles posed by the aberrations surveyed here have at least offered a chance to experience the satisfaction of seeing Gordian knots unraveled, one after another. Recreational embryology indeed.

References

1. Abdelilah, S., Solnica-Krezel, L., Stainier, D.Y.R., and Driever, W. (1994). Implications for dorsoventral axis determination from the zebrafish mutation *janus*. *Nature* 370, 468–471.

2. Abe, M. and Kuroda, R. (2019). The development of CRISPR for a mollusc establishes the formin *Lsdia1* as the long-sought gene for snail dextral/sinistral coiling. *Development* 146, dev175976.

3. Aboitiz, F. and Montiel, J.F. (2019). Morphological evolution of the vertebrate forebrain: from mechanical to cellular processes. *Evol. Dev.* 21, 330–341.

4. Abu-Shaar, M. and Mann, R.S. (1998). Generation of multiple antagonistic domains along the proximodistal axis during *Drosophila* leg development. *Development* 125, 3821–3830.

5. Abzhanov, A. (2017). The old and new faces of morphology: the legacy of D'Arcy Thompson's "theory of transformations" and "laws of growth". *Development* 144, 4284–4297.

6. Acurio, A.E., Rhebergen, F.T., Paulus, S., Courtier-Orgogozo, V., and Lang, M. (2019). Repeated evolution of asymmetric genitalia and right-sided mating behavior in the *Drosophila nannoptera* species group. *BMC Evol. Biol.* 19, 109.

7. Adhikari, K., Fontanil, T., Cal, S., Mendoza-Revilla, J., Fuentes-Guajardo, M., Chacón-Duque, J.C., Al-Saadi, F., Johansson, J.A., Quinto-Sánchez, M., Acuña-Alonzo, V., Jaramillo, C., Arias, W., Lozano, R.B., Pérez, G.M., Gómez-Valdés, J., Villamil-Ramírez, H., Hunemeier, T., Ramallo, V., Silva de Cerqueira, C.C., Hurtado, M., Villegas, V., Granja, V., Gallo, C., Poletti, G., Schuler-Faccini, L., Salzano, F.M., Cátira Bortolinia, M., Canizales-Quinteros, S., Rothhammer, F., Bedoya, G., González-José, R., Headon, D., López-Otín, C., Tobin, D.J., Balding, D., and Ruiz-Linares, A. (2016). A genome-wide association scan in admixed Latin Americans identifies loci influencing facial and scalp hair features. *Nat. Commun.* 7, 10815.

8. Agi, E., Langen, M., Altschuler, S.J., Wu, L.F., Zimmermann, T., and Hiesinger, P.R. (2014). The evolution and development of neural superposition. *J. Neurogenet.* 28, 216–232.

9. Aiello, D. and Lasagna, E. (2018). The *myostatin* gene: an overview of mechanisms of action and its relevance to livestock animals. *Anim. Genet.* 49, 505–519.

10. Akam, M. (1998). *Hox* genes: from master genes to micromanagers. *Curr. Biol.* 8, R676–R678.

11. Akey, J.M., Ruhe, A.L., Akey, D.T., Wong, A.K., Connelly, C.F., Madeoy, J., Nicholas, T.J., and Neff, M.W. (2010). Tracking footprints of artificial selection in the dog genome. *PNAS* 107, 1160–1165.

12. Alberch, P. (1982). Developmental constraints in evolutionary processes. In *Evolution and Development*, Bonner, J.T., editor. Springer-Verlag, Berlin, pp. 313–332.

13. Alberch, P. (1985). Developmental constraints: why St. Bernards often have an extra digit and poodles never do. *Am. Nat.* 126, 430–433.

14. Alberch, P. (1986). Possible dogs. *Nat. Hist.* 95(12), 4–8.

15. Alberch, P. (1989). The logic of monsters: evidence for internal constraint in development and evolution. *Geobios* 22(Suppl. 2), 21–57.

16. Alberch, P., Gould, S.J., Oster, G.F., and Wake, D.B. (1979). Size and shape in ontogeny and phylogeny. *Paleobiology* 5, 296–317.

17. Alberts, B., Johnson, A., Lewis, J., Raff, M., Roberts, K., and Walter, P. (2002). *Molecular Biology of the Cell*, 4th ed. Garland, New York.

18. Alexander, R.M. (1995). Big flies have bigger cells. *Nature* 375, 20.

19. Alibardi, L. (2018). Appendage regeneration in amphibians and some reptiles derived from specific evolutionary histories. *J. Exp. Zool. B Mol. Dev. Evol.* 330, 396–405.

20. Alibardi, L. (2018). Limb regeneration in humans: dream or reality? *Ann. Anat.* 217, 1–6.

21. Allchin, D. (2019). How the tiger changed its stripes. *Am. Biol. Teacher* 81, 599–604.

22. Allen, W.L., Cuthill, I.C., Scott-Samuel, N.E., and Baddeley, R. (2011). Why the leopard got its spots: relating pattern development to ecology in felids. *Proc. R. Soc. B* 278, 1373–1380.

23. Alpert, B.O. (2013). The meaning of the dots on the horses of Pech Merle. *Arts* 2, 476–490.

24. Alsina, B. and Whitfield, T.T. (2017). Sculpting the labyrinth: morphogenesis of the developing inner ear. *Semin. Cell Dev. Biol.* 65, 47–59.

25. Ambegaonkar, A.A. and Irvine, K.D. (2015). Coordination of planar cell polarity pathways through Spiny legs. *eLife* 4, e09946.

26. Ambrosi, D., Ben Amar, M., Cyron, C.J., DeSimone, A., Goriely, A., Humphrey, J.D., and Kuhl, E. (2019). Growth and remodelling of living tissues: perspectives, challenges and opportunities. *J. R. Soc. Interface* 16, 20190233.

27. Anderson, D. and Brenner, S. (2008). Seymour Benzer (1921–2007): restless spirit, and pioneer in molecular genetics. *Nature* 451, 139.

28. Andl, T., Reddy, S.T., Gaddapara, T., and Millar, S.E. (2002). WNT signals are required for the initiation of hair follicle development. *Dev. Cell* 2, 643–653.

29. Andrew, D.J. and Ewald, A.J. (2010). Morphogenesis of epithelial tubes: insights into tube formation, elongation, and elaboration. *Dev. Biol.* 341, 34–55.

30. Angelini, D.R. and Kaufman, T.C. (2005). Insect appendages and comparative ontogenetics. *Dev. Biol.* 286, 57–77.

31. Anon. (1957). *The Science of Fingerprints*. Federal Bureau of Investigation, Washington, DC.

32. Arendt, J. (2007). Ecological correlates of body size in relation to cell size and cell number: patterns in flies, fish, fruits and foliage. *Biol. Rev.* 82, 241–256.

33. Argyriou, T., Clauss, M., Maxwell, E.E., Furrer, H., and Sánchez-Villagra, M.R. (2016). Exceptional preservation reveals gastrointestinal anatomy and evolution in early actinopterygian fishes. *Sci. Rep.* 6, 18758.

34. Arnone, M.I. and Davidson, E.H. (1997). The hardwiring of development: organization and function of genomic regulatory systems. *Development* 124, 1851–1864.

35. Arnosti, D.N. and Kulkarni, M.M. (2005). Transcriptional enhancers: intelligent enhanceosomes or flexible billboards? *J. Cell. Biochem.* 94, 890–898.

36. Artavanis-Tsakonas, S. and Muskavitch, M.A.T. (2010). Notch: the past, the present, and the future. *Curr. Top. Dev. Biol.* 92, 1–29.

37. Artavanis-Tsakonas, S., Rand, M.D., and Lake, R.J. (1999). Notch signaling: cell fate control and signal integration in development. *Science* 284, 770–776.

38. Arthur, W. (2006). D'Arcy Thompson and the theory of transformations. *Nat. Rev. Genet.* 7, 401–406.

39. Atallah, J. and Larsen, E. (2009). Genotype–phenotype mapping: developmental biology confronts the toolkit paradox. *Int. Rev. Cell Mol. Biol.* 278, 119–148.

40. Atallah, J., Liu, N.H., Dennis, P., Hon, A., Godt, D., and Larsen, E.W. (2009). Cell dynamics and developmental bias in the ontogeny of a complex sexually dimorphic trait in *Drosophila melanogaster*. *Evol. Dev.* 11, 191–204.

41. Atallah, J., Watabe, H., and Kopp, A. (2012). Many ways to make a novel structure: a new mode of sex comb development in Drosophilidae. *Evol. Dev.* 14, 476–483.

42. Athanasiadis, A.P., Tzannatos, C., Mikos, T., Zafrakas, M., and Bontis, J.N. (2005). A unique case of conjoined triplets. *Am. J. Obstet. Gynecol.* 192, 2084–2087.

43. Ather, S., Proudlock, F.A., Welton, T., Morgan, P.S., Sheth, V., Gottlob, I., and Dineen, R.A. (2018). Aberrant visual pathway development in albinism: from retina to cortex. *Hum. Brain Mapp.* 40, 777–788.

44. Audette, D.S., Scheiblin, D.A., and Duncan, M.K. (2017). The molecular mechanisms underlying lens fiber elongation. *Exp. Eye Res.* 156, 41–49.

45. Auerbach, C. (1936). The development of the legs, wings, and halteres in wild type and some mutant strains of *Drosophila melanogaster*. *Trans. R. Soc. Edinb.* 58, 787–815.

46. Averbach, B. and Chein, O. (1999). *Problem Solving Through Recreational Mathematics*. Dover, New York.

47. Aw, S. and Levin, M. (2008). What's left in asymmetry? *Dev. Dynamics* 237, 3453–3463.

48. Aw, W.Y. and Devenport, D. (2016). Planar cell polarity: global inputs establishing cellular asymmetry. *Curr. Opin. Cell Biol.* 44, 110–116.

49. Axelrod, J. (2008). Bad hair days for mouse PCP mutants. *Nat. Cell Biol.* 10, 1251–1252.

50. Ayala-Carmago, A., Ekas, L.A., Flaherty, M.S., Baeg, G.-H., and Bach, E.A. (2007). The JAK/STAT pathway regulates proximo-distal patterning in *Drosophila*. *Dev. Dynamics* 236, 2721–2730.

51. Ayukawa, T., Akiyama, M., Mummery-Widmer, J.L., Stoeger, T., Sasaki, J., Knoblich, J.A., Senoo, H., Sasaki, T., and Yamazaki, M. (2014). Dachsous-dependent asymmetric localization of Spiny-legs determines planar cell polarity orientation in *Drosophila*. *Cell Rep.* 8, 610–621.

52. Babler, W.J. (1991). Embryologic development of epidermal ridges and their configurations. *Birth Defects Orig. Artic. Ser.* 27, 95–112.

53. Baer, M.M., Chanut-Delalande, H., and Affolter, M. (2009). Cellular and molecular mechanisms underlying the formation of biological tubes. *Curr. Top. Dev. Biol.* 89, 137–162.

54. Baker, N.E. (2011). Proximodistal patterning in the *Drosophila* leg: models and mutations. *Genetics* 187, 1003–1010.

55. Baker, N.E. and Brown, N.L. (2018). All in the family: proneural bHLH genes and neuronal diversity. *Development* 145, 159426.

56. Baker, R.E., Schnell, S., and Maini, P.K. (2006). A clock and wavefront mechanism for somite formation. *Dev. Biol.* 293, 116–126.

57. Ball, P. (1999). *The Self-Made Tapestry: Pattern Formation in Nature*. Oxford University Press, New York.

58. Ball, P. (2015). Forging patterns and making waves from biology to geology: a commentary on Turing (1952) "The chemical basis of morphogenesis". *Philos. Trans. R. Soc. Lond. B* 370, 20140218.

59. Bando, T., Mito, T., Hamada, Y., Ishimaru, Y., Noji, S., and Ohuchi, H. (2018). Molecular mechanisms of limb regeneration: insights from regenerating legs of the cricket *Gryllus bimaculatus*. *Int. J. Dev. Biol.* 62, 559–569.

60. Bando, T., Mito, T., Maeda, Y., Nakamura, T., Ito, F., Watanabe, T., Ohuchi, H., and Noji, S. (2009). Regulation of leg size and shape by the Dachsous/Fat signalling pathway during regeneration. *Development* 136, 2235–2245.

61. Bangru, S. and Kalsotra, A. (2020). Cellular and molecular basis of liver regeneration. *Semin. Cell Dev. Biol.* 100, 74–87.

62. Bao, R., Dia, S.E., Issa, H.A., Alhusein, D., and Friedrich, M. (2018). Comparative evidence of an exceptional impact of gene duplication on the developmental evolution of *Drosophila* and the higher Diptera. *Front. Ecol. Evol.* 6, 63.

63. Barad, O., Hornstein, E., and Barkai, N. (2011). Robust selection of sensory organ precursors by the Notch–Delta pathway. *Curr. Opin. Cell Biol.* 23, 663–667.

64. Bard, J. (2011). A systems biology representation of developmental anatomy. *J. Anat.* 218, 591–599.

65. Bard, J.B.L. (1977). A unity underlying the different zebra striping patterns. *J. Zool. Lond.* 183, 527–539.

66. Bard, J.B.L. (1981). A model for generating aspects of zebra and other mammalian coat patterns. *J. Theor. Biol.* 93, 363–385.

67. Bard, J.B.L. (2008). Waddington's legacy to developmental and theoretical biology. *Biol. Theory* 3, 188–197.

68. Bard, J.B.L. (2018). Tinkering and the origins of heritable anatomical variation in vertebrates. *Biology* 7, 7010020.

69. Barham, G. and Clarke, N.M.P. (2008). Genetic regulation of embryological limb development with relation to congenital limb deformity in humans. *J. Child. Orthop.* 2, 1–9.

70. Barmina, O. and Kopp, A. (2007). Sex-specific expression of a HOX gene associated with rapid morphological evolution. *Dev. Biol.* 311, 277–286.

71. Barnes, J., ed. (1984). *The Complete Works of Aristotle: The Revised Oxford Translation*, Vol. 1. Princeton University Press, Princeton, NJ.

72. Barolo, S. and Posakony, J.W. (2002). Three habits of highly effective signaling pathways: principles of transcriptional control by developmental cell signaling. *Genes Dev.* 16, 1167–1181.

73. Barriga, E.H. and Mayor, R. (2015). Embryonic cell–cell adhesion: a key player in collective neural crest migration. *Curr. Top. Dev. Biol.* 112, 301–323.

74. Bartos, L., Bubenik, G.A., and Kuzmova, E. (2012). Endocrine relationships between rank-related behavior and antler growth in deer. *Front. Biosci.* E4, 1111–1126.

75. Bassnett, S. and Costello, M.J. (2017). The cause and consequence of fiber cell compaction in the vertebrate lens. *Exp. Eye Res.* 156, 50–57.

76. Bassnett, S., Shi, Y., and Vrensen, G.F.J.M. (2011). Biological glass: structural determinants of eye lens transparency. *Philos. Trans. R. Soc. Lond. B* 366, 1250–1264.

77. Bastida, M.F., Pérez-Gómez, R., Trofka, A., Zhu, J., Rada-Iglesias, A., Sheth, R., Stadler, H.S., Mackem, S., and Ros, M.A. (2020). The formation of the thumb

requires direct modulation of *Gli3* transcription by *Hoxa13*. *PNAS* 117, 1090–1096.

78. Bate, M. and Martinez Arias, A. (1991). The embryonic origin of imaginal discs in *Drosophila*. *Development* 112, 755–761.

79. Bateson, G. (1971). A re-examination of "Bateson's rule". *J. Genet.* 60, 230–240.

80. Bateson, W. (1894). *Materials for the Study of Variation Treated with Especial Regard to Discontinuity in the Origin of Species*. Macmillan, London.

81. Bateson, W. (1909). *Mendel's Principles of Heredity*. Cambridge University Press, Cambridge.

82. Baudouin-Gonzalez, L., Santos, M.A., Tempesta, C., Sucena, E., Roch, F., and Tanaka, K. (2017). Diverse *cis*-regulatory mechanisms contribute to expression evolution of tandem gene duplicates. *Mol. Biol. Evol.* 34, 3132–3147.

83. Baumeister, F.A.M., Egger, J., Schildhauer, M.T., and Stengel-Rutkowski, S. (1993). Ambras syndrome: delineation of a unique hypertrichosis universalis congenita and association with a balanced pericentric inversion (8) (p11.2; q22). *Clin. Genet.* 44, 121–128.

84. Bazin-Lopez, N., Valdivia, L.E., Wilson, S.W., and Gestri, G. (2015). Watching eyes take shape. *Curr. Opin. Genet. Dev.* 32, 73–79.

85. Bazopoulou-Kyrkanidou, E. (2001). Chimeric creatures in Greek mythology and reflections in science. *Am. J. Med. Genet.* 100, 66–80.

86. Beadle, G.W. (1970). Alfred Henry Sturtevant (1891–1970). *Am. Philos. Soc. Yearbook* 1970, 166–171.

87. Bear, J.E. and Haugh, J.M. (2014). Directed migration of mesenchymal cells: where signaling and the cytoskeleton meet. *Curr. Opin. Cell Biol.* 30, 74–82.

88. Bechtel, H.B. (1995). *Reptile and Amphibian Variants: Colors, Patterns, and Scales*. Krieger, Malabar, FL.

89. Beebe, D.C., Vasiliev, O., Guo, J., Shui, Y.-B., and Bassnett, S. (2001). Changes in adhesion complexes define stages in the differentiation of lens fiber cells. *Invest. Ophthalmol. Vis. Sci.* 42, 727–734.

90. Beermann, F., Orlow, S.J., and Lamoreux, M.L. (2004). The *Tyr* (albino) locus of the laboratory mouse. *Mamm. Genome* 15, 749–758.

91. Begley, S. (1982). How life begins. *Newsweek* (Jan. 11, 1982), 38–43.

92. Beira, J.V. and Paro, R. (2016). The legacy of *Drosophila* imaginal discs. *Chromosoma* 125, 573–592.

93. Bell, M.L., Earl, J.B., and Britt, S.G. (2007). Two types of *Drosophila* R7 photoreceptor cells are arranged randomly: a model for stochastic cell-fate determination. *J. Comp. Neurol.* 502, 75–85.

94. Bellaïche, Y., Gho, M., Kaltschmidt, J., Brand, A., and Schweisguth, F. (2001). Frizzled regulates localization of cell-fate determinants and mitotic spindle rotation during asymmetric cell division. *Nat. Cell Biol.* 3, 50–57.

95. Bellone, R.R., Forsyth, G., Leeb, T., Archer, S., Sigurdsson, S., Imsland, F., Mauceli, E., Engensteiner, M., Bailey, E., Sandmeyer, L., Grahn, B., Lindblad-Toh, K., and Wade, C.M. (2010). Fine-mapping and mutation analysis of *TRPM1*: a candidate gene for leopard complex (*LP*) spotting and congenital stationary night blindness in horses. *Brief. Funct. Genomics* 9, 193–207.

96. Beloussov, L.V., Opitz, J.M., and Gilbert, S.F. (1997). Life of Alexander G. Gurwitsch and his relevant contribution to the theory of morphogenetic fields. *Int. J. Dev. Biol.* 41, 771–779.

97. Bénazéraf, B. and Pourquie, O. (2013). Formation and segmentation of the vertebrate body axis. *Annu. Rev. Cell Dev. Biol.* 29, 1–26.

98. Benkel, B.F., Rouvinen-Watt, K., Farid, H., and Anistoroaei, R. (2009). Molecular characterization of the Himalayan mink. *Mamm. Genome* 20, 256–259.

99. Benzer, S. (1971). From the gene to behavior. *JAMA* 218, 1015–1022.

100. Benzer, S. (1973). Genetic dissection of behavior. *Sci. Am.* 229(6), 24–37.

101. Bercovitch, F.B. (2019). Giraffe taxonomy, geographic distribution and conservation. *Afr. J. Ecol.* 58, 150–158.

102. Bergman, J. (2002). Darwin's ape-men and the exploitation of deformed humans. *Technical Journal* 16, 116–122.

103. Bergmann, P., Richter, S., Glöckner, N., and Betz, O. (2018). Morphology of hind-wing veins in the shield bug *Graphosoma italicum* (Heteroptera: Pentatomidae). *Arthropod Struct. Dev.* 47, 375–390.

104. Bernard, B.A. (2017). The hair follicle enigma. *Exp. Dermatol.* 26, 472–477.

105. Bernays, M.E. and Smith, R. (1999). Convergent strabismus in a white Bengal tiger. *Aust. Vet. J.* 77, 152–155.

106. Berton, P. (1977). *The Dionne Years: A Thirties Melodrama*. W. W. Norton, New York.

107. Beverdam, A., Merlo, G.R., Paleari, L., Mantero, S., Genova, F., Barbieri, O., Janvier, P., and Levi, G. (2002). Jaw transformation with gain of symmetry after *Dlx5/Dlx6* inactivation: mirror of the past? *Genesis* 34, 221–227.

108. Bhalla, U.S. and Iyengar, R. (1999). Emergent properties of networks of biological signaling pathways. *Science* 283, 381–387.

109. Biesecker, L.G. (2011). Polydactyly: how many disorders and how many genes? 2010 update. *Dev. Dynamics* 240, 931–942.

110. Biesecker, L.G. and Spinner, N.B. (2013). A genomic view of mosaicism and human disease. *Nat. Rev. Genet.* 14, 307–320.

111. Bigas, A. and Espinosa, L. (2016). Notch signaling in cell-cell communication pathways. *Curr. Stem Cell Rep.* 2, 349–355.

112. Binns, W., James, L.F., Shupe, J.L., and Everett, G. (1963). A congenital cyclopian-type malformation in lambs induced by maternal ingestion of a range plant, *Veratrum californicum*. *Am. J. Vet. Res.* 24, 1164–1175.

113. Biosa, G., Bastianoni, S., and Rustici, M. (2006). Chemical waves. *Chem. Eur. J.* 12, 3430–3437.

114. Bishop, S.A., Klein, T., Martinez Arias, A., and Couso, J.P. (1999). Composite signalling from *Serrate* and *Delta* establishes leg segments in *Drosophila* through *Notch*. *Development* 126, 2993–3003.

115. Bizzarri, M., Giuliani, A., Minini, M., Monti, N., and Cucina, A. (2020). Constraints shape cell function and morphology by canalizing the developmental path along the Waddington's landscape. *BioEssays* 42, 1900108.

116. Black, S.D. and Gerhart, J.C. (1986). High frequency twinning in *Xenopus* eggs centrifuged before first cleavage. *Dev. Biol.* 116, 228–240.

117. Blair, S.S. (2004). Developmental biology: Notching the hindbrain. *Curr. Biol.* 14, R570–R572.

118. Blair, S.S., Brower, D.L., Thomas, J.B., and Zavortink, M. (1994). The role of *apterous* in the control of dorsoventral compartmentalization and PS integrin gene expression in the developing wing of *Drosophila*. *Development* 120, 1805–1815.

119. Blanco, J., Girard, F., Kamachi, Y., Kondoh, H., and Gehring, W. (2005). Functional analysis of the chicken d*1-crystallin* enhancer activity in *Drosophila* reveals remarkable evolutionary conservation between chicken and fly. *Development* 132, 1895–1905.

120. Blanco, M.J., Misof, B.Y., and Wagner, G.P. (1998). Heterochronic differences of *Hoxa–11* expression in *Xenopus* fore- and hind limb development: evidence for lower limb identity of the anural ankle bones. *Dev. Genes Evol.* 208, 175–187.

121. Blaustein, A.R. and Johnson, P.T.J. (2003). Explaining frog deformities. *Sci. Am.* 288 (2), 60–65.

122. Blaustein, A.R. and Johnson, P.T.J. (2003). The complexity of deformed amphibians. *Front. Ecol. Environ.* 1, 87–94.

123. Blum, M. and Ott, T. (2018). Animal left–right asymmetry. *Curr. Biol.* 28, R301–R304.

124. Blum, M. and Ott, T. (2019). Mechanical strain, novel genes and evolutionary insights: news from the frog left–right organizer. *Curr. Biol.* 56, 8–14.

125. Blum, M., Feistel, K., Thumberger, T., and Schweickert, A. (2014). The evolution and conservation of left–right patterning mechanisms. *Development* 141, 1603–1613.

126. Blum, M., Schweickert, A., Vick, P., Wright, C.V.E., and Danilchik, M.V. (2014). Symmetry breakage in the vertebrate embryo: when does it happen and how does it work? *Dev. Biol.* 393, 109–123.

127. Blumberg, M.S. (2009). *Freaks of Nature: What Anomalies Tell Us about Development.* Oxford University Press, New York.

128. Boareto, M. (2020). Patterning via local cell–cell interactions in developing systems. *Dev. Biol.* 460, 77–85.

129. Boer, L.L., Schepens-Franke, A.N., and Oostra, R.J. (2019). Two is a crowd: on the enigmatic etiopathogenesis of conjoined twinning. *Clin. Anat.* 32, 722–741.

130. Bohn, H. (1965). Analyse der Regenerationfähigkeit der Insektenextremität durch Amputations- und Transplantationsversuche an Larven der Afrikanischen Schabe (*Leucophaea maderae* Fabr.). II. Achsendetermination. *Roux Arch. Entw.-Mech.* 156, 449–503.

131. Böhni, R., Riesgo-Escovar, J., Oldham, S., Brogiolo, W., Stocker, H., Andruss, B.F., Beckingham, K., and Hafen, E. (1999). Autonomous control of cell and organ size by CHICO, a *Drosophila* homolog of vertebrate IRS1–4. *Cell* 97, 865–875.

132. Bohring, A., Stamm, T., Spaich, C., Haase, C., Spree, K., Hehr, U., Hoffmann, M., Ledig, S., Sel, S., Wieacker, P., and Röpke, A. (2009). *WNT10A* mutations are a frequent cause of a broad spectrum of ectodermal dysplasias with sex-biased manifestation pattern in heterozygotes. *Am. J. Hum. Genet.* 85, 97–105.

133. Bökel, C. and Brand, M. (2014). Endocytosis and signaling during development. *Cold Spring Harb. Perspect. Biol.* 6, a017020.

134. Boklage, C.E. (2006). Embryogenesis of chimeras, twins and anterior midline asymmetries. *Hum. Reprod.* 21, 579–591.

135. Bolk, L. (1926). *Das Problem der Menschwerdung.* Gustav Fischer, Jena.

136. Borok, M.J., Tran, D.A., Ho, M.C.W., and Drewell, R.A. (2010). Dissecting the regulatory switches of development: lessons from enhancer evolution in *Drosophila*. *Development* 137, 5–13.

137. Bosch, M., Bishop, S.-A., Baguña, J., and Couso, J.-P. (2010). Leg regeneration in *Drosophila* abridges the normal developmental program. *Int. J. Dev. Biol.* 54, 1241–1250.

138. Botelho, J.F., Smith-Paredes, D., Soto-Acuña, S., Núñez-León, D., Palma, V., and Vargas, A.O. (2016). Greater growth of proximal metatarsals in bird embryos and the evolution of hallux position in the grasping foot. *J. Exp. Zool. B Mol. Dev. Evol.* 328, 106–118.

139. Botstein, D. and Maurer, R. (1982). Genetic approaches to the analysis of microbial development. *Annu. Rev. Genet.* 16, 61–83.

140. Bower, M.A., McGivney, B.A., Campana, M.G., Gu, J., Andersson, L.S., Barrett, E., Davis, C.R., Mikko, S., Stock, F., Voronkova, V., Bradley, D.G., Fahey, A.G., Lindgren, G., MacHugh, D.E., Sulimova, G., and Hill, E.W. (2012). The genetic origin and history of speed in the Thoroughbred racehorse. *Nat. Commun.* 3, 643.

141. Bownes, M. and Seiler, M. (1977). Developmental effects of exposing *Drosophila* embryos to ether vapour. *J. Exp. Zool.* 199, 9–23.

142. Boyce, D. (1992). A sight to behold: Deidre finds a toad with eyes in its mouth. *The Hamilton Spectator* (Sept. 3, 1992).

143. Bozorgmehr, J.E.H. (2014). The role of self-organization in developmental evolution. *Theory Biosci.* 133, 145–163.

144. Brakefield, P.M. (1999). Butterfly wings: the evolution of development of colour patterns. *BioEssays* 21, 391–401.

145. Brakefield, P.M., French, V., and Zwaan, B.J. (2003). Development and the genetics of evolutionary change within insect species. *Annu. Rev. Ecol. Evol. Syst.* 34, 633–660.

146. Bray, S. (1998). Notch signalling in *Drosophila*: three ways to use a pathway. *Semin. Cell Dev. Biol.* 9, 591–597.

147. Bredov, D. and Volodyaev, I. (2018). Increasing complexity: mechanical guidance and feedback loops as a basis for self-organization in morphogenesis. *BioSystems* 173, 133–156.

148. Brewer, A.A. (2009). Visual maps: to merge or not to merge. *Curr. Biol.* 19, R945–R947.

149. Bridges, C.B. and Brehme, K.S. (1944). *The Mutants of* Drosophila melanogaster. Carnegie Institution of Washington, Washington, DC.

150. Brigham, P.A., Cappas, A., and Uno, H. (1988). The stumptailed macaque as a model for androgenetic alopecia: effects of topical minoxidil analyzed by use of the folliculogram. *Clin. Dermatol.* 6(4), 177–187.

151. Briscoe, J. and Kicheva, A. (2017). The physics of development 100 years after D'Arcy Thompson's "On Growth and Form". *Mech. Dev.* 145, 26–31.

152. Brison, N., Debeer, P., and Tylzanowski, P. (2013). Joining the fingers: a *HOXD13* story. *Dev. Dynamics* 243, 37–48.

153. Britton, N.F. (1986). *Reaction–Diffusion Equations and Their Applications to Biology.* Academic Press, New York.

154. Brogiolo, W., Stocker, H., Ikeya, T., Rintelen, F., Fernandez, R., and Hafen, E. (2001). An evolutionarily conserved function of the *Drosophila* insulin receptor and insulin-like peptides in growth control. *Curr. Biol.* 11, 213–221.

155. Brommage, R., Powell, D.R., and Vogel, P. (2019). Predicting human disease mutations and identifying drug targets from mouse gene knockout phenotyping campaigns. *Dis. Model. Mech.* 12, dmm038224.

156. Bronner, M.E. and LeDouarin, N.M. (2012). Development and evolution of the neural crest: an overview. *Dev. Biol.* 366, 2–9.

157. Browd, S.R., Goodrich, J.T., and Walker, M.L. (2008). Craniopagus twins. *J. Neurosurg. Pediatr.* 1, 1–20.

158. Brower, J.S., Wootton-Gorges, S.L., Costouros, J.G., Boakes, J., and Greenspan, A. (2003). Congenital diplopia. *Pediatr. Radiol.* 33, 797–799.

159. Brown, D.M., Brenneman, R.A., Koepfli, K.-P., Pollinger, J.P., Milá, B., Georgiadis, N.J., Louis, E.E., Jr., Grether, G.F., Jacobs, D.K., and Wayne, R.K. (2007). Extensive population genetic structure in the giraffe. *BMC Biol.* 5, 57.

160. Brückner, K., Perez, L., Clausen, H., and Cohen, S. (2000). Glycosyltransferase activity of Fringe modulates Notch–Delta interactions. *Nature* 406, 411–415.

161. Brunet, T., Larson, B.T., Linden, T.A., Vermeij, M.J.A., McDonald, K., and King, N. (2019). Light-regulated collective contractility in a multicellular choanoflagellate. *Science* 366, 326–334.

162. Bryant, P.J. (1971). Regeneration and duplication following operations *in situ* on the imaginal discs of *Drosophila melanogaster*. *Dev. Biol.* 26, 637–651.

163. Bryant, S.V., French, V., and Bryant, P.J. (1981). Distal regeneration and symmetry. *Science* 212, 993–1002.

164. Budday, S., Nay, R., de Rooij, R., Steinmann, P., Wyrobek, T., Ovaert, T.C., and Kuhl, E. Mechanical properties of gray and white matter brain tissue by indentation. *J. Mech. Behav. Biomed. Mater.* 46, 318–330.

165. Buffry, A.D., Mendes, C.C., and McGregor, A.P. (2016). The functionality and evolution of eukaryotic transcriptional enhancers. *Adv. Genet.* 96, 143–206.

166. Bulger, M. and Groudine, M. (2010). Enhancers: the abundance and function of regulatory sequences beyond promoters. *Dev. Biol.* 339, 250–257.

167. Bullough, W.S. (1962). The control of mitotic activity in adult mammalian tissues. *Biol. Rev.* 37, 307–342.

168. Burger, B., Fuchs, D., Sprecher, E., and Itin, P. (2011). The immigration delay disease: adermatoglyphia – inherited absence of epidermal ridges. *J. Am. Acad. Dermatol.* 64, 974–980.

169. Butler, M.T. and Wallingford, J.B. (2017). Planar cell polarity in development and disease. *Nat. Rev. Mol. Cell Biol.* 18, 375–388.

170. Cadieu, E., Neff, M.W., Quignon, P., Walsh, K., Chase, K., Parker, H.G., VonHoldt, B.M., Rhue, A., Boyko, A., Byers, A., Wong, A., Mosher, D.S., Elkahloun, A.G., Spady, T.C., André, C., Lark, K.G., Cargill, M., Bustamante, C.D., Wayne, R.K., and Ostrander, E.A. (2009). Coat variation in the domestic dog is governed by variants in three genes. *Science* 326, 150–153.

171. Cai, J., Townsend, J.P., Dodson, T.C., Heiney, P.A., and Sweeney, A.M. (2017). Eye patches: protein assembly of index-gradient squid lenses. *Science* 357, 564–569.

172. Campbell, G. (2002). Distalization of the *Drosophila* leg by graded EGF-receptor activity. *Nature* 418, 781–785.

173. Campbell, G. and Tomlinson, A. (1995). Initiation of the proximodistal axis in insect legs. *Development* 121, 619–628.

174. Campbell, G. and Tomlinson, A. (1998). The roles of the homeobox genes *aristaless* and *Distal-less* in patterning the legs and wings of *Drosophila*. *Development* 125, 4483–4493.

175. Campbell, G., Weaver, T., and Tomlinson, A. (1993). Axis specification in the developing *Drosophila* appendage: the role of *wingless*, *decapentaplegic*, and the homeobox gene *aristaless*. *Cell* 74, 1113–1123.

176. Campos-Ortega, J.A. (1998). The genetics of the *Drosophila achaete-scute* gene complex: a historical appraisal. *Int. J. Dev. Biol.* 42, 291–297.

177. Cañestro, C., Albalat, R., Irimia, M., and Garcia-Fernàndez, J. (2013). Impact of gene gains, losses and duplication modes on the origin and diversification of vertebrates. *Semin. Cell Dev. Biol.* 24, 83–94.

178. Capdevila, M.P. and García-Bellido, A. (1978). Phenocopies of *bithorax* mutants. *W. Roux Arch. Dev. Biol.* 185, 105–126.

179. Capek, D. and Müller, P. (2019). Positional information and tissue scaling during development and regeneration. *Development* 146, dev177709.

180. Capilla, A., Johnson, R., Daniels, M., Benavente, M., Bray, S.J., and Galindo, M.I. (2012). Planar cell polarity controls directional Notch signaling in the *Drosophila* leg. *Development* 139, 2584–2593.

181. Caro, T. (2009). Contrasting coloration in terrestrial mammals. *Philos. Trans. R. Soc. Lond. B* 364, 537–548.

182. Caro, T. and Mallarino, R. (2020). Coloration in mammals. *Trends Ecol. Evol.* 35, 357–366.

183. Carpenter, A.C., Smith, A.N., Wagner, H., Cohen-Tayar, Y., Rao, S., Wallace, V., Ashery-Padan, R., and Lang, R.A. (2015). Wnt ligands from the embryonic surface ectoderm regulate "bimetallic strip" optic cup morphogenesis in mouse. *Development* 142, 972–982.

184. Carroll, L. and Gardner, M. (1960). *The Annotated Alice: Alice's Adventures in Wonderland & Through the Looking Glass*. Meridian, New York.

185. Carroll, R.L. and Holmes, R.B. (2007). Evolution of the appendicular skeleton of amphibians. In *Fins into Limbs: Evolution, Development, and Transformation*, Hall, B.K., editor. University of Chicago Press, Chicago, IL, pp. 185–224.

186. Carroll, S.B. (2000). Endless forms: the evolution of gene regulation and morphological diversity. *Cell* 101, 577–580.

187. Carroll, S.B. (2005). *Endless Forms Most Beautiful: The New Science of Evo Devo and the Making of the Animal Kingdom*. Norton, New York.

188. Carroll, S.B., Grenier, J.K., and Weatherbee, S.D. (2005). *From DNA to Diversity: Molecular Genetics and the Evolution of Animal Design*, 2nd ed. Blackwell, Malden, MA.

189. Carson, H.L. and Kaneshiro, K.Y. (1976). *Drosophila* of Hawaii: systematics and ecological genetics. *Annu. Rev. Ecol. Syst.* 7, 311–345.

190. Casares, F. and Mann, R.S. (2001). The ground state of the ventral appendage in *Drosophila*. *Science* 293, 1477–1480.

191. Casas, E. and Kehrli, M.E., Jr. (2016). A review of selected genes with known effects on performance and health of cattle. *Front. Vet. Sci.* 3, 113.

192. Cassina, M., Cagnoli, G.A., Zuccarello, D., Di Gianantonio, E., and Clementi, M. (2017). Human teratogens and genetic phenocopies. Understanding pathogenesis through human genes mutation. *Eur. J. Med. Genet.* 60, 22–31.

193. Castelli-Gair, J. (1998). Implications of the spatial and temporal regulation of *Hox* genes on development and evolution. *Int. J. Dev. Biol.* 42, 437–444.

194. Castelli-Gair Hombría, J. and Lovegrove, B. (2003). Beyond homeosis: HOX function in morphogenesis and organogenesis. *Differentiation* 71, 461–476.

195. Cavodeassi, F. and Houart, C. (2011). Brain regionalization: of signaling centers and boundaries. *Dev. Neurobiol.* 72, 218–233.

196. Cavodeassi, F., del Corral, R.D., Campuzano, S., and Domínguez, M. (1999). Compartments and organising boundaries in the *Drosophila* eye: the role of the homeodomain Iroquois proteins. *Development* 126, 4933–4942.

197. Chan, C.J., Heisenberg, C.-P., and Hiiragi, T. (2017). Coordination of morphogenesis and cell-fate specification in development. *Curr. Biol.* 27, R1024–R1035.

198. Chang, H.Y. (2009). Anatomic demarcation of cells: genes to patterns. *Science* 326, 1206–1207.

199. Chang, S. (2017). How squid build their graded-index spherical lenses. *Physics Today* 70, 26–28.

200. Chang, X., Li, D., Tian, L., Liu, Y., March, M., Wang, T., Hou, C., Pellegrino, R., Levy, R., Jen, M., Soccio, R., Sleiman, P., Hakonarson, H., and Castelo-Soccio, L. (2018). Heterozygous deletion impacting *SMARCAD1* in the original kindred with absent dermatoglyphs and associated features (Baird, 1964). *J. Pediatr.* 194, 248–252.

201. Charlton-Perkins, M., Brown, N.L., and Cook, T.A. (2011). The lens in focus: a comparison of lens development in *Drosophila* and vertebrates. *Mol. Genet. Genomics* 286, 189–213.

202. Chauhan, B., Plageman, T., Lou, M., and Lang, R. (2015). Epithelial morphogenesis: the mouse eye as a model system. *Curr. Top. Dev. Biol.* 111, 375–399.

203. Chauhan, B.K., Lou, M., Zheng, Y., and Lang, R.A. (2011). Balanced Rac1 and RhoA activities regulate cell shape and drive invagination morphogenesis in epithelia. *PNAS* 108, 18289–18294.

204. Chen, J. and Chuong, C.-M. (2011). Patterning skin by planar cell polarity: the multi-talented hair designer. *Exp. Dermatol.* 21, 81–85.

205. Chen, J., Jacox, L.A., Saldanha, F., and Sive, H. (2017). Mouth development. *Wiley Interdiscip. Rev. Dev. Biol.* 6, e275.

206. Chew, K.Y., Yu, H., Pask, A.J., Shaw, G., and Renfree, M.B. (2012). *HOXA13* and *HOXD13* expression during development of the syndactylous digits in the marsupial *Macropus eugenii. BMC Dev. Biol.* 12, 2.

207. Chiang, C., Litingtung, Y., Lee, E., Young, K.E., Corden, J.L., Westphal, H., and Beachy, P.A. (1996). Cyclopia and defective axial patterning in mice lacking Sonic hedgehog gene function. *Nature* 383, 407–413.

208. Choe, C.P. and Crump, J.G. (2015). Dynamic epithelia of the developing vertebrate face. *Curr. Opin. Genet. Dev.* 32, 66–72.

209. Chouard, T. (2010). Revenge of the hopeful monster. *Nature* 463, 864–867.

210. Ciechanska, E., Dansereau, D.A., Svendsen, P.C., Heslip, T.R., and Brook, W.J. (2007). *dAP-2* and *defective proventriculus* regulate *Serrate* and *Delta* expression in the tarsus of *Drosophila melanogaster. Genome* 50, 693–705.

211. Cieslak, J., Borowska, A., Wodas, L., and Mackowski, M. (2018). Interbreed distribution of the myostatin (*MSTN*) gene 5′-flanking variants and their relationship with horse biometric traits. *J. Equine Vet. Sci.* 60, 83–89.

212. Cieslak, M., Reissmann, M., Hofreiter, M., and Ludwig, A. (2011). Colours of domestication. *Biol. Rev.* 86, 885–899.

213. Clark, D.A., Mitra, P.P., and Wang, S.S.-H. (2001). Scalable architecture in mammalian brains. *Nature* 411, 189–193.

214. Clark, E. (2017). Dynamic patterning by the *Drosophila* pair-rule network reconciles long-germ and short-germ segmentation. *PLoS Biol.* 15(9), e2002439.

215. Clark, W.C. and Russell, M.A. (1977). The correlation of lysosomal activity and adult phenotype in a cell-lethal mutant of *Drosophila. Dev. Biol.* 57, 160–173.

216. Cline, T.W. (1993). The *Drosophila* sex determination signal: how do flies count to two? *Trends Genet.* 9, 385–390.

217. Cloutier, R., Clement, A.M., Lee, M.S.Y., Noël, R., Béchard, I., Roy, V., and Long, J.A. (2020). *Elpistostege* and the origin of the vertebrate hand. *Nature* 579, 549–554.

218. Cobourne, M.T. and Sharpe, P.T. (2003). Tooth and jaw: molecular mechanisms of patterning in the first branchial arch. *Arch. Oral Biol.* 48, 1–14.

219. Cock, A.G. and Forsdyke, D.R. (2008). *Treasure Your Exceptions: The Science and Life of William Bateson*. Springer, New York.

220. Coen, E. (1999). *The Art of Genes: How Organisms Make Themselves*. Oxford University Press, New York.

221. Cohen, M.M., Jr. (2006). Holoprosencephaly: clinical, anatomic, and molecular dimensions. *Birth Defects Res. A Clin. Mol. Teratol.* 76, 658–673.

222. Cohen, S.M. (1993). Imaginal disc development. In *The Development of* Drosophila melanogaster, Bate, M. and Martinez Arias, A., editors. Cold Spring Harbor Laboratory Press, Plainview, NY, pp. 747–841.

223. Cohen, S.M. (2003). Long-range signalling by touch. *Nature* 426, 503–504.

224. Coile, D.C. (2005). *Encyclopedia of Dog Breeds*. Barron's Educational Series, Hauppauge, NY.

225. Colas, J.-F. and Schoenwolf, G.C. (2001). Towards a cellular and molecular understanding of neurulation. *Dev. Dynamics* 221, 117–145.

226. Coletti, S.M., Ide, C.F., Blankenau, A.J., and Meyer, R.L. (1990). Ocular dominance stripe formation by regenerated isogenic double temporal retina in *Xenopus laevis*. *J. Neurobiol.* 21, 276–282.

227. Collins, T.N., Mao, Y., Li, H., Bouaziz, M., Hong, A., Feng, G.-S., Wang, F., Quilliam, L.A., Chen, L., Park, T., Curran, T., and Zhang, X. (2018). Crk proteins transduce FGF signaling to promote lens fiber cell elongation. *eLife* 7, e32586.

228. Condic, M.L., Fristrom, D., and Fristrom, J.W. (1991). Apical cell shape changes during *Drosophila* imaginal leg disc elongation: a novel morphogenetic mechanism. *Development* 111, 23–33.

229. Conlin, L.K., Thiel, B.D., Bonnemann, C.G., Medne, L., Ernst, L.M., Zackai, E.H., Deardorff, M.A., Krantz, I.D., Hakonarson, H., and Spinner, N.B. (2010). Mechanisms of mosaicism, chimerism and uniparental disomy identified by single nucleotide polymorphism array analysis. *Hum. Mol. Genet.* 19, 1263–1275.

230. Constantine-Paton, M. and Law, M.I. (1978). Eye-specific termination bands in tecta of three-eyed frogs. *Science* 202, 639–641.

231. Cook, T.A. (1914). *The Curves of Life*. Constable, London.

232. Cooke, J. (1975). The emergence and regulation of spatial organization in early animal development. *Annu. Rev. Biophys. Bioeng.* 4, 185–217.

233. Cooke, J. and Zeeman, E.C. (1976). A clock and wavefront model for control of the number of repeated structures during animal morphogenesis. *J. Theor. Biol.* 58, 455–476.

234. Cooper, K.L. (2019). Developmental and evolutionary allometry of the mammalian limb skeleton. *Integr. Comp. Biol.* 59, 1356–1368.

235. Cooper, S.B. and Van Leeuwen, J., eds. (2013). *Alan Turing: His Work and Impact*. Elsevier, New York.

236. Corallo, D., Trapani, V., and Bonaldo, P. (2015). The notochord: structure and functions. *Cell. Mol. Life Sci.* 72, 2989–3008.

237. Cordero, R.J.B. and Casadevall, A. (2020). Melanin. *Curr. Biol.* 30, R142–R143.

238. Cordingley, J.E., Sundaresan, S.R., Fischhoff, I.R., Shapiro, B., Ruskey, J., and Rubenstein, D.I. (2009). Is the endangered Grevy's zebra threatened by hybridization? *Anim. Conserv.* 12, 505–513.

239. Córdoba, S. and Estella, C. (2018). The transcription factor Dysfusion promotes fold and joint morphogenesis through regulation of Rho1. *PLoS Genet.* 14(8), e1007584.

240. Corona, M., Libbrecht, R., and Wheeler, D.E. (2016). Molecular mechanisms of phenotypic plasticity in social insects. *Curr. Opin. Insect Sci.* 13, 55–60.

241. Corson, F., Couturier, L., Rouault, H., Mazouni, K., and Schweisguth, F. (2017). Self-organized Notch dynamics generate stereotyped sensory organ patterns in *Drosophila*. *Science* 356, 501.

242. Cortes, C., Francou, A., De Bono, C., and Kelly, R.G. (2018). Epithelial properties of the second heart field. *Circ. Res.* 122, 142–154.

243. Coulombre, J.L. and Coulombre, A.J. (1963). Lens development: fiber elongation and lens orientation. *Science* 142, 1489–1490.

244. Courgeon, M. and Desplan, C. (2019). Coordination between stochastic and deterministic specification in the *Drosophila* visual system. *Science* 366, 325.

245. Couso, J.P., Bishop, S.A., and Martinez Arias, A. (1994). The wingless signalling pathway and the patterning of the wing margin in *Drosophila*. *Development* 120, 621–636.

246. Coutelis, J.-B., González-Morales, N., Géminard, C., and Noselli, S. (2014). Diversity and convergence in the mechanisms establishing L/R asymmetry in metazoa. *EMBO Rep.* 15, 926–937.

247. Cozzitorto, C. and Spagnoli, F.M. (2019). Pancreas organogenesis: the interplay between surrounding microenvironment(s) and epithelium-intrinsic factors. *Curr. Top. Dev. Biol.* 132, 221–256.

248. Cranford, T.W., Amundin, M., and Norris, K.S. (1996). Functional morphology and homology in the odontocete nasal complex: implications for sound generation. *J. Morphol.* 228, 223–285.

249. Creel, D., Garber, S.R., King, R.A., and Witkop, C.J., Jr. (1980). Auditory brainstem anomalies in human albinos. *Science* 209, 1253–1255.

250. Cretekos, C.J., Wang, Y., Green, E.D., Martin, J.F., Rasweiler, J.J., IV, and Behringer, R.R. (2008). Regulatory divergence modifies limb length between mammals. *Genes Dev.* 22, 141–151.

251. Crews, D. (2003). Sex determination: where environment and genetics meet. *Evol. Dev.* 5, 50–55.

252. Crispo, E. (2007). The Baldwin effect and genetic assimilation: revisiting two mechanisms of evolutionary change mediated by phenotypic plasticity. *Evolution* 61, 2469–2479.

253. Crow, J.F. and Bender, W. (2004). Edward B. Lewis, 1918–2004. *Genetics* 168, 1773–1783.

254. Cubas, P., de Celis, J.-F., Campuzano, S., and Modolell, J. (1991). Proneural clusters of *achaete-scute* expression and the generation of sensory organs in the *Drosophila* imaginal wing disc. *Genes Dev.* 5, 996–1008.

255. Cubeñas-Potts, C. and Corces, V.G. (2015). Architectural proteins, transcription, and the three-dimensional organization of the genome. *FEBS Lett.* 589, 2923–2930.

256. Cummins, H. and Midlo, C. (1943). *Finger Prints, Palms and Soles: An Introduction to Dermatoglyphics*. Dover, New York.

257. Currie, A. (2013). Convergence as evidence. *Br. J. Philos. Sci.* 64, 763–786.

258. Curtis, A.S.G. (1960). Cortical grafting in *Xenopus laevis. J. Embryol. Exp. Morphol.* 8, 163–173.

259. Curtis, A.S.G. (1962). Morphogenetic interactions before gastrulation in the amphibian, *Xenopus laevis*: the cortical field. *J. Embryol. Exp. Morphol.* 10, 410–422.

260. Curtiss, J., Halder, G., and Mlodzik, M. (2002). Selector and signalling molecules cooperate in organ patterning. *Nat. Cell Biol.* 4, E48–E51.

261. Cvekl, A. and Ashery-Padan, R. (2014). The cellular and molecular mechanisms of vertebrate lens development. *Development* 141, 4432–4447.

262. Cvekl, A. and Zhang, X. (2017). Signaling and gene regulatory networks in mammalian lens development. *Trends Genet.* 33, 677–702.

263. D'Souza, B., Meloty-Kapella, C., and Weinmaster, G. (2010). Canonical and non-canonical Notch ligands. *Curr. Top. Dev. Biol.* 92, 73–129.

264. Dall'Olio, S., Fontanesi, L., Costa, L.N., Tassinari, M., Minieri, L., and Falaschini, A. (2010). Analysis of horse myostatin gene and identification of single nucleotide polymorphisms in breeds of different morphological types. *J. Biomed. Biotech.* 2010, 542945.

265. Dall'Olio, S., Wang, Y., Sartori, C., Fontanesi, L., and Mantovani, R. (2014). Association of myostatin (MSTN) gene polymorphisms with morphological traits in the Italian Heavy Draft Horse breed. *Livestock Sci.* 160, 29–36.

266. Darbellay, F. and Duboule, D. (2016). Topological domains, metagenes, and the emergence of pleiotropic regulations at *Hox* loci. *Curr. Top. Dev. Biol.* 116, 299–314.

267. Darwin, C. (1859). *On the Origin of Species by Means of Natural Selection, or the Preservation of Favoured Races in the Struggle for Life.* John Murray, London.

268. Darwin, C. (1872). *The Expression of the Emotions in Man and Animals.* John Murray, London.

269. Dasgupta, A. and Amack, J.D. (2016). Cilia in left–right patterning. *Philos. Trans. R. Soc. Lond. B* 371, 20150410.

270. DasGupta, R. and Fuchs, E. (1999). Multiple roles for activated LEF/TCF transcription complexes during hair follicle development and differentiation. *Development* 126, 4557–4568.

271. Davidson, L.A. (2012). Epithelial machines that shape the embryo. *Trends Cell Biol.* 22, 82–87.

272. Davidson, L.A. (2017). Mechanical design in embryos: mechanical signalling, robustness and developmental defects. *Philos. Trans. R. Soc. Lond. B* 372, 20150516.

273. Davies-Thompson, J., Scheel, M., Lanyon, L.J., and Barton, J.J.S. (2013). Functional organisation of visual pathways in a patient with no optic chiasm. *Neuropsychologia* 51, 1260–1272.

274. Davis, A.P., Witte, D.P., Hsieh-Li, H.M., Potter, S.S., and Capecchi, M.R. (1995). Absence of radius and ulna in mice lacking *hoxa-11* and *hoxd-11. Nature* 375, 791–795.

275. Davis, D.D. (1964). *The Giant Panda: A Morphological Study of Evolutionary Mechanisms.* Fieldiana: Zoology Memoirs, Vol. 3. Chicago Natural History Museum, Chicago, IL.

276. Dawkins, R. (1996). *Climbing Mount Improbable.* Norton, New York.

277. Day, S.J. and Lawrence, P.A. (2000). Measuring dimensions: the regulation of size and shape. *Development* 127, 2977–2987.

278. de Beer, G. (1958). *Embryos and Ancestors*, 3rd ed. Clarendon Press, Oxford.

279. de Celis, J.F., García-Bellido, A., and Bray, S.J. (1996). Activation and function of *Notch* at the dorsal–ventral boundary of the wing imaginal disc. *Development* 122, 359–369.

280. de Celis, J.F., Tyler, D.M., de Celis, J., and Bray, S.J. (1998). Notch signalling mediates segmentation of the *Drosophila* leg. *Development* 125, 4617–4626.

281. de Joussineau, C., Soulé, J., Martin, M., Anguille, C., Montcourrier, P., and Alexandre, D. (2003). Delta-promoted filopodia mediate long-range lateral inhibition in *Drosophila*. *Nature* 426, 555–559.

282. de Juan Romero, C. and Borrell, V. (2017). Genetic maps and patterns of cerebral cortex folding. *Curr. Opin. Cell Biol.* 49, 31–37.

283. De Pascalis, C. and Etienne-Manneville, S. (2017). Single and collective cell migration: the mechanics of adhesions. *Mol. Biol. Cell* 28, 1833–1846.

284. De Robertis, E.M. (2009). Spemann's organizer and the self-regulation of embryonic fields. *Mech. Dev.* 126, 925–941.

285. De Robertis, E.M., Morita, E.A., and Cho, K.W.Y. (1991). Gradient fields and homeobox genes. *Development* 112, 669–678.

286. De Robertis, E.M., Moriyama, Y., and Colozza, G. (2017). Generation of animal form by the Chordin/Tolloid/BMP gradient: 100 years after D'Arcy Thompson. *Dev. Growth Differ.* 59, 580–592.

287. Deane-Coe, P.E., Chu, E.T., Slavney, A., Boyko, A.R., and Sams, A.J. (2018). Direct-to-consumer DNA testing of 6,000 dogs reveals 98.6-kb duplication associated with blue eyes and heterochromia in Siberian Huskies. *PLoS Genet.* 14(10), e1007648

288. Degabriele, R. (1980). The physiology of the koala. *Sci. Am.* 243(1), 110–117.

289. del Álamo, D., Terriente, J., and Díaz-Benjumea, F.J. (2002). Spitz/EGFr signalling via the Ras/MAPK pathway mediates the induction of bract cells in *Drosophila* legs. *Development* 129, 1975–1982.

290. Delgado, I. and Torres, M. (2016). Gradients, waves and timers: an overview of limb patterning models. *Semin. Cell Dev. Biol.* 49, 109–115.

291. Delgado, I. and Torres, M. (2017). Coordination of limb development by crosstalk among axial patterning pathways. *Dev. Biol.* 429, 382–386.

292. Delpretti, S., Zakany, J., and Duboule, D. (2012). A function for all posterior *Hoxd* genes during digit development? *Dev. Dynamics* 241, 792–802.

293. Deng-Lobnig, M. and Martin, A.C. (2020). Divergent and combinatorial mechanical strategies that promote epithelial folding during morphogenesis. *Curr. Opin. Genet. Dev.* 63, 24–29.

294. Depew, M.J., Lufkin, T., and Rubenstein, J.L.R. (2002). Specification of jaw subdivisions by *Dlx* genes. *Science* 298, 381–385.

295. Deschamps, J. (2008). Tailored *Hox* gene transcription and the making of the thumb. *Genes Dev.* 22, 293–296.

296. Deutsch, J.S., ed. (2010). *Hox Genes: Studies from the 20th to the 21st Century*. Advances in Experimental Medicine and Biology. Landes Bioscience, New York.

297. Devenport, D. (2016). Tissue morphodynamics: translating planar polarity cues into polarized cell behaviors. *Semin. Cell Dev. Biol.* 55, 99–110.

298. Diaz de la Loza, M.C. and Thompson, B.J. (2017). Forces shaping the *Drosophila* wing. *Mech. Dev.* 144, 23–32.

299. Diaz de la Loza, M.C., Loker, R., Mann, R.S., and Thompson, B.J. (2020). Control of tissue morphogenesis by the HOX gene *Ultrabithorax*. *Development* 147, dev184564.

300. Diaz de la Loza, M.C., Ray, R.P., Ganguly, P.S., Alt, S., Davis, J.R., Hoppe, A., Tapon, N., Salbreux, G., and Thompson, B.J. (2018). Apical and basal remodeling control epithelial morphogenesis. *Dev. Cell* 46, 23–39.

301. Dickerson, B.H., de Souza, A.M., Huda, A., and Dickinson, M.H. (2019). Flies regulate wing motion via active control of a dual-function gyroscope. *Curr. Biol.* 29, 3517–3524.

302. Dickinson, M.H. (1999). Haltere-mediated equilibrium reflexes of the fruit fly, *Drosophila melanogaster*. *Philos. Trans. R. Soc. Lond. B* 354, 903–916.

303. Dieters, J., Kowalczyk, W., and Seidl, T. (2016). Simultaneous optimisation of earwig hindwings for flight and folding. *Biol. Open* 5, 638–644.

304. Dietrich, M.R. (2000). From hopeful monsters to homeotic effects: Richard Goldschmidt's integration of development, evolution, and genetics. *Am. Zool.* 40, 738–747.

305. Dietrich, M.R. (2003). Richard Goldschmidt: hopeful monsters and other "heresies". *Nat. Rev. Genet.* 4, 68–74.

306. Diogo, R. (2017). *Evolution Driven by Organismal Behavior: A Unifying View of Life, Function, Form, Mismatches, and Trends*. Springer Nature, Cham, Switzerland.

307. Diogo, R., Guinard, G., and Diaz, R.E., Jr. (2017). Dinosaurs, chameleons, humans, and evo-devo path: linking Étienne Geoffroy's teratology, Waddington's homeorhesis, Alberch's logic of "monsters," and Goldschmidt hopeful "monsters". *J. Exp. Zool. B Mol. Dev. Evol.* 328, 207–229.

308. Diogo, R., Linde-Medina, M., Abdala, V., and Ashley-Ross, M.A. (2013). New, puzzling insights from comparative myological studies on the old and unsolved fore-limb/hindlimb enigma. *Biol. Rev.* 88, 196–214.

309. Diogo, R., Smith, C.M., and Ziermann, J.M. (2015). Evolutionary developmental pathology and anthropology: a new field linking development, comparative anatomy, human evolution, morphological variations and defects, and medicine. *Dev. Dynamics* 244, 1357–1374.

310. Dittrich-Reed, D.R. and Fitzpatrick, B.M. (2013). Transgressive hybrids as hopeful monsters. *Evol. Biol.* 40, 310–315.

311. Docampo, M.J., Zanna, G., Fondevila, D., Cabrera, J., López-Iglesias, C., Carvalho, A., Cerrato, S., Ferrer, L., and Bassols, A. (2011). Increased HAS2-driven hyaluronic acid synthesis in shar-pei dogs with hereditary cutaneous hyaluronosis (mucinosis). *Vet. Dermatol.* 22, 535–545.

312. Doe, C.Q. and Spana, E.P. (1995). A collection of cortical crescents: asymmetric protein localization in CNS precursor cells. *Neuron* 15, 991–995.

313. Doherty, D., Feger, G., Younger-Shepherd, S., Jan, L.Y., and Jan, Y.N. (1996). Delta is a ventral to dorsal signal complementary to Serrate, another Notch ligand, in *Drosophila* wing formation. *Genes Dev.* 10, 421–434.

314. Doiguchi, M., Nakagawa, T., Imamura, Y., Yoneda, M., Higashi, M., Kubota, K., Yamashita, S., Asahara, H., Iida, M., Fujii, S., Ikura, T., Liu, Z., Nandu, T., Kraus, W.L., Ueda, H., and Ito, T. (2016). SMARCAD1 is an ATP-dependent stimulator of nucleosomal H2A acetylation via CBP, resulting in transcriptional regulation. *Sci. Rep.* 6, 20179.

315. Domingos, P.M., Jenny, A., Combie, K.F., del Alamo, D., Mlodzik, M., Steller, H., and Mollereau, B. (2019). Regulation of Numb during planar cell polarity establishment in the *Drosophila* eye. *Mech. Dev.* 160, 103583.

316. Donahue, C.J., Glasser, M.F., Preuss, T.M., Rilling, J.K., and Van Essen, D.C. (2018). Quantitative assessment of prefrontal cortex in humans relative to nonhuman primates. *PNAS* 115, E5183–E5192.

317. Donahue, C.J., Glasser, M.F., Preuss, T.M., Rilling, J.K., and Van Essen, D.C. (2019). Reply to Barton and Montgomery: a case for preferential prefrontal cortical expansion. *PNAS* 116, 5–6.

318. Dongen, S.V. (2006). Fluctuating asymmetry and developmental instability in evolutionary biology: past, present and future. *J. Evol. Biol.* 19, 1727–1743.

319. Donnelly, D.E. and Morrison, P.J. (2014). Hereditary gigantism: the biblical giant Goliath and his brothers. *Ulster Med. J.* 83, 86–88.

320. Dougoud, M., Mazza, C., Schwaller, B., and Pecze, L. (2019). Extending the mathematical palette for developmental pattern formation: piebaldism. *Bull. Math. Biol.* 81, 1461–1478.

321. Dover, G. (2000). How genomic and developmental dynamics affect evolutionary processes. *BioEssays* 22, 1153–1159.

322. Drimmer, F. (1973). *Very Special People*. Amjon Publishers, New York.

323. Driscoll, C.A., Clutton-Brock, J., Kitchener, A.C., and O'Brien, S.J. (2009). The taming of the cat. *Sci. Am.* 300(6), 68–75.

324. Drögemüller, C., Karlsson, E.K., Hytönen, M.K., Perloski, M., Dolf, G., Sainio, K., Lohi, H., Lindblad-Toh, K., and Leeb, T. (2008). A mutation in hairless dogs implicates *FOXI3* in ectodermal development. *Science* 321, 1462.

325. Duan, D., Xia, S., Rekik, I., Wu, Z., Wang, L., Lin, W., Gilmore, J.H., Shen, D., and Li, G. (2020). Individual identification and individual variability analysis based on cortical folding features in developing infant singletons and twins. *Hum. Brain Mapp.* 41, 1985–2003.

326. Duboule, D. (2007). The rise and fall of Hox gene clusters. *Development* 134, 2549–2560.

327. Duncan, I. and Montgomery, G. (2002). E. B. Lewis and the bithorax complex: part I. *Genetics* 160, 1265–1272.

328. Duncan, I. and Montgomery, G. (2002). E. B. Lewis and the bithorax complex: part II. From *cis–trans* test to the genetic control of development. *Genetics* 161, 1–10.

329. Dworkin, I. (2005). Canalization, cryptic variation, and developmental buffering: a critical examination and analytical perspective. In *Variation: A Central Concept in Biology*, Hallgrímsson, B. and Hall, B.K., editors. Elsevier Academic Press, New York, pp. 131–158.

330. Dybus, A., Proskura, W.S., Sadkowski, S., and Pawlina, E. (2013). A single nucleotide polymorphism in exon 3 of the myostatin gene in different breeds of domestic pigeon (*Columba livia* var. *domestica*). *Vet. Med. (Praha)* 58, 32–38.

331. Ebisuya, M. and Briscoe, J. (2018). What does time mean in development? *Development* 145, dev164368.

332. Economou, A.D., Ohazama, A., Porntaveetus, T., Sharpe, P.T., Kondo, S., Basson, M.A., Gritli-Linde, A., Cobourne, M.T., and Green, J.B.A. (2012). Periodic stripe formation by a Turing mechanism operating at growth zones in the mammalian palate. *Nat. Genet.* 44, 348–352.

333. Ede, D.A. (1972). Cell behaviour and embryonic development. *Int. J. Neurosci.* 3, 165–174.

334. Eder, D., Aegerter, C., and Basler, K. (2017). Forces controlling organ growth and size. *Mech. Dev.* 114, 53–61.

335. Edgar, B.A. (2006). How flies get their size: genetics meets physiology. *Nat. Rev. Genet.* 7, 907–916.

336. Edgar, B.A. and Orr-Weaver, T.L. (2001). Endoreplication cell cycles: more for less. *Cell* 105, 297–306.

337. Edwards, J.S. (1994). In memoriam. Sir Vincent Brian Wigglesworth (1899–1994). *Dev. Biol.* 166, 361–362.

338. Edwards, J.S. (1998). Sir Vincent Wigglesworth and the coming of age of insect development. *Int. J. Dev. Biol.* 42, 471–473.

339. Efstratiadis, A. (1998). Genetics of mouse growth. *Int. J. Dev. Biol.* 42, 955–976.

340. Eizirik, E., David, V.A., Buckley-Beason, V., Roelke, M.E., Schäffer, A.A., Hannah, S.S., Narfström, K., O'Brien, S.J., and Menotti-Raymond, M. (2010). Defining and mapping mammalian coat pattern genes: multiple genomic regions implicated in domestic cat stripes and spots. *Genetics* 184, 267–275.

341. Elgjo, K. and Reichelt, K.L. (2004). Chalones: from aqueous extracts to oligopeptides. *Cell Cycle* 3, 1208–1211.

342. Elliott, K.L., Houston, D.W., and Fritzsch, B. (2015). Sensory afferent segregation in three-eared frogs resemble the dominance columns observed in three-eyed frogs. *Sci. Rep.* 5, 8338.

343. Elsdale, T. and Wasoff, F. (1976). Fibroblast cultures and dermatoglyphics: the topology of two planar patterns. *W. Roux Arch. Dev. Biol.* 180, 121–147.

344. Emerson, S.B. (1985). Jumping and leaping. In *Functional Vertebrate Morphology*, Hildebrand, M., Bramble, D.M., Liem, K.F., and Wake, D.B., editors. Harvard University Press, Cambridge, MA, pp. 58–72.

345. Emlen, D.J. (2008). The evolution of animal weapons. *Annu. Rev. Ecol. Evol. Syst.* 39, 387–413.

346. Emlen, D.J. (2014). *Animal Weapons: The Evolution of Battle*. Henry Holt, New York.

347. Emlen, D.J., Warren, I.A., Johns, A., Dworkin, I., and Lavine, L.C. (2012). A mechanism of extreme growth and reliable signaling in sexually selected ornaments and weapons. *Science* 337, 860–864.

348. Enard, D., Depaulis, F., and Crollius, H.R. (2010). Human and non-human primate genomes share hotspots of positive selection. *PLoS Genet.* 6(2), e1000840.

349. Eom, D.S., Bain, E.J., Patterson, L.B., Grout, M.E., and Parichy, D.M. (2015). Long-distance communication by specialized cellular projections during pigment pattern development and evolution. *eLife* 4, e12401.

350. Erickson, J.R. and Echeverri, K. (2018). Learning from regeneration research organisms: the circuitous road to scar free wound healing. *Dev. Biol.* 433, 144–154.

351. Estella, C., Voutev, R., and Mann, R.S. (2012). A dynamic network of morphogens and transcription factors patterns the fly leg. *Curr. Top. Dev. Biol.* 98, 173–198.

352. Estrellas, K.M., Chung, L., Cheu, L.A., Sadtler, K., Majumdar, S., Mula, J., Wolf, M.T., Elisseeff, J.H., and Wagner, K.R. (2018). Biological scaffold-mediated delivery of myostatin inhibitor promotes a regenerative immune response in an animal model of Duchenne muscular dystrophy. *J. Biol. Chem.* 293, 15594–15605.

353. Etienne-Manneville, S. (2011). Control of polarized cell morphology and motility by adherens junctions. *Semin. Cell Dev. Biol.* 22, 850–857.

354. Etienne-Manneville, S. (2014). Neighborly relations during collective migration. *Curr. Opin. Cell Biol.* 30, 51–59.

355. Falk, D., Lepore, F.E., and Noe, A. (2013). The cerebral cortex of Albert Einstein: a description and preliminary analysis of unpublished photographs. *Brain* 136, 1304–1327.

356. Fankhauser, G. (1945). The effects of changes in chromosome number on amphibian development. *Q. Rev. Biol.* 20, 20–78.

357. Fantauzzo, K.A., Tadin-Strapps, M., You, Y., Mentzer, S.E., Baumeister, F.A.M., Cianfarani, S., Van Maldergem, L., Warburton, D., Sundberg, J.P., and Christiano, A.M. (2008). A position effect on *TRPS1* is associated with Ambras syndrome in humans and the Koala phenotype in mice. *Hum. Mol. Genet.* 17, 3539–3551.

358. Fehilly, C.B., Willadsen, S.M., and Tucker, E.M. (1984). Interspecific chimaerism between sheep and goat. *Nature* 307, 634–636.

359. Feigin, C.Y. and Mallarino, R. (2018). Setting the bar: analyzing the genomes of rock pigeons demonstrates that genetic variation comes in many forms and can have unexpected origins. *eLife* 7, e39068.

360. Fernandes, J., Celniker, S.E., Lewis, E.B., and VijayRaghavan, K. (1994). Muscle development in the four-winged *Drosophila* and the role of the *Ultrabithorax* gene. *Curr. Biol.* 4, 957–964.

361. Fernández, V., Llinares-Benadero, C., and Borrell, V. (2016). Cerebral cortex expansion and folding: what have we learned? *EMBO J.* 35, 1021–1044.

362. Ferree, P.L., Deneke, V.E., and Di Talia, S. (2016). Measuring time during early embryonic development. *Semin. Cell Dev. Biol.* 55, 80–88.

363. Ferreira, R.R., Fukui, H., Chow, R., Vilfan, A., and Vermot, J. (2019). The cilium as a force sensor: myth versus reality. *J. Cell Sci.* 132, jcs213496.

364. Figuera, L.E., Pandolfo, M., Dunne, P.W., Cantú, J.M., and Patel, P.I. (1995). Mapping of the congenital generalized hypertrichosis locus to chromosome Xq24–q27.1. *Nat. Genet.* 10, 202–207.

365. Findlay, G.H. and Harris, W.F. (1977). The topology of hair streams and whorls in man, with an observation on their relationship to epidermal ridge patterns. *Am. J. Phys. Anthrop.* 46, 427–438.

366. Finet, C., Decaras, A., Armisén, D., and Khila, A. (2018). The *achaete-scute* complex contains a single gene that controls bristle development in the semi-aquatic bugs. *Proc. R. Soc. B* 285, 20182387.

367. Finlay, B.L. and Huang, K. (2020). Developmental duration as an organizer of the evolving mammalian brain: scaling, adaptations, and exceptions. *Evol. Dev.* 22, 181–195.

368. Finlay, B.L., Darlington, R.B., and Nicastro, N. (2001). Developmental structure in brain evolution. *Behav. Brain Sci.* 24, 263–308.

369. Fisher, A. and Caudy, M. (1998). The function of hairy-related bHLH repressor proteins in cell fate decisions. *BioEssays* 20, 298–306.

370. Fitch, C.L., Girton, J.R., and Girton, L. (1992). The *suppressor of forked* locus in *Drosophila melanogaster*: genetic and molecular analyses. *Genetica* 85, 185–203.

371. Fondon, J.W., III and Garner, H.R. (2004). Molecular origins of rapid and continuous morphological evolution. *PNAS* 101, 18058–18063.

372. Fouilloux, C., Ringler, E., and Rojas, B. (2019). Cannibalism. *Curr. Biol.* 29, R1295–R1297.

373. Francavilla, A., Ove, P., Polimeno, L., Coetzee, M., Makowka, L., Barone, M., Vanthiel, D.H., and Starzl, T.E. (1988). Regulation of liver size and regeneration: importance in liver-transplantation. *Transplant. Proc.* 20, 494–497.

374. François, L., Fegraeus, K.J., Eriksson, S., Andersson, L.S., Tesfayonas, Y.G., Viluma, A., Imsland, F., Buys, N., Mikko, S., Lindgren, G., and Velie, B.D. (2016).

Conformation traits and gaits in the Icelandic horse are associated with genetic variants in *myostatin* (*MSTN*). *J. Hered.* 107, 431–437.

375. Frank, S.A. (2014). Somatic mosaicism and disease. *Curr. Biol.* 24, R577–R581.

376. Frantsevich, L. (2016). A Houdini's trick in a fly: leg unfolding with the aid of transient hinges in an extricating *Calliphora vicina* (Diptera: Calliphoridae). *Arthropod Struct. Dev.* 45, 2–13.

377. French, V., Bryant, P.J., and Bryant, S.V. (1976). Pattern regulation in epimorphic fields. *Science* 193, 969–981.

378. Freytes, D.O., Wan, L.Q., and Vunjak-Novakovic, G. (2009). Geometry and force control of cell function. *J. Cell. Biochem.* 108, 1047–1058.

379. Fristrom, D. (1988). The cellular basis of epithelial morphogenesis: a review. *Tissue Cell* 20, 645–690.

380. Fristrom, D. and Fristrom, J.W. (1993). The metamorphic development of the adult epidermis. In *The Development of* Drosophila melanogaster, Bate, M. and Martinez Arias, A., editors. Cold Spring Harbor Laboratory Press, Plainview, NY, pp. 843–897.

381. Fristrom, D.K., Fekete, E., and Fristrom, J.W. (1981). Imaginal disc development in a non-pupariating lethal mutant in *Drosophila melanogaster*. *W. Roux Arch. Dev. Biol.* 190, 11–21.

382. Fromental-Ramain, C., Warot, X., Lakkaraju, S., Favier, B., Haack, H., Birling, C., Dierich, A., Dollé, P., and Chambon, P. (1996). Specific and redundant functions of the paralogous *Hoxa-9* and *Hoxd-9* genes in forelimb and axial skeleton patterning. *Development* 122, 461–472.

383. Fuchs, E. (2007). Scratching the surface of skin development. *Nature* 445, 834–842.

384. Fujisawa, Y., Kosakamoto, H., Chihara, T., and Miura, M. (2019). Non-apoptotic function of *Drosophila* caspase activation in epithelial thorax closure and wound healing. *Development* 146, dev169037.

385. Furman, D.P. and Bukharina, T.A. (2009). The gene network determining development of *Drosophila melanogaster* mechanoreceptors. *Comp. Biol. Chem.* 33, 231–234.

386. Furman, D.P. and Bukharina, T.A. (2018). The bristle pattern development in *Drosophila melanogaster*: the prepattern and *achaete-scute* genes. *Vavilov J. Genet. Breed.* 22, 1046–1054.

387. Fusco, G. and Minelli, A. (2010). Phenotypic plasticity in development and evolution: facts and concepts. *Philos. Trans. R. Soc. Lond. B* 365, 547–556.

388. Galant, R. and Carroll, S.B. (2002). Evolution of a transcriptional repression domain in an insect Hox protein. *Nature* 415, 910–913.

389. Galindo, M.I., Bishop, S.A., and Couso, J.P. (2005). Dynamic EGFR-Ras signalling in *Drosophila* leg development. *Dev. Dynamics* 233, 1496–1508.

390. Galindo, M.I., Bishop, S.A., Greig, S., and Couso, J.P. (2002). Leg patterning driven by proximal–distal interactions and EGFR signaling. *Science* 297, 256–259.

391. Galis, F., van Alphen, J.J.M., and Metz, J.A.J. (2001). Why five fingers? Evolutionary constraints on digit numbers. *Trends Ecol. Evol.* 16, 637–646.

392. Galloni, M., Gyurkovics, H., Schedl, P., and Karch, F. (1993). The bluetail transposon: evidence for independent *cis*-regulatory domains and domain boundaries in the bithorax complex. *EMBO J.* 12, 1087–1097.

393. Gandolfi, B., Outerbridge, C.A., Beresford, L.G., Myers, J.A., Pimentel, M., Alhaddad, H., Grahn, J.C., Grahn, R.A., and Lyons, L.A. (2010). The naked truth: Sphynx and Devon Rex cat breed mutations in *KRT71*. *Mamm. Genome* 21, 509–515.

394. Gao, B., Song, H., Bishop, K., Elliot, G., Garrett, L., English, M.A., Andre, P., Robinson, J., Sood, R., Minami, Y., Economides, A.N., and Yang, Y. (2011). Wnt signaling gradients establish planar cell polarity by inducing Vangl2 phosphorylation through Ror2. *Dev. Cell* 20, 163–176.

395. Garcia, K.E., Kroenke, C.D., and Bayly, P.V. (2018). Mechanics of cortical folding: stress, growth and stability. *Philos. Trans. R. Soc. Lond. B* 373, 20170321.

396. García-Bellido, A. (1975). Genetic control of wing disc development in *Drosophila*. In *Cell Patterning*, Porter, R. and Rivers, J., editors. Elsevier, Amsterdam, pp. 161–182.

397. García-Bellido, A. and de Celis, J.F. (2009). The complex tale of the *achaete-scute* complex: a paradigmatic case in the analysis of gene organization and function during development. *Genetics* 182, 631–639.

398. García-Bellido, A. and Merriam, J.R. (1969). Cell lineage of the imaginal discs in *Drosophila* gynandromorphs. *J. Exp. Zool.* 170, 61–76.

399. García-Bellido, A., Lawrence, P.A., and Morata, G. (1979). Compartments in animal development. *Sci. Am.* 241(1), 102–110.

400. Garcia-Cruz, D., Figuera, L.E., and Cantu, J.M. (2002). Inherited hypertrichoses. *Clin. Genet.* 61, 321–329.

401. Gardner, E.W., Miller, H.M., and Lowney, E.D. (1979). Folded skin associated with underlying nevus lipomatosus. *Arch. Derm.* 115, 978–979.

402. Gardner, S. (2015). #ThrowBackThursday: the toad and I. *TheSpec.com* (Jan. 15, 2015).

403. Garzón-Alvarado, D.A. and Ramirez Martinez, A.M. (2011). A biochemical hypothesis on the formation of fingerprints using a Turing patterns approach. *Theor. Biol. Med. Model.* 8, 24.

404. Gawne, R., McKenna, K.Z., and Nijhout, H.F. (2018). Unmodern synthesis: developmental hierarchies and the origin of phenotypes. *BioEssays* 40, 1600265.

405. Gayon, J. (2000). History of the concept of allometry. *Am. Zool.* 40, 748–758.

406. Gebo, D.L. (1987). Functional anatomy of the tarsier foot. *Am. J. Phys. Anthrop.* 73, 9–31.

407. Gee, H. (2013). *The Accidental Species: Misunderstandings of Human Evolution.* University of Chicago Press, Chicago, IL.

408. Gehring, W.J. (2012). The animal body plan, the prototypic body segment, and eye evolution. *Evol. Dev.* 14, 34–46.

409. Géminard, C., González-Morales, N., Coutelis, J.B., and Noselli, S. (2014). The Myosin ID pathway and left–right asymmetry in *Drosophila. Genesis* 52, 471–480.

410. Gerhart, J. (1999). Signaling pathways in development (1998 Warkany lecture). *Teratology* 60, 226–239.

411. Gerhart, J. (2001). Evolution of the organizer and the chordate body plan. *Int. J. Dev. Biol.* 45, 133–153.

412. Gerhart, J. (2002). Changing the axis changes the perspective. *Dev. Dynamics* 225, 380–383.

413. Gerhart, J. (2010). Enzymes, embryos, and ancestors. *Annu. Rev. Cell Dev. Biol.* 26, 1–20.

414. Gerhart, J. and Kirschner, M. (1997). *Cells, Embryos, and Evolution.* Blackwell Science, Malden, MA.

415. Gerhart, J. and Kirschner, M. (2007). The theory of facilitated variation. *PNAS* 104 (Suppl.1), 8582–8589.

416. Gerhart, J., Ubbels, G., Black, S., Hara, K., and Kirschner, M. (1981). A reinvestigation of the role of the grey crescent in axis formation in *Xenopus laevis*. *Nature* 292, 511–516.

417. Gerhart, J.C. (1987). Determinants of early amphibian development. *Am. Zool.* 27, 593–605.

418. Germani, F., Bergantinos, C., and Johnston, L.A. (2018). Mosaic analysis in *Drosophila*. *Genetics* 208, 473–490.

419. Geyer, P.K. and Corces, V.G. (1987). Separate regulatory elements are responsible for the complex pattern of tissue-specific and developmental transcription of the *yellow* locus in *Drosophila melanogaster*. *Genes Dev.* 1, 996–1004.

420. Gho, M., Bellaïche, Y., and Schweisguth, F. (1999). Revisiting the *Drosophila* microchaete lineage: a novel intrinsically asymmetric cell division generates a glial cell. *Development* 126, 3573–3584.

421. Ghysen, A. (2009). Ontogeny of an adventurous mind: the origin of Antonio García-Bellido's contributions to developmental genetics. *Int. J. Dev. Biol.* 53, 1277–1290.

422. Ghysen, A. and Dambly-Chaudière, C. (1988). From DNA to form: the *achaete-scute* complex. *Genes Dev.* 2, 495–501.

423. Ghysen, A. and Dambly-Chaudière, C. (1989). Genesis of the *Drosophila* peripheral nervous system. *Trends Genet.* 5, 251–255.

424. Gibert, J.-M. and Simpson, P. (2003). Evolution of *cis*-regulation of the proneural genes. *Int. J. Dev. Biol.* 47, 643–651.

425. Gibson, G. and Hogness, D.S. (1996). Effect of polymorphism in the *Drosophila* regulatory gene *Ultrabithorax* on homeotic stability. *Science* 271, 200–203.

426. Gibson, M.C. (2019). Commentary on "Regeneration, duplication and transdetermination in fragments of the leg disc of *Drosophila melanogaster*": Schubiger, G. (1971). *Dev. Biol.* 449, 63–82.

427. Giebel, B. and Wodarz, A. (2012). Notch signaling: Numb makes the difference. *Curr. Biol.* 22, R133–R135.

428. Giebel, L.B., Tripathi, R.K., King, R.A., and Spritz, R.A. (1991). A tyrosinase gene missense mutation in temperature-sensitive Type I oculocutaneous albinism: a human homologue to the Siamese cat and the Himalayan mouse. *J. Clin. Invest.* 87, 1119–1122.

429. Gierer, A. and Meinhardt, H. (1974). Biological pattern formation involving lateral inhibition. In *Lectures on Mathematics in the Life Sciences*, Vol. 7. American Mathematical Society, Providence, RI, pp. 163–183.

430. Gilbert, S.F. (2001). Ecological developmental biology: developmental biology meets the real world. *Dev. Biol.* 233, 1–12.

431. Gilbert, S.F. (2014). *Developmental Biology*, 10th ed. Sinauer, Sunderland, MA.

432. Gilbert, S.F. (2016). Developmental plasticity and developmental symbiosis: the return of eco-devo. *Curr. Top. Dev. Biol.* 116, 415–433.

433. Gilbert, S.F. and Barresi, M.J.F. (2019). *Developmental Biology*, 11th ed. Sinauer, Sunderland, MA.

434. Gilgenkrantz, H. and de l'Hortet, A.C. (2018). Understanding liver regeneration: from mechanisms to regenerative medicine. *Am J. Pathol.* 188, 1316–1327.

435. Gillham, N.W. (2001). Evolution by jumps: Francis Galton and William Bateson and the mechanism of evolutionary change. *Genetics* 159, 1383–1392.

436. Gilmour, D., Rembold, M., and Leptin, M. (2017). From morphogen to morphogenesis and back. *Nature* 541, 311–320.

437. Girton, J.R. (1981). Pattern triplications produced by a cell-lethal mutation in *Drosophila*. *Dev. Biol.* 84, 164–172.

438. Girton, J.R. (1982). Genetically induced abnormalities in *Drosophila*: two or three patterns? *Am. Zool.* 22, 65–77.

439. Girton, J.R. (1983). Morphological and somatic clonal analyses of pattern triplications. *Dev. Biol.* 99, 202–209.

440. Girton, J.R. and Berns, M.W. (1982). Pattern abnormalities induced in *Drosophila* imaginal discs by an ultraviolet laser microbeam. *Dev. Biol.* 91, 73–77.

441. Girton, J.R. and Bryant, P.J. (1980). The use of cell lethal mutations in the study of *Drosophila* development. *Dev. Biol.* 77, 233–243.

442. Girton, J.R. and Kumor, A.L. (1985). The role of cell death in the induction of pattern abnormalities in a cell-lethal mutation of *Drosophila*. *Dev. Genet.* 5, 93–102.

443. Girton, J.R. and Russell, M.A. (1980). A clonal analysis of pattern duplication in a temperature-sensitive cell-lethal mutant of *Drosophila melanogaster*. *Dev. Biol.* 77, 1–21.

444. Girton, J.R. and Russell, M.A. (1981). An analysis of compartmentalization in pattern duplications induced by a cell-lethal mutation in *Drosophila*. *Dev. Biol.* 85, 55–64.

445. Gloor, H. (1947). Phänokopie-Versuche mit Äther an *Drosophila*. *Rev. Suisse Zool.* 54, 637–712.

446. Gnatzy, W., Grünert, U., and Bender, M. (1987). Campaniform sensilla of *Calliphora vicina* (Insecta, Diptera). I. Typography. *Zoomorphology* 106, 312–319.

447. Goldschmidt, R. (1940). *The Material Basis of Evolution*. Yale University Press, New Haven, CT.

448. Goldschmidt, R.B. (1949). Phenocopies. *Sci. Am.* 181(10), 46–49.

449. Goldschmidt, R.B. (1952). Homoeotic mutants and evolution. *Acta Biotheor.* 10, 87–104.

450. Goldstein, B. and Freeman, G. (1997). Axis specification in animal development. *BioEssays* 19, 105–116.

451. Golovnin, A., Gause, M., Georgieva, S., Gracheva, E., and Georgiev, P. (1999). The su(Hw) insulator can disrupt enhancer–promoter interactions when located more than 20 kilobases away from the *Drosophila* achaete-scute complex. *Mol. Cell. Biol.* 19, 3443–3456.

452. Gómez, J.A., Ceacero, F., Landete-Castillejos, T., Gaspar-Lopez, E., García, A.J., and Gallego, L. (2012). Factors affecting antler investment in Iberian red deer. *Anim. Prod. Sci.* 52, 867–873.

453. Gómez-Skarmeta, J.L., Campuzano, S., and Modolell, J. (2003). Half a century of neural prepatterning: the story of a few bristles and many genes. *Nat. Rev. Neurosci.* 4, 587–598.

454. Gómez-Skarmeta, J.L., Rodríguez, I., Martínez, C., Culí, J., Ferrés-Marcó, D., Beamonte, D., and Modolell, J. (1995). *Cis*-regulation of *achaete* and *scute*: shared enhancer-like elements drive their coexpression in proneural clusters of the imaginal discs. *Genes Dev.* 9, 1869–1882.

455. Gönczy, P. (2008). Mechanisms of asymmetric cell division: flies and worms pave the way. *Nat. Rev. Mol. Cell Biol.* 9, 355–366.

456. González-Forero, M. and Gardner, A. (2018). Inference of ecological and social drivers of human brain-size evolution. *Nature* 557, 554–557.

457. González-Méndez, L., Gradilla, A.-C., and Guerreiro, I. (2019). The cytoneme connection: direct long-distance signal transfer during development. *Development* 146, dev174607.

458. Goodman, B.A. and Johnson, P.T.J. (2011). Disease and the extended phenotype: parasites control host performance and survival through induced changes in body plan. *PLoS ONE* 6(5), e20193.

459. Goodrich, L.V. and Strutt, D. (2011). Principles of planar polarity in animal development. *Development* 138, 1877–1892.

460. Goodwin, B.C. (1985). Developing organisms as self-organizing fields. In *Mathematical Essays on Growth and the Emergence of Form*, Antonelli, P.L., editor. University of Alberta Press, Edmonton, pp. 185–200.

461. Gotoh, H., Hust, J.A., Miura, T., Niimi, T., Emlen, D.J., and Lavine, L.C. (2015). The Fat/Hippo signaling pathway links within-disc morphogen patterning to whole-animal signals during phenotypically plastic growth in insects. *Dev. Dynamics* 244, 1039–1045.

462. Gou, J., Stotsky, J.A., and Othmer, H.G. (2020). Growth control in the *Drosophila* wing disk. *Wiley Interdiscip. Rev. Syst. Biol. Med.* 2020, e1478.

463. Gould, G.M. and Pyle, W.L. (1896). *Anomalies and Curiosities of Medicine*. Julian Press, New York.

464. Gould, S.J. (1966). Allometry and size in ontogeny and phylogeny. *Biol. Rev.* 41, 587–640.

465. Gould, S.J. (1971). D'Arcy Thompson and the science of form. *New Lit. Hist.* 2, 229–258.

466. Gould, S.J. (1974). The origin and function of "bizarre" structures: antler size and skull size in the "Irish Elk," *Megaloceros giganteus. Evolution* 28, 191–220.

467. Gould, S.J. (1977). *Ontogeny and Phylogeny*. Harvard University Press, Cambridge, MA.

468. Gould, S.J. (1977). The return of hopeful monsters. *Nat. Hist.* 86(6), 22–30.

469. Gould, S.J. (1980). *The Panda's Thumb: More Reflections in Natural History*. W. W. Norton, New York.

470. Gould, S.J. (1981). What, if anything, is a zebra? *Nat. Hist.* 90(7), 6–12.

471. Gould, S.J. (1981). What color is a zebra? *Nat. Hist.* 90(8), 16–22.

472. Gould, S.J. (1982). Living with connections: are Siamese twins one person or two? *Nat. Hist.* 91(11), 18–22.

473. Gould, S.J. (1986). The egg-a-day barrier. *Nat. Hist.* 95(7), 16–24.

474. Gould, S.J. (1990). *Wonderful Life: The Burgess Shale and the Nature of History*. Norton, New York.

475. Gould, S.J. (1994). Cabinet museums revisited. *Nat. Hist.* 103(1), 12–20.

476. Govind, C.K. (1989). Asymmetry in lobster claws. *Am. Sci.* 77, 468–474.

477. Graff, J.M. (1997). Embryonic patterning: to BMP or not to BMP, that is the question. *Cell* 89, 171–174.

478. Grall, E. and Tschopp, P. (2019). A sense of place, many times over: pattern formation and evolution of repetitive morphological structures. *Dev. Dynamics* 249, 313–327.

479. Grantham, M.E., Shingleton, A.W., Dudley, E., and Brisson, J.A. (2020). Expression profiling of winged- and wingless-destined pea aphid embryos implicates insulin/insulin growth factor signaling in morph differences. *Evol. Dev.* 22, 257–268.

480. Graván, C.P. and Lahoz-Beltra, R. (2004). Evolving morphogenetic fields in the zebra skin pattern based on Turing's morphogen hypothesis. *Int. J. Appl. Math. Comput. Sci.* 14, 351–361.

481. Gray, G.W. (1948). The great ravelled knot. *Sci. Am.* 179(10), 26–39.

482. Green, H. and Thomas, J. (1978). Pattern formation by cultured human epidermal cells: development of curved ridges resembling dermatoglyphics. *Science* 200, 1385–1388.

483. Green, J.B.A. and Sharpe, J. (2015). Positional information and reaction–diffusion: two big ideas in developmental biology combine. *Development* 142, 1203–1211.

484. Green, M.C. (1961). Himalayan, a new allele of albino in the mouse. *J. Hered.* 52, 73–75.

485. Greenberg, L. and Hatini, V. (2011). Systematic expression and loss-of-function analysis defines spatially restricted requirements for *Drosophila RhoGEFs* and *RhoGAPs* in leg morphogenesis. *Mech. Dev.* 128, 5–17.

486. Greenwald, I. (2012). *Notch* and the awesome power of genetics. *Genetics* 191, 655–669.

487. Greiling, T.M.S. and Clark, J.I. (2012). New insights into the mechanism of lens development using zebra fish. *Int. Rev. Cell Mol. Biol.* 296, 1–61.

488. Grimaldi, D.A. (1987). Amber fossil Drosophilidae (Diptera), with particular reference to the Hispaniolan taxa. *Am. Mus. Novitates* 2880, 1–23.

489. Grimes, D.T. (2019). Making and breaking symmetry in development, growth and disease. *Development* 146, dev170985.

490. Grimes, D.T. and Burdine, R.D. (2017). Left–right patterning: breaking symmetry to asymmetric morphogenesis. *Trends Genet.* 33, 616–628.

491. Grobet, L., Martin, L.J.R., Poncelet, D., Pirottin, D., Brouwers, B., Riquet, J., Schoeberlein, A., Dunner, S., Ménissier, F., Massabanda, J., Fries, R., Hanset, R., and Georges, M. (1997). A deletion in the bovine myostatin gene causes the double-muscled phenotype in cattle. *Nat. Genet.* 17, 71–74.

492. Grochowska, E., Borys, B., Lisiak, D., and Mroczkowski, S. (2019). Genotypic and allelic effects of the myostatin gene (MSTN) on carcass, meat quality, and biometric traits in Colored Polish Merino sheep. *Meat Sci.* 151, 4–17.

493. Gross, J.B., Kerney, R., Hanken, J., and Tabin, C.J. (2011). Molecular anatomy of the developing limb in the coquí frog, *Eleutherodactylus coqui*. *Evol. Dev.* 13, 415–426.

494. Groves, A.K. and Fekete, D.M. (2012). Shaping sound in space: the regulation of inner ear patterning. *Development* 139, 245–257.

495. Gu, L., Mo, E., Yang, Z., Zhu, X., Fang, Z., Sun, B., Wang, C., Bao, J., and Sung, C. (2007). Expression and localization of insulin-like growth factor-I in four parts of the red deer antler. *Growth Factors* 25, 264–279.

496. Gubb, D. and García-Bellido, A. (1982). A genetic analysis of the determination of cuticular polarity during development in *Drosophila melanogaster*. *J. Embryol. Exp. Morphol.* 68, 37–57.

497. Guerreiro, I. and Duboule, D. (2014). Snakes: hatching a model system for Evo-Devo? *Int. J. Dev. Biol.* 58, 727–732.

498. Guinard, G. (2015). Introduction to evolutionary teratology, with an application to the forelimbs of Tyrannosauridae and Carnotaurinae (Dinosauria: Theropoda). *Evol. Biol.* 42, 20–41.

499. Gumbiner, B.M. and Kim, N.-G. (2014). The Hippo–YAP signaling pathway and contact inhibition of growth. *J. Cell Sci.* 127, 709–717.

500. Gunhaga, L. (2011). The lens: a classical model of embryonic induction providing new insights into cell determination in early development. *Philos. Trans. R. Soc. Lond. B* 366, 1193–1203.

501. Haas, B.J. and Whited, J.L. (2017). Advances in decoding axolotl limb regeneration. *Trends Genet.* 33, 553–565.

502. Hadorn, E. (1961). *Developmental Genetics and Lethal Factors.* Methuen, London [translated from 1955 German original, Thieme Verlag, Stuttgart, by U. Mittwoch].

503. Hadorn, E. (1978). Transdetermination. In *The Genetics and Biology of Drosophila*, Ashburner, M. and Wright, T.R.F., editors. Academic Press, New York, pp. 555–617.

504. Halder, G. and Johnson, R.L. (2011). Hippo signaling: growth control and beyond. *Development* 138, 9–22.

505. Hall, B.K. (2008). EvoDevo concepts in the work of Waddington. *Biol. Theory* 3, 198–203.

506. Hall, B.K. (2018). Germ layers, the neural crest and emergent organization in development and evolution. *Genesis* 56, e23103.

507. Hall, J.C., Gelbart, W.M., and Kankel, D.R. (1976). Mosaic systems. In *The Genetics and Biology of Drosophila*, Ashburner, M. and Novitski, E., editors. Academic Press, New York, pp. 265–314.

508. Hallgrímsson, B., Green, R.M., Katz, D.C., Fish, J.L., Bernier, F.P., Roseman, C.C., Young, N.M., Cheverud, J.M., and Marcucio, R.S. (2019). The developmental-genetics of canalization. *Semin. Cell Dev. Biol.* 88, 67–79.

509. Hallgrímsson, B., Jamniczky, H., Young, N.M., Rolian, C., Parsons, T.E., Boughner, J.C., and Marcucio, R.S. (2009). Deciphering the palimpsest: studying the relationship between morphological integration and phenotypic covariation. *Evol. Biol.* 36, 355–376.

510. Hamant, O. (2017). Mechano-devo. *Mech. Dev.* 145, 2–9.

511. Hamburger, V. (1988). *The Heritage of Experimental Embryology: Hans Spemann and the Organizer.* Oxford University Press, New York.

512. Hamburger, V. (2001). Induction of embryonic primordia by implantation of organizers from a different species. [English translation of 1924 German paper by Hans Spemann and Hilde Mangold.] *Int. J. Dev. Biol.* 45, 13–38.

513. Hamelin, A., Conchou, F., Fusellier, M., Duchenij, B., Vieira, I., Filhol, E., de Citres, C.D., Tiret, L., Gache, V., and Abitbol, M. (2020). Genetic heterogeneity of polydactyly in Maine Coon cats. *J. Feline Med. Surg.* 1098612X20905061 [published online, 18 Feb 2020].

514. Handrigan, G.R. and Wassersug, R.J. (2007). The anuran *Bauplan*: a review of the adaptive, developmental, and genetic underpinnings of frog and tadpole morphology. *Biol. Rev.* 82, 1–25.

515. Hannah-Alava, A. (1960). Genetic mosaics. *Sci. Am.* 202(5), 118–130.

516. Hannezo, E. and Heisenberg, C.-P. (2019). Mechanochemical feedback loops in development and disease. *Cell* 178, 12–25.

517. Hanset, R. and Michaux, C. (1985). On the genetic determinism of muscular hypertrophy in the Belgian White and Blue cattle breed. II. Population data. *Génét. Sél. Evol.* 17, 369–386.

518. Happle, R. (2015). The categories of cutaneous mosaicism: a proposed classification. *Am. J. Med. Genet. A* 170A, 452–459.

519. Hardie, R.C. (1985). Functional organization of the fly retina. In *Progress in Sensory Physiology*, Ottoson, D., editor. Springer-Verlag, Berlin, pp. 1–79.

520. Harfe, B.D., Scherz, P.J., Nissim, S., Tian, H., McMahon, A.P., and Tabin, C.J. (2004). Evidence for an expansion-based temporal Shh gradient in specifying vertebrate digit identities. *Cell* 118, 517–528.

521. Hariharan, I.K. (2015). Organ size control: lessons from *Drosophila*. *Dev. Cell* 34, 255–265.

522. Hariharan, I.K. and Bilder, D. (2006). Regulation of imaginal disc growth by tumor-suppressor genes in *Drosophila*. *Annu. Rev. Genet.* 40, 335–361.

523. Hariharan, I.K. and Serras, F. (2017). Imaginal disc regeneration takes flight. *Curr. Opin. Cell Biol.* 48, 10–16.

524. Harris, M.L., Chora, L., Bishop, C.A., and Bogart, J.P. (2000). Species- and age-related differences in susceptibility to pesticide exposure for two amphibians, *Rana pipiens* and *Bufo americanus*. *Bull. Environ. Contam. Toxicol.* 64, 263–270.

525. Harris, R.E., Setiawan, L., Saul, J., and Hariharan, I.K. (2016). Localized epigenetic silencing of a damage-activated WNT enhancer limits regeneration in mature *Drosophila* imaginal discs. *eLife* 5, e11588.

526. Harrison, R.G. (1918). Experiments on the development of the fore limb of *Amblystoma*, a self-differentiating equipotential system. *J. Exp. Zool.* 25, 413–461.

527. Harrison, R.G. (1921). On relations of symmetry in transplanted limbs. *J. Exp. Zool.* 32, 1–136.

528. Hartfelder, K., Guidugli-Lazzarini, K.R., Cervoni, M.S., Santos, D.E., and Humann, F.C. (2015). Old threads make new tapestry: rewiring of signalling pathways underlies caste phenotypic plasticity in the honey bee, *Apis mellifera* L. *Adv. Insect Physiol.* 48, 1–36.

529. Hartwell, L.H. (1967). Macromolecule synthesis in temperature-sensitive mutants of yeast. *J. Bacteriol.* 93, 1662–1670.

530. Harvey, P.H. and Krebs, J.R. (1990). Comparing brains. *Science* 249, 140–146.

531. Harzsch, S., Benton, J., and Beltz, B.S. (2000). An unusual case of a mutant lobster embryo with double brain and double ventral nerve cord. *Arthropod Struct. Dev.* 29, 95–99.

532. Hashimoto, H., Mizuta, A., Okada, N., Suzuki, T., Tagawa, M., Tabata, K., Yokoyama, Y., Sakaguchi, M., Tanaka, M., and Toyohara, H. (2002). Isolation and characterization of a Japanese flounder clonal line, *reversed*, which exhibits reversal of metamorphic left–right asymmetry. *Mech. Dev.* 111, 17–24.

533. Hassan, B.A. and Hiesinger, P.R. (2015). Beyond molecular codes: simple rules to wire complex brains. *Cell* 163, 285–291.

534. Hassanpour, M. and Joss, J. (2009). Anatomy and histology of the spiral valve intestine in juvenile Australian lungfish, *Neoceratodus forsteri*. *Open Zool. J.* 2, 62–85.

535. Hatchwell, E. and Dennis, N. (1996). Mirror hands and feet: a further case of Laurin–Sandrow syndrome. *J. Med. Genet.* 33, 426–428.

536. Hattori, A., Sugime, Y., Sasa, C., Miyakaya, H., Ishikawa, Y., Miyazaki, S., Okada, Y., Cornette, R., Lavine, L.C., Emlen, D.J., Koshikawa, S., and Miura, T. (2013). Soldier morphogenesis in the damp-wood termite is regulated by the insulin signaling pathway. *J. Exp. Zool. B Mol. Dev. Evol.* 320, 295–306.

537. Haupaix, N. and Manceau, M. (2020). The embryonic origin of periodic color patterns. *Dev. Biol.* 460, 70–76.

538. Hauswirth, R., Haase, B., Blatter, M., Brooks, S.A., Burger, D., Drögemüller, C., Gerber, V., Henke, D., Janda, J., Jude, R., Magdesian, K.G., Matthews, J.M., Poncet, P.-A., Svansson, V., Tozaki, T., Wilkinson-White, L., Penedo, M.C.T., Rieder, S., and Leeb, T. (2012). Mutations in *MITF* and *PAX3* cause "splashed white" and other white spotting phenotypes in horses. *PLoS Genet.* 8(4), e1002653.

539. Haynie, J.L. and Bryant, P.J. (1986). Development of the eye-antenna imaginal disc and morphogenesis of the adult head in *Drosophila melanogaster*. *J. Exp. Zool.* 237, 293–308.

540. He, X.J., Zhou, L.B., Pan, Q.Z., Barron, A.B., Yan, W.Y., and Zeng, Z.J. (2017). Making a queen: an epigenetic analysis of the robustness of the honeybee (*Apis mellifera*) queen developmental pathway. *Mol. Ecol.* 26, 1598–1607.

541. Heer, N.C. and Martin, A.C. (2017). Tension, contraction and tissue morphogenesis. *Development* 144, 4249–4260.

542. Heimeier, R.A., Das, B., Buchholz, D.R., Fiorentino, M., and Shi, Y.-B. (2010). Studies on *Xenopus laevis* intestine reveal biological pathways underlying vertebrate gut adaptation from embryo to adult. *Genome Biol.* 11, R55.

543. Heingard, M., Turetzek, N., Prpic, N.-M., and Janssen, R. (2019). FoxB, a new and highly conserved key factor in arthropod dorsal–ventral (DV) limb patterning. *EvoDevo* 10, 28.

544. Heintzman, N.D. and Ren, B. (2009). Finding distal regulatory elements in the human genome. *Curr. Opin. Genet. Dev.* 19, 541–549.

545. Heisenberg, C.-P. and Bellaïche, Y. (2013). Forces in tissue morphogenesis and patterning. *Cell* 153, 948–962.

546. Hejtmancik, J.F., Riazuddin, S.A., McGreal, R., Liu, W., Cvekl, A., and Shiels, A. (2015). Lens biology and biochemistry. *Prog. Mol. Biol. Transl. Sci.* 134, 169–201.

547. Held, L.I., Jr. (1977). Analysis of bristle-pattern formation in *Drosophila*. PhD thesis, Department of Molecular Biology, University of California, Berkeley, CA.

548. Held, L.I., Jr. (1979). Pattern as a function of cell number and cell size on the second-leg basitarsus of *Drosophila*. *W. Roux Arch. Dev. Biol.* 187, 105–127.

549. Held, L.I., Jr. (1979). A high-resolution morphogenetic map of the second-leg basitarsus in *Drosophila melanogaster*. *W. Roux Arch. Dev. Biol.* 187, 129–150.

550. Held, L.I., Jr. (1990). Sensitive periods for abnormal patterning on a leg segment in *Drosophila melanogaster*. *W. Roux Arch. Dev. Biol.* 199, 31–47.

551. Held, L.I., Jr. (1990). Arrangement of bristles as a function of bristle number on a leg segment in *Drosophila melanogaster*. *W. Roux Arch. Dev. Biol.* 199, 48–62.

552. Held, L.I., Jr. (1991). Bristle patterning in *Drosophila*. *BioEssays* 13, 633–640.

553. Held, L.I., Jr. (1992). *Models for Embryonic Periodicity*. Monographs in Developmental Biology, Vol. 24. Karger, Basel.

554. Held, L.I., Jr. (1995). Axes, boundaries and coordinates: the ABCs of fly leg development. *BioEssays* 17, 721–732.

555. Held, L.I., Jr. (2002). *Imaginal Discs: The Genetic and Cellular Logic of Pattern Formation*. Cambridge University Press, New York.

556. Held, L.I., Jr. (2002). Bristles induce bracts via the EGFR pathway on *Drosophila* legs. *Mech. Dev.* 117, 225–234.

557. Held, L.I., Jr. (2009). *Quirks of Human Anatomy: An Evo-Devo Look at the Human Body*. Cambridge University Press, New York.

558. Held, L.I., Jr. (2010). The evo-devo puzzle of human hair patterning. *Evol. Biol.* 37, 113–122.

559. Held, L.I., Jr. (2010). How does *Scr* cause first legs to deviate from second legs? *Dros. Info. Serv.* 93, 132–146.

560. Held, L.I., Jr. (2014). *How the Snake Lost Its Legs: Curious Tales from the Frontier of Evo-Devo*. Cambridge University Press, New York.

561. Held, L.I., Jr. (2017). *Deep Homology? Uncanny Similarities of Humans and Flies.* Cambridge University Press, New York.

562. Held, L.I., Jr. and Bryant, P.J. (1984). Cell interactions controlling the formation of bristle patterns in *Drosophila.* In *Pattern Formation: A Primer in Developmental Biology,* Malacinski, G.M. and Bryant, S.V., editors. Macmillan, New York, pp. 291–322.

563. Held, L.I., Jr. and Heup, M. (1996). Genetic mosaic analysis of *decapentaplegic* and *wingless* gene function in the *Drosophila* leg. *Dev. Genes Evol.* 206, 180–194.

564. Held, L.I., Jr. and Sessions, S.K. (2019). Reflections on Bateson's rule: solving an old riddle about why extra legs are mirror-symmetric. *J. Exp. Zool. B Mol. Dev. Evol.* 332, 219–237.

565. Held, L.I., Jr., Davis, A.L., and Aybar, R.S. (2017). Instigating an "identity crisis" to investigate how a *Hox* gene acts on fly legs. *Dros. Info. Serv.* 100, 75–89.

566. Held, L.I., Jr., Duarte, C.M., and Derakhshanian, K. (1986). Extra tarsal joints and abnormal cuticular polarities in various mutants of *Drosophila melanogaster. W. Roux Arch. Dev. Biol.* 195, 145–157.

567. Held, L.I., Jr., Grimson, M.J., and Du, Z. (2004). Proving an old prediction: the sex comb rotates at 16 to 24 hours after pupariation. *Dros. Info. Serv.* 87, 76–78.

568. Held, L.I., Jr., McNeme, S.C., and Hernandez, D. (2018). Induction of ectopic transverse rows by *Ubx* on fly legs. *Dros. Info. Serv.* 101, 25–32.

569. Heller, E. and Fuchs, E. (2015). Tissue patterning and cellular mechanics. *J. Cell Biol.* 211, 219–231.

570. Henderson, D.J., Long, D.A., and Dean, C.H. (2018). Planar cell polarity in organ formation. *Curr. Opin. Cell Biol.* 55, 96–103.

571. Hendrikse, J.L., Parsons, T.E., and Hallgrímsson, B. (2007). Evolvability as the proper focus of evolutionary developmental biology. *Evol. Dev.* 9, 393–401.

572. Henke, K. and Maas, H. (1946). Über sensible Perioden der allgemeinen Körpergliederung von *Drosophila. Nachr. Akad. Wiss. Göttingen. Math.-Phys. Kl. IIb, Biol.-Physiol.-Chem. Abt.* 1, 3–4.

573. Henneberg, M., Lambert, K.M., and Leigh, C.M. (1998). Fingerprinting a chimpanzee and a koala: animal dermatoglyphics can resemble human ones. In *Conference of the Australian and New Zealand International Symposium on the Forensic Sciences, 1996.* Sydney.

574. Henrique, D. and Schweisguth, F. (2019). Mechanisms of Notch signaling: a simple logic deployed in time and space. *Development* 146, dev172148.

575. Henry, J.J. and Hamilton, P.W. (2018). Diverse evolutionary origins and mechanisms of lens regeneration. *Mol. Biol. Evol.* 35, 1563–1575.

576. Herndon, J.G., Tigges, J., Anderson, D.C., Klumpp, S.A., and McClure, H.M. (1999). Brain weight throughout the life span of the chimpanzee. *J. Comp. Neurol.* 409, 567–572.

577. Hersh, B.M., Nelson, C.E., Stoll, S.J., Norton, J.E., Albert, T.J., and Carroll, S.B. (2007). The UBX-regulated network in the haltere imaginal disc of *D. melanogaster. Dev. Biol.* 302, 717–727.

578. Hershkovitz, P. (1977). *Living New World Monkeys (Platyrrhini) With an Introduction to Primates.* Vol. 1. University of Chicago Press, Chicago, IL.

579. Heyning, J.E. and Lento, G.M. (2002). The evolution of marine mammals. In *Marine Mammal Biology: An Evolutionary Approach,* Hoelzel, A.R., editor. Blackwell Science, Malden, MA, pp. 38–72.

580. Hill, E.W., McGivney, B.A., Rooney, M.F., Katz, L.M., Parnell, A., and MacHugh, D.E. (2019). The contribution of myostatin (MSTN) and additional modifying genetic loci to race distance aptitude in Thoroughbred horses racing in different geographic regions. *Equine Vet. J.* 51, 625–633.

581. Hillmer, A.M., Flaquer, A., Hanneken, S., Eigelshoven, S., Kortüm, A.-K., Brockschmidt, F.F., Golla, A., Metzen, C., Thiele, H., Kolberg, S., Reinartz, R., Betz, R.C., Ruzicka, T., Hennies, H.C., Kruse, R., and Nöthen, M.M. (2008). Genome-wide scan and fine-mapping linkage study of androgenetic alopecia reveals a locus on chromosome 3q26. *Am. J. Hum. Genet.* 82, 737–743.

582. Hintz, M., Bartholmes, C., Nutt, P., Ziermann, J., Hameister, S., Neuffer, B., and Theissen, G. (2007). Catching a "hopeful monster": shepherd's purse (*Capsella bursa-pastoris*) as a model system to study the evolution of flower development. *J. Exp. Botany* 57, 3531–3542.

583. Hiscock, T.W. and Megason, S.G. (2015). Mathematically guided approaches to distinguish models of periodic patterning. *Development* 142, 409–419.

584. Hiscock, T.W. and Megason, S.G. (2015). Orientation of Turing-like patterns by morphogen gradients and tissue anisotropies. *Cell Syst.* 1, 408–416.

585. Hiscock, T.W., Tschopp, P., and Tabin, C.J. (2017). On the formation of digits and joints during limb development. *Dev. Cell* 41, 459–465.

586. Hjorth, M., Pourteymour, S., Görgens, S.W., Langleite, T.M., Lee, S., Holen, T., Gulseth, H.L., Birkeland, K.I., Jensen, J., Drevon, C.A., and Norheim, F. (2016). Myostatin in relation to physical activity and dysglycaemia and its effect on energy metabolism in human skeletal muscle cells. *Acta Physiol.* 217, 45–60.

587. Ho, E.C.Y., Malagón, J.N., Ahuja, A., Singh, R., and Larsen, E. (2018). Rotation of sex combs in *Drosophila melanogaster* requires precise and coordinated spatio-temporal dynamics from forces generated by epithelial cells. *PLoS Comput. Biol.* 14(10), e1006455.

588. Ho, M.-W., Bolton, E., and Saunders, P.T. (1983). Bithorax phenocopy and pattern formation. I. Spatiotemporal characteristics of the phenocopy response. *Exp. Cell Biol.* 51, 282–290.

589. Ho, M.-W., Saunders, P.T., and Bolton, E. (1983). Bithorax phenocopy and pattern formation. II. A model of prepattern formation. *Exp. Cell Biol.* 51, 291–299.

590. Hodges, A. (1983). *Alan Turing: The Enigma.* Simon & Schuster, New York.

591. Hoekstra, H.E. (2006). Genetics, development and evolution of adaptive pigmentation in vertebrates. *Heredity* 97, 222–234.

592. Hofer, H., Carroll, J., Neitz, J., Neitz, M., and Williams, D.R. (2005). Organization of the human trichromatic cone mosaic. *J. Neurosci.* 25, 9669–9679.

593. Hoffman, B.D. and Yap, A.S. (2015). Towards a dynamic understanding of cadherin-based mechanobiology. *Trends Cell Biol.* 25, 803–814.

594. Hoffmann, M.B. and Dumoulin, S.O. (2015). Congenital visual pathway abnormalities: a window onto cortical stability and plasticity. *Trends Neurosci.* 38, 55–65.

595. Holland, N.D. (2003). Early central nervous system evolution: an era of skin brains? *Nat. Rev. Neurosci.* 4, 1–11.

596. Holland, P.W.H., Marlétaz, F., Maeso, I., Dunwell, T.L., and Paps, J. (2016). New genes from old: asymmetric divergence of gene duplicates and the evolution of development. *Philos. Trans. R. Soc. Lond. B* 372, 20150480.

597. Holloway, R.L. (2001). Does allometry mask important brain structure residuals relevant to species-specific behavioral evolution? *Behav. Brain Sci.* 24, 286–287.

598. Holtfreter, J. and Hamburger, V. (1955). Amphibians. In *Analysis of Development*, Willier, B.H., Weiss, P.A., and Hamburger, V., editors. Hafner, New York, pp. 230–296.

599. Holtzer, H. (1978). Cell lineages, stem cells and the "quantal" cell cycle concept. In *Stem Cells and Tissue Homeostasis*, Lord, B.I., Potten, C.S., and Cole, R.J., editors. Cambridge University Press, Cambridge, pp. 1–27.

600. Hombria, J.C.-G. and Sotillos, S. (2020). Evo-devo: when four became two plus two. *Curr. Biol.* 30, R655–R657.

601. Honeycutt, R.L. (2010). Unraveling the mysteries of dog evolution. *BMC Biol.* 8, 20.

602. Hoopes, B.C., Rimbault, M., Liebers, D., Ostrander, E.A., and Sutter, N.B. (2012). The insulin-like growth factor 1 receptor (IGF1R) contributes to reduced size in dogs. *Mamm. Genome* 23, 780–790.

603. Horton, J.C. and Adams, D.L. (2005). The cortical column: a structure without a function. *Philos. Trans. R. Soc. Lond. B* 360, 837–862.

604. Hosseini, H.S. and Taber, L.A. (2018). How mechanical forces shape the developing eye. *Prog. Biophys. Mol. Biol.* 137, 25–36.

605. Hotta, Y. and Benzer, S. (1970). Genetic dissection of the *Drosophila* nervous system by means of mosaics. *PNAS* 67, 1156–1163.

606. Hotta, Y. and Benzer, S. (1972). Mapping of behaviour in *Drosophila* mosaics. *Nature* 240, 527–535.

607. Houle, D., Jones, L.T., Fortune, R., and Sztepanacz, J.L. (2019). Why does allometry evolve so slowly? *Integr. Comp. Biol.* 59, 1429–1440.

608. Houston, D.W. (2012). Cortical rotation and messenger RNA localization in *Xenopus* axis formation. *Wiley Interdiscip. Rev. Dev. Biol.* 1, 371–388.

609. Huang, T., Zhang, M., Yan, G., Huang, X., Chen, H., Zhou, L., Deng, W., Zhang, Z., Qiu, H., Ai, H., and Huang, L. (2019). Genome-wide association and evolutionary analyses reveal the formation of swine facial wrinkles in Chinese Erhualian pigs. *Aging* 11, 4672–4687.

610. Huelsken, J., Vogel, R., Erdmann, B., Cotsarelis, G., and Birchmeier, W. (2001). β-Catenin controls hair follicle morphogenesis and stem cell differentiation in the skin. *Cell* 105, 533–545.

611. Hughes, M.W., Wu, P., Jiang, T.-X., Lin, S.-J., Dong, C.-Y., Li, A., Hsieh, F.-J., Widelitz, R.B., and Chuong, C.M. (2011). In search of the Golden Fleece: unraveling principles of morphogenesis by studying the integrative biology of skin appendages. *Integr. Biol.* 3, 388–407.

612. Hunter, G.L., Hadjivasiliou, Z., Bonin, H., He, L., Perrimon, N., Charras, G., and Baum, B. (2016). Coordinated control of Notch/Delta signalling and cell cycle progression drives lateral inhibition-mediated tissue patterning. *Development* 143, 2305–2310.

613. Hunter, M.V. and Fernandez-Gonzalez, R. (2017). Coordinating cell movements *in vivo*: junctional and cytoskeletal dynamics lead the way. *Curr. Opin. Cell Biol.* 48, 54–62.

614. Huxley, J.S. (1932). *Problems of Relative Growth*, 2nd ed. Methuen, London.

615. Huxley, J.S. and Teissier, G. (1936). Terminology of relative growth. *Nature* 137, 780–781.

616. Ichikawa, M. and Bui, K.H. (2018). Microtubule inner proteins: a meshwork of luminal proteins stabilizing the doublet microtubule. *BioEssays* 40, 1700209.

617. Iljin, N.A. and Iljin, V.N. (1930). Temperature effects on the color of the Siamese cat. *J. Hered.* 21, 309–318.

618. Im, K. and Grant, P.E. (2019). Sulcal pits and patterns in developing human brains. *NeuroImage* 185, 881–890.

619. Im, K., Pienaar, R., Lee, J.-M., Seong, J.-K., Choi, Y.Y., Lee, K.H., and Grant, P.E. (2011). Quantitative comparison and analysis of sulcal patterns using sulcal graph matching: a twin study. *NeuroImage* 57, 1077–1086.

620. Imes, D.L., Geary, L.A., Grahn, R.A., and Lyons, L.A. (2006). Albinism in the domestic cat (*Felis catus*) is associated with a *tyrosinase* (*TYR*) mutation. *Anim. Genet.* 37, 175–178.

621. Ingham, P.W. and McMahon, A.P. (2001). Hedgehog signaling in animal development: paradigms and principles. *Genes Dev.* 15, 3059–3087.

622. Irimia, M., Maeso, I., Roy, S.W., and Fraser, H.B. (2013). Ancient *cis*-regulatory constraints and the evolution of genome architecture. *Trends Genet.* 29, 521–528.

623. Irvine, K.D. (1999). Fringe, Notch, and making developmental boundaries. *Curr. Opin. Genet. Dev.* 9, 434–441.

624. Irvine, K.D. and Rauskolb, C. (2001). Boundaries in development: formation and function. *Annu. Rev. Cell Dev. Biol.* 17, 189–214.

625. Irvine, K.D. and Shraiman, B.I. (2017). Mechanical control of growth: ideas, facts and challenges. *Development* 144, 4238–4248.

626. Ishibashi, S., Saldanha, F.Y.L., and Amaya, E. (2017). *Xenopus* as a model organism for biomedical research. In *Basic Science Methods for Clinical Researchers*, Jalali, M., Saldanha, F., and Jalali, M., editors. Academic Press, New York, pp. 263–290.

627. Iten, L.E. and Bryant, S.V. (1975). The interaction between the blastema and stump in the establishment of the anterior–posterior and proximal–distal organization of the limb regenerate. *Dev. Biol.* 44, 119–147.

628. Ito, M., Yang, Z., Andl, T., Cui, C., Kim, N., Millar, S.E., and Cotsarelis, G. (2007). Wnt-dependent *de novo* hair follicle regeneration in adult mouse skin after wounding. *Nature* 447, 316–320.

629. Jablonski, N.G. (2006). *Skin: A Natural History*. University of California Press, Berkeley, CA.

630. Jablonski, N.G. (2010). The naked truth. *Sci. Am.* 302(2), 42–49.

631. Jablonski, N.G. and Chaplain, G. (2017). The colours of humanity: the evolution of pigmentation in the human lineage. *Proc. R. Soc. B* 372, 20160349.

632. Jackson, I.J. (2013). How the leopard gets its spots: a transmembrane peptidase specifies feline pigmentation patterns. *Pigment Cell Melanoma Res.* 26, 438–439.

633. Jacobs, M.D. (2009). Multiscale systems integration in the eye. *Wiley Interdiscip. Rev. Syst. Biol. Med.* 1, 15–27.

634. Jaeger, J. and Verd, B. (2020). Dynamic positional information: patterning mechanism versus precision in gradient-driven systems. *Curr. Top. Dev. Biol.* 137, 219–246.

635. Jahner, J.P., Lucas, L.K., Wilson, J.S., and Forister, M.L. (2015). Morphological outcomes of gynandromorphism in *Lycaeides* butterflies (Lepidoptera: Lycaenidae). *J. Insect Sci.* 15, 38.

636. Jahoda, C.A.B. (1998). Cellular and developmental aspects of androgenetic alopecia. *Exp. Dermatol.* 7, 235–248.

637. Jansen, A.G., Mous, S.E., White, T., Posthuma, D., and Polderman, T.J.C. (2015). What twin studies tell us about the heritability of brain development, morphology, and function: a review. *Neuropsychol. Rev.* 25, 27–46.

638. Janzen, F.J. and Paukstis, G.L. (1991). Environmental sex determination in reptiles: ecology, evolution, and experimental design. *Q. Rev. Biol.* 66, 149–179.

639. Jaraud, A., Bossé, P., de Citres, C.D., Tiret, L., Gache, V., and Abitbol, M. (2020). Feline chimerism revealed by DNA profiling. *Anim. Genet.* 51, 631–633.

640. Jarvik, J. and Botstein, D. (1973). A genetic method for determining the order of events in a biological pathway. *PNAS* 70, 2046–2050.

641. Jattiot, R., Fara, E., Brayard, A., Urdy, S., and Goudemand, N. (2019). Learning from beautiful monsters: phylogenetic and morphogenetic implications of left–right asymmetry in ammonoid shells. *BMC Evol. Biol.* 19, 210.

642. Jeong, D., Li, Y., Choi, Y., Yoo, M., Kang, D., Park, J., Choi, J., and Kim, J. (2017). Numerical simulation of the zebra pattern formation on a three-dimensional model. *Physica A* 475, 106–116.

643. Jernvall, J. and Salazar-Ciudad, I. (2007). The economy of tinkering mammalian teeth. In *Tinkering: The Microevolution of Development*, Bock, G. and Goode, J., editors. Wiley, Chichester, pp. 207–224.

644. Ji, S., Liu, Q., Zhang, S., Chen, Q., Wang, C., Zhang, W., Xiao, C., Li, Y., Nian, C., Li, J., Li, J., Geng, J., Hong, L., Xie, C., He, Y., Chen, X., Li, X., Yin, Z.-Y., You, H., Lin, K.-H., Wu, Q., Yu, C., Johnson, R.L., Wang, L., Chen, L., Wang, F., and Zhou, D. (2018). FGF15 activates Hippo signaling to suppress bile acid metabolism and liver tumorigenesis. *Dev. Cell* 48, 460–474.

645. Jiang, J. and Struhl, G. (1996). Complementary and mutually exclusive activities of Decapentaplegic and Wingless organize axial patterning during *Drosophila* leg development. *Cell* 86, 401–409.

646. Jidigam, V.K., Srinivasan, R.C., Patthey, C., and Gunhaga, L. (2015). Apical constriction and epithelial invagination are regulated by BMP activity. *Biol. Open* 4, 1782–1791.

647. Jin, H., Fisher, M., and Grainger, R.M. (2012). Defining progressive stages in the commitment process leading to embryonic lens formation. *Genesis* 50, 728–740.

648. Johnson, M.R., Barsh, G.S., and Mallarino, R. (2018). Periodic patterns in Rodentia: development and evolution. *Exp. Dermatol.* 28, 509–513.

649. Johnson, N.A.N., Wang, Y., Zeng, Z., Wang, G.-D., Yao, Q., and Chen, K.-P. (2019). Phylogenetic analysis and classification of insect achaete-scute complex genes. *J. Asia-Pacific Entomol.* 22, 398–403.

650. Johnson, P.T.J. and Sutherland, D.R. (2003). Amphibian deformities and *Ribeiroia* infection: an emerging helminthiasis. *Trends Parasitol.* 19, 332–335.

651. Johnson, P.T.J., Lunde, K.B., Ritchie, E.G., and Launer, A.E. (1999). The effect of trematode infection on amphibian limb development and survivorship. *Science* 284, 802–804.

652. Johnson, P.T.J., Lunde, K.B., Zelmer, D.A., and Werner, J.K. (2003). Limb deformities as an emerging parasitic disease in amphibians: evidence from museum specimens and resurvey data. *Conserv. Biol.* 17, 1724–1737.

653. Johnson, P.T.J., Preu, E.R., Sutherland, D.R., Romansic, J.M., Han, B., and Blaustein, A.R. (2006). Adding infection to injury: synergistic effects of predation and parasitism on amphibian malformations. *Ecology* 87, 2227–2235.

654. Johnston, M. (2020). Model organisms: nature's gift to disease research. *Genetics* 214, 233–234.

655. Johnston, R.J., Jr. and Desplan, C. (2010). Stochastic mechanisms of cell fate specification that yield random or robust outcomes. *Annu. Rev. Cell Dev. Biol.* 26, 689–719.

656. Jordan, W., III, Rieder, L.E., and Larschan, E. (2019). Diverse genome topologies characterize dosage compensation across species. *Trends Genet.* 35, 308–315.

657. Joshi, M., Buchanan, K.T., Shroff, S., and Orenic, T.V. (2006). Delta and Hairy establish a periodic prepattern that positions sensory bristles in *Drosophila* legs. *Dev. Biol.* 293, 64–76.

658. Joshi, S.D. and Davidson, L.A. (2012). Epithelial machines of morphogenesis and their potential application in organ assembly and tissue engineering. *Biomech. Model. Mechanobiol.* 11, 1109–1121.

659. Jülicher, F. and Eaton, S. (2017). Emergence of tissue shape changes from collective cell behaviours. *Semin. Cell Dev. Biol.* 67, 103–112.

660. Jürgens, H., Peitgen, H.-O., and Saupe, D. (1990). The language of fractals. *Sci. Am.* 263(2), 60–67.

661. Kaas, J.H. (2000). Organizing principles of sensory representations. In *Evolutionary Developmental Biology of the Cerebral Cortex*, Bock, G.R. and Cardew, G., editors. J. Wiley & Sons, New York, pp. 188–205.

662. Kaas, J.H. (2005). Serendipity and the Siamese cat: the discovery that genes for coat and eye pigment affect the brain. *ILAR J.* 46, 357–363.

663. Kaelin, C.B. and Barsh, G. (2013). Genetics of pigmentation in dogs and cats. *Annu. Rev. Anim. Biosci.* 1, 125–156.

664. Kaelin, C.B., Xu, X., Hong, L.Z., David, V.A., McGowan, K.A., Schmidt-Küntzel, A., Roelke, M.E., Pino, J., Pontius, J., Cooper, G.M., Manuel, H., Swanson, W.F., Marker, L., Harper, C.K., van Dyk, A., Yue, B., Mullikin, J.C., Warren, W.C., Eizirik, E., Kos, L., O'Brien, S.J., Barsh, G.S., and Menotti-Raymond, M. (2012). Specifying and sustaining pigmentation patterns in domestic and wild cats. *Science* 337, 1536–1541.

665. Kageura, H. (1997). Activation of dorsal development by contact between the cortical dorsal determinant and the equatorial core cytoplasm in eggs of *Xenopus laevis*. *Development* 124, 1543–1551.

666. Kalay, G., Lachowiec, J., Rosas, U., Dome, M.R., and Wittkopp, P.J. (2019). Redundant and cryptic enhancer activities of the *Drosophila yellow* gene. *Genetics* 212, 343–360.

667. Kalcheim, C. (2016). Epithelial–mesenchymal transitions during neural crest and somite development. *J. Clinical Med.* 5, 5010001.

668. Kalcheim, C. (2018). Neural crest emigration: from start to stop. *Genesis* 56, e23090.

669. Kandachar, V. and Roegiers, F. (2012). Endocytosis and control of Notch signaling. *Curr. Opin. Cell Biol.* 24, 534–540.

670. Kaneshiro, K.Y. (1988). Speciation in the Hawaiian *Drosophila*: sexual selection appears to play an important role. *BioScience* 38, 258–263.

671. Kango-Singh, M. and Singh, A. (2009). Regulation of organ size: insights from the *Drosophila* Hippo signaling pathway. *Dev. Dynamics* 238, 1627–1637.

672. Kango-Singh, M., Nolo, R., Tao, C., Verstreken, P., Hiesinger, P.R., Bellen, H.J., and Halder, G. (2002). Shar-pei mediates cell proliferation arrest during imaginal disc growth in *Drosophila*. *Development* 129, 5719–5730.

673. Kapanidou, M., Lee, S., and Bolanos-Garcia, V.M. (2015). BubR1 kinase: protection against aneuploidy and premature aging. *Trends Mol. Med.* 21, 364–372.

674. Kapoor, P. and Gonsalves, W.I. (2020). Of lions, shar-pei, and doughnuts: a tale retold. *Blood* 135, 1074–1076.

675. Karlsson, E.K., Baranowska, I., Wade, C.M., Salmon Hillbertz, N.H.C., Zody, M.C., Anderson, N., Biagi, T.M., Patterson, N., Pielberg, G.R., Kulbokas, E.J., III, Comstock, K.E., Keller, E.T., Mesirov, J.P., von Euler, H., Kämpe, O., Hedhammar, A., Lander, E.S., Andersson, G., Andersson, L., and Lindblad-Toh, K. (2007). Efficient mapping of mendelian traits in dogs through genome-wide association. *Nat. Genet.* 39, 1321–1328.

676. Karzbrun, E., Kshirsagar, A., Cohen, S.R., Hanna, J.H., and Reiner, O. (2018). Human brain organoids on a chip reveal the physics of folding. *Nat. Phys.* 14, 515–522.

677. Katanaev, V.L., Egger-Adam, D., and Tomlinson, A. (2018). Antagonistic PCP signaling pathways in the developing *Drosophila* eye. *Sci. Rep.* 8, 5741.

678. Katz, L.C. and Crowley, J.C. (2002). Development of cortical circuits: lessons from ocular dominance columns. *Nat. Rev. Neurosci.* 3, 34–42.

679. Kauffman, S.A. (1983). Developmental constraints: internal factors in evolution. In *Development and Evolution*, Goodwin, B.C., Holder, N., and Wylie, C.C., editors. Cambridge University Press, Cambridge, pp. 195–225.

680. Kaufman, L. (1925). An experimental study on the partial albinism in Himalayan rabbits. *Biol. Generalis Vienna* 1, 7–21.

681. Kaufman, M.H. (2004). The embryology of conjoined twins. *Childs Nerv. Syst.* 20, 508–525.

682. Kaufman, T.C., Seeger, M.A., and Olsen, G. (1990). Molecular and genetic organization of the Antennapedia gene complex of *Drosophila melanogaster*. *Adv. Genet.* 27, 309–362.

683. Kawahira, N., Ohtsuka, D., Kida, N., Hironaka, K.-i., and Morishita, Y. (2020). Quantitative analysis of 3D tissue deformation reveals key cellular mechanism associated with initial heart looping. *Cell Rep.* 30, 3889–3903.

684. Kawamori, A. and Yamaguchi, M. (2011). DREF is critical for *Drosophila* bristle development by regulating endoreplication in shaft cells. *Cell Struct. Funct.* 36, 103–119.

685. Kawasaki, S., Makuuchi, M., Ishizone, S., Matsunami, H., Terada, M., and Kawarazaki, H. (1992). Liver regeneration in recipients and donors after transplantation. *Lancet* 339, 580–581.

686. Keeler, R.F. and Binns, W. (1968). Teratogenic compounds of *Veratrum californicum* (Durand). V. Comparison of cyclopian effects of steroidal alkaloids from the plant and structurally related compounds from other sources. *Teratology* 1, 5–10.

687. Kemp, P.R., Griffiths, M., and Polkey, M.I. (2019). Muscle wasting in the presence of disease, why is it so variable? *Biol. Rev.* 94, 1038–1055.

688. Kenward, B., Wachtmeister, C.-A., Ghirlanda, S., and Enquist, M. (2004). Spots and stripes: the evolution of repetition in visual signal form. *J. Theor. Biol.* 230, 407–419.

689. Keyte, A.L. and Smith, K.K. (2014). Heterochrony and developmental timing mechanisms: changing ontogenies in evolution. *Semin. Cell Dev. Biol.* 34, 99–107.

690. Kidd, W. (1920). *Initiative in Evolution*. H. F. & G. Witherby, London.

691. Kidson, S. and Fabian, B. (1979). Pigment synthesis in the Himalayan mouse. *J. Exp. Zool.* 210, 145–152.

692. Kiecker, C. and Lumsden, A. (2012). The role of organizers in patterning the nervous system. *Annu. Rev. Neurosci.* 35, 347–367.

693. Kim, J., Sebring, A., Esch, J.J., Kraus, M.E., Vorwerk, K., Magee, J., and Carroll, S.B. (1996). Integration of positional signals and regulation of wing formation and identity by *Drosophila vestigial* gene. *Nature* 382, 133–138.

694. Kim, J., Williams, F.J., Dreger, D.L., Plassais, J., Davis, B.W., Parker, H.G., and Ostrander, E.A. (2018). Genetic selection of athletic success in sport-hunting dogs. *PNAS* 115, E7212–E7221.

695. Kim, M.J., Oh, H.J., Kim, G.A., Park, J.E., Park, E.J., Jang, G., Ra, J.C., Kang, S.K., and Lee, B.C. (2012). Lessons learned from cloning dogs. *Reprod. Dom. Anim.* 47 (Suppl. 4), 115–119.

696. King, R.A., Townsend, D., Oetting, W., Summers, C.G., Olds, D.P., White, J.G., and Spritz, R.A. (1991). Temperature-sensitive tyrosinase associated with peripheral pigmentation in oculocutaneous albinism. *J. Clin. Invest.* 87, 1046–1053.

697. Kingsley, M.C.S. and Ramsay, M.A. (1988). The spiral in the tusk of the narwhal. *Arctic* 41, 236–238.

698. Kirikoshi, H., Sekihara, H., and Katoh, M. (2001). *WNT10A* and *WNT6*, clustered in human chromosome 2q35 region with head-to-tail manner, are strongly coexpressed in SW480 cells. *Biochem. Biophys. Res. Comm.* 283, 798–805.

699. Kirschner, M.W. and Gerhart, J.C. (2005). *The Plausibility of Life: Resolving Darwin's Dilemma*. Yale University Press, New Haven, CT.

700. Kiskowski, M., Glimm, T., Moreno, N., Gamble, T., and Chiari, Y. (2019). Isolating and quantifying the role of developmental noise in generating phenotypic variation. *PLoS Comput. Biol.* 15(4), e1006943.

701. Kivell, T.L., Lemelin, P., Richmond, B.G., and Schmitt, D., eds. (2016). *The Evolution of the Primate Hand*. Springer, New York.

702. Klaassen, Z., Shoja, M.M., Tubbs, R.S., and Loukas, M. (2011). Supernumerary and absent limbs and digits of the lower limb: a review of the literature. *Clin. Anat.* 24, 570–575.

703. Klar, A.J.S. (2003). Human handedness and scalp hair-whorl direction develop from a common genetic mechanism. *Genetics* 165, 269–276.

704. Klar, A.J.S. (2005). A 1927 study supports a current genetic model for inheritance of human scalp hair-whorl orientation and hand-use preference traits. *Genetics* 170, 2027–2030.

705. Klein, T. and Martinez Arias, A. (1999). The Vestigial gene product provides a molecular context for the interpretation of signals during the development of the wing in *Drosophila*. *Development* 126, 913–925.

706. Kley, N.J. and Kearney, M. (2007). Adaptations for digging and burrowing. In *Fins into Limbs: Evolution, Development, and Transformation*, Hall, B.K., editor. University of Chicago Press, Chicago, IL, pp. 284–309.

707. Klimczewska, K., Kasperczuk, A., and Suwinska, A. (2018). The regulative nature of mammalian embryos. *Curr. Top. Dev. Biol.* 128, 105–149.

708. Klingenberg, C.P. (1998). Heterochrony and allometry: the analysis of evolutionary change in ontogeny. *Biol. Rev.* 73, 79–123.

709. Kmita, M. and Duboule, D. (2003). Organizing axes in time and space: 25 years of colinear tinkering. *Science* 301, 331–333.

710. Kmita, M., Fraudeau, N., Hérault, Y., and Duboule, D. (2002). Serial deletions and duplications suggest a mechanism for the collinearity of *Hoxd* genes in limbs. *Nature* 420, 145–150.

711. Knebel, D., Rillich, J., Ayali, A., Pflüger, H.-J., and Rigosi, E. (2018). *Ex vivo* recordings reveal desert locust forelimb control is asymmetric. *Curr. Biol.* 28, R1283–R1295.

712. Knezevic, V., De Santo, R., Schughart, K., Huffstadt, U., Chiang, C., Mahon, K.A., and Mackem, S. (1997). *Hoxd-12* differentially affects preaxial and postaxial chondrogenic branches in the limb and regulates *Sonic hedgehog* in a positive feedback loop. *Development* 124, 4523–4536.

713. Koca, Y., Housden, B.E., Gault, W.J., Bray, S.J., and Mlodzik, M. (2019). Notch signaling coordinates ommatidial rotation in the *Drosophila* eye via transcriptional regulation of the EGF-Receptor ligand Argos. *Sci. Rep.* 9, 18628.

714. Koch, S.L., Tridico, S.R., Bernard, B.A., Shriver, M.D., and Jablonski, N.G. (2019). The biology of human hair: a multidisciplinary review. *Am. J. Hum. Biol.* 2019, e23316.

715. Koenig, K.M., Sun, P., Meyer, E., and Gross, J.M. (2016). Eye development and photoreceptor differentiation in the cephalopod *Doryteuthis pealeii*. *Development* 143, 3168–3181.

716. Koga, A., Hisakawa, C., and Yoshizawa, M. (2020). Baboon bearing resemblance in pigmentation pattern to Siamese cat carries a missense mutation in the tyrosinase gene. *Genome* 63, 275–279.

717. Kohler, R.E. (1994). *Lords of the Fly:* Drosophila *Genetics and the Experimental Life*. University of Chicago Press, Chicago, IL.

718. Kojima, T. (2004). The mechanism of *Drosophila* leg development along the proximo-distal axis. *Dev. Growth Differ.* 46, 115–129.

719. Kojima, T. (2017). Developmental mechanism of the tarsus in insect legs. *Curr. Opin. Insect Sci.* 19, 36–42.

720. Kondo, S. and Asai, R. (1995). A reaction–diffusion wave on the skin of the marine angelfish *Pomacanthus*. *Nature* 376, 765–768.

721. Kondo, S. and Miura, T. (2010). Reaction–diffusion model as a framework for understanding biological pattern formation. *Science* 329, 1616–1620.

722. Kondo, S. and Shirota, H. (2009). Theoretical analysis of mechanisms that generate the pigmentation pattern of animals. *Semin. Cell Dev. Biol.* 20, 82–89.

723. Kondo, S., Iwashita, M., and Yamaguchi, M. (2009). How animals get their skin patterns: fish pigment pattern as a live Turing wave. *Int. J. Dev. Biol.* 53, 851–856.

724. Kondo, T. and Hayashi, S. (2015). Mechanisms of cell height changes that mediate epithelial invagination. *Dev. Growth Differ.* 57, 313–323.

725. Kong, X.-Z., Mathias, S.R., Guadalupe, T., Group, E.L.W., Glahn, D.C., Franke, B., Crivello, F., Tzourio-Mazoyer, N., Fisher, S.E., Thompson, P.M., and Francks, C. (2018). Mapping cortical brain asymmetry in 17,141 healthy individuals worldwide via the ENIGMA Consortium. *PNAS* 115, ES154–ES163.

726. Kopp, A. (2009). Metamodels and phylogenetic replication: a systematic approach to the evolution of developmental pathways. *Evolution* 63, 2771–2789.

727. Kopp, A. (2011). *Drosophila* sex combs as a model of evolutionary innovations. *Evol. Dev.* 13, 504–522.

728. Kopp, A. (2012). *Dmrt* genes in the development and evolution of sexual dimorphism. *Trends Genet.* 28, 175–184.

729. Kopp, A. and True, J.R. (2002). Evolution of male sexual characters in the Oriental *Drosophila melanogaster* species group. *Evol. Dev.* 4, 278–291.

730. Koshikawa, S. (2015). Enhancer modularity and the evolution of new traits. *Fly* 9, 155–159.

731. Krapp, H.G. (2009). Ocelli. *Curr. Biol.* 19, R435–R437.

732. Kruggel, F. and Solodkin, A. (2019). Determinants of structural segregation and patterning in the human cortex. *NeuroImage* 196, 248–260.

733. Krumlauf, R. (2016). Hox genes and the hindbrain: a study in segments. *Curr. Top. Dev. Biol.* 116, 581–596.

734. Kücken, M. (2007). Models for fingerprint pattern formation. *Forensic Sci. Int.* 171, 85–96.

735. Kücken, M. and Newell, A.C. (2005). Fingerprint formation. *J. Theor. Biol.* 235, 71–83.

736. Kumar, J.P. (2012). Building an ommatidium one cell at a time. *Dev. Dynamics* 241, 136–149.

737. Kunhardt, P.B., Jr., Kunhardt, P.B., III, and Kunhardt, P.W. (1995). *P. T. Barnum: America's Greatest Showman.* Knopf, New York.

738. Kuratani, S. (2012). Evolution of the vertebrate jaw from developmental perspectives. *Evol. Dev.* 14, 76–92.

739. Kutsarova, E., Munz, M., and Ruthazer, E.S. (2017). Rules for shaping neural connections in the developing brain. *Front. Neural Circuits* 10, 11.

740. Kwak, J.Y. and Kwon, K.-S. (2019). Pharmacological interventions for treatment of sarcopenia: current status of drug development for sarcopenia. *Ann. Geriatr. Med. Res.* 23, 98–104.

741. Kwon, B.S., Halaban, R., and Chintamaneni, C. (1989). Molecular basis of mouse Himalayan mutation. *Biochem. Biophys. Res. Comm.* 161, 252–260.

742. Ladher, R.K. (2017). Changing shape and shaping change: inducing the inner ear. *Dev. Biol.* 65, 39–46.

743. Ladoux, B., Mège, R.-M., and Trepat, X. (2016). Front–rear polarization by mechanical cues: from single cells to tissues. *Trends Cell Biol.* 26, 420–433.

744. Lafuente, E. and Beldade, P. (2019). Genomics of developmental plasticity in animals. *Front. Genet.* 10, 720.

745. Lamas, J.A., Rueda-Ruzafa, L., and Herrera-Pérez, S. (2019). Ion channels and thermosensitivity: TRP, TREK, or both? *Int. J. Mol. Sci.* 20, 2371.

746. Lancaster, M.A. (2018). Crinkle-cut brain organoids. *Cell Stem Cell* 22, 616–618.

747. Lander, A.D. (2011). Pattern, growth, and control. *Cell* 144, 955–969.

748. Landge, A.N., Jordan, B.M., Diego, X., and Müller, P. (2020). Pattern formation mechanisms of self-organizing reaction–diffusion systems. *Dev. Biol.* 460, 2–11.

749. Lange, A. and Müller, G.B. (2017). Polydactyly in development, inheritance, and evolution. *Q. Rev. Biol.* 92, 1–38.

750. Lange, A., Nemeschkal, H.L., and Müller, G.B. (2014). Biased polyphenism in polydactylous cats carrying a single point mutation: the Hemingway model for digit novelty. *Evol. Biol.* 41, 262–275.

751. Larsen, E.W. (1997). Evolution of development: the shuffling of ancient modules by ubiquitous bureaucracies. In *Physical Theory in Biology: Foundations and Explorations*, Lumsden, C.J., Brandts, W.A., and Trainor, L.E.H., editors. World Scientific, Singapore, pp. 431–441.

752. Larue, L., de Vuyst, F., and Delmas, V. (2013). Modeling melanoblast development. *Cell. Mol. Life Sci.* 70, 1067–1079.

753. Laubichler, M.D. and Hall, B.K. (2008). Conrad Hal Waddington: forefather of theoretical EvoDevo. *Biol. Theory* 3, 185–187.

754. Laufer, E., Dahn, R., Orozco, O.E., Yeo, C.-Y., Pisenti, J., Henrique, D., Abbott, U.K., Fallon, J.F., and Tabin, C. (1997). Expression of *Radical fringe* in limb-bud ectoderm regulates apical ectodermal ridge formation. *Nature* 386, 366–373.

755. Laurent-Gengoux, P., Petit, V., and Larue, L. (2019). Modeling and analysis of melanoblast motion. *J. Math. Biol.* 79, 2111–2132.

756. Lavine, L., Gotoh, H., Brent, C.S., Dworkin, I., and Emlen, D.J. (2015). Exaggerated trait growth in insects. *Annu. Rev. Entomol.* 60, 453–472.

757. Lawrence, P. (2020). Practice makes perfect: Sir Michael J. Berridge. *Curr. Biol.* 20, R377.

758. Lawrence, P.A. (2004). A Wigglesworth classic: how cells make patterns. *J. Exp. Biol.* 207, 192–193.

759. Lawrence, P.A. (2019). Sydney Brenner: a master of science and wit. *Development* 146, dev179879.

760. Lawrence, P.A. and Casal, J. (2018). Planar cell polarity: two genetic systems use one mechanism to read gradients. *Development* 145, dev168229.

761. Lawrence, P.A. and Locke, M. (1997). A man for our season. *Nature* 386, 757–758.

762. Lawrence, P.A., Struhl, G., and Casal, J. (2008). Do the protocadherins Fat and Dachsous link up to determine both planar cell polarity and the dimensions of organs? *Nat. Cell Biol.* 10, 1379–1382.

763. Lawrence, P.A., Struhl, G., and Morata, G. (1979). Bristle patterns and compartment boundaries in the tarsi of *Drosophila*. *J. Embryol. Exp. Morphol.* 51, 195–208.

764. Lawton, A.K., Engstrom, T., Rohrbach, D., Omura, M., Turnbull, D.H., Mamou, J., Zhang, T., Schwarz, J.M., and Joyner, A.L. (2019). Cerebellar folding is initiated by mechanical constraints on a fluid-like layer without a cellular pre-pattern. *eLife* 8, e45019.

765. Le Garrec, J.-F. and Kerszberg, M. (2008). Modeling polarity buildup and cell fate decision in the fly eye: insight into the connection between the PCP and Notch pathways. *Dev. Genes Evol.* 218, 413–426.

766. Le Gros Clark, W.E. (1945). Deformation patterns in the cerebral cortex. In *Essays on Growth and Form*, Le Gros Clark, W.E. and Medawar, P.B., editors. Clarendon Press, Oxford, pp. 1–22.

767. Lebreton, G., Géminard, C., Lapraz, F., Pyrpassopoulos, S., Cerezo, D., Spéder, P., Ostap, E.M., and Noselli, S. (2018). Molecular to organismal chirality is induced by the conserved myosin 1D. *Science* 362, 949–952.

768. Lee, D.E., Cavener, D.R., and Bond, M.L. (2018). Seeing spots: quantifying mother–offspring similarity and assessing fitness consequences of coat pattern traits in a wild population of giraffes (*Giraffa camelopardalis*). *PeerJ* 6, e5690.

769. Lee, L.A. and Orr-Weaver, T.L. (2003). Regulation of cell cycles in *Drosophila* development: intrinsic and extrinsic cues. *Annu. Rev. Genet.* 37, 545–578.

770. Lee, S.-J. (2007). Quadrupling muscle mass in mice by targeting TGF-β signaling pathways. *PLoS ONE* 2(8), e789.

771. Lee, S.-J. and McPherron, A.C. (2001). Regulation of myostatin activity and muscle growth. *PNAS* 98, 9306–9311.

772. Leevers, S.J., Weinkove, D., MacDougall, L.K., Hafen, E., and Waterfield, M.D. (1996). The *Drosophila* phosphoinositide 3-kinase Dp110 promotes cell growth. *EMBO J.* 15, 6584–6594.

773. Lejeune, E., Dortdivanlioglu, B., Kuhl, E., and Linder, C. (2019). Understanding the mechanical link between oriented cell division and cerebellar morphogenesis. *Soft Matter* 15, 2204–2215.

774. Lemons, D. and McGinnis, W. (2006). Genomic evolution of Hox gene clusters. *Science* 313, 1918–1922.

775. Lenhoff, H.M., Wang, P.P., Greenberg, F., and Bellugi, U. (1997). Williams Syndrome and the brain. *Sci. Am.* 277(6), 68–73.

776. Lerner, A.B. and Fitzpatrick, T.B. (1950). Biochemistry of melanin formation. *Physiol. Rev.* 30, 91–126.

777. Leroi, A.M. (2003). *Mutants: On Genetic Variety and the Human Body.* Viking Press, New York.

778. Lettice, L.A., Williamson, I., Devenney, P.S., Kilanowski, F., Dorin, J., and Hill, R.E. (2014). Development of five digits is controlled by a bipartite long-range *cis*-regulator. *Development* 141, 1715–1725.

779. Leventhal, A.G., Vitek, D.J., and Creel, D.J. (1985). Abnormal visual pathways in normally pigmented cats that are heterozygous for albinism. *Science* 229, 1395–1397.

780. Levin, M. (1999). Twinning and embryonic left–right asymmetry. *Laterality* 4, 197–208.

781. Levin, M. (2012). Morphogenetic fields in embryogenesis, regeneration, and cancer: non-local control of complex patterning. *BioSystems* 109, 243–261.

782. Levin, M., Klar, A.J.S., and Ramsdell, A.F. (2016). Introduction to provocative questions in left–right asymmetry. *Philos. Trans. R. Soc. Lond. B* 371, 20150399.

783. Levin, M., Roberts, D.J., Holmes, L.B., and Tabin, C. (1996). Laterality defects in conjoined twins. *Nature* 384, 321.

784. Levis, N.A. and Pfennig, D.W. (2020). Plasticity-led evolution: a survey of developmental mechanisms and empirical tests. *Evol. Dev.* 22, 71–87.

785. Lewis, E.B. (1976). Alfred Henry Sturtevant. In *Dictionary of Scientific Biography.* C. Scribner's Sons, New York, pp. 133–138.

786. Lewis, E.B. (1978). A gene complex controlling segmentation in *Drosophila. Nature* 276, 565–570.

787. Lewis, E.B. (1994). Homeosis: the first 100 years. *Trends Genet.* 10, 341–343.

788. Lewis, E.B. (1998). The *bithorax* complex: the first fifty years. *Int. J. Dev. Biol.* 42, 403–415.

789. Lewis, E.B., Pfeiffer, B.D., Mathog, D.R., and Celniker, S.E. (2003). Evolution of the homeobox complex in the Diptera. *Curr. Biol.* 13, R587–R588.

790. Lewis, J. (1982). Continuity and discontinuity in pattern formation. In *Developmental Order: Its Origin and Regulation*, Subtelny, S. and Green, P.B., editors. Liss, New York, pp. 511–531.

791. Li, A., Xue, J., and Peterson, E.H. (2008). Architecture of the mouse utricle: macular organization and hair bundle heights. *J. Neurophysiol.* 99, 718–733.

792. Li, C., Littlejohn, R.P., and Suttie, J.M. (1999). Effects of insulin-like growth factor 1 and testosterone on the proliferation of antlerogenic cells in vitro. *J. Exp. Zool.* 284, 82–90.

793. Li, C., Zhao, H., Liu, Z., and McMahon, C. (2014). Deer antler: a novel model for studying organ regeneration in mammals. *Int. J. Biochem. Cell Biol.* 56, 111–122.

794. Li, H., Mao, Y., Bouaziz, M., Yu, H., Qu, X., Wang, F., Feng, G.-S., Shawber, C., and Zhang, X. (2019). Lens differentiation is controlled by the balance between PDGF and FGF signaling. *PLoS Biol.* 17(2), e3000133.

795. Li, P. and Elowitz, M.B. (2019). Communication codes in developmental signaling pathways. *Development* 146, dev170977.

796. Li, X., Guo, C., and Li, L. (2019). Functional morphology and structural characteristics of the hind wings of the bamboo weevil *Cyrtotrachelus buqueti* (Coleoptera, Curculionidae). *Anim. Cells Syst.* 23, 143–153.

797. Lichtwark, G.A. and Kelly, L.A. (2020). Ahead of the curve in the evolution of human feet. *Nature* 579, 31–32.

798. Lim, Y.H., Moscato, Z., and Choate, K.A. (2017). Mosaicism in cutaneous disorders. *Annu. Rev. Genet.* 51, 123–141.

799. Lin, A.Y. and Wang, L.H. (2018). Molecular therapies for muscular dystrophies. *Curr. Treat. Options Neurol.* 20, 27.

800. Lin, J.Y. and Fisher, D.E. (2007). Melanocyte biology and skin pigmentation. *Nature* 445, 843–850.

801. Lincoln, G.A., Clarke, I.J., Hut, R.A., and Hazelrigg, D.G. (2006). Characterizing a mammalian circannual pacemaker. *Science* 314, 1941–1944.

802. Lindsley, D.L. and Grell, E.H. (1968). *Genetic Variations of* Drosophila melanogaster. Carnegie Institution of Washington, Washington, DC.

803. Lipshitz, H.D., ed. (2004). *Genes, Development and Cancer: The Life and Work of Edward B. Lewis*. Springer, New York.

804. Lipshitz, H.D. (2005). From fruit flies to fallout: Ed Lewis and his science. *Dev. Dynamics* 232, 529–546.

805. Lister, A.M., Edwards, C.J., Nock, D.A.W., Bunce, M., van Pijlen, I.A., Bradley, D.G., Thomas, M.G., and Barnes, I. (2005). The phylogenetic position of the "giant deer" *Megaloceros giganteus*. *Nature* 438, 850–853.

806. Litingtung, Y., Dahn, R.D., Li, Y., Fallon, J.F., and Chiang, C. (2002). *Shh* and *Gli3* are dispensable for limb skeleton formation but regulate digit number and identity. *Nature* 418, 979–983.

807. Liu, F., Chen, Y., Zhu, G., Hysi, P.G., Wu, S., Adhikari, K., Breslin, K., Pospiech, E., Hamer, M.A., Peng, F., Muralidharan, C., Acuna-Alonzo, V., Canizales-Quinteros, S., Bedoya, G., Gallo, C., Poletti, G., Rothhammer, F., Bortolini, M.C., Gonzalez-Jose, R., Zeng, C., Xu, S., Jin, L., Uitterlinden, A.G., Ikram, M.A., van Duijn, C.M., Nijsten, T., Walsh, S., Branicki, W., Wang, S., Ruiz-Linares, A., Spector, T.D., Martin, N.G., Medland, S.E., and Kayser, M. (2018). Meta-analysis of genome-wide association studies identifies 8 novel loci involved in shape variation of human head hair. *Hum. Mol. Genet.* 27, 559–575.

808. Liu, F., Hamer, M.A., Deelen, J., Lall, J.S., Jacobs, L., van Heemst, D., Murray, P.G., Wollstein, A., de Craen, A.J.M., Uh, H.-W., Zeng, C., Hofman, A., Uitterlinden, A.G., Houwing-Duistermaat, J.J., Pardo, L.M., Beekman, M., Slagboom, P.E., Nijsten, T., Kayser, M., and Gunn, D.A. (2016). The *MC1R* gene and youthful looks. *Curr. Biol.* 26, 1213–1220.

809. Liu, R.T., Liaw, S.S., and Maini, P.K. (2006). Two-stage Turing model for generating pigment patterns on the leopard and the jaguar. *Phys. Rev. E* 74, 011914.

810. Liu, W., Selever, J., Lu, M.-F., and Martin, J.F. (2003). Genetic dissection of *Pitx2* in craniofacial development uncovers new functions in branchial arch morphogenesis, late aspects of tooth morphogenesis and cell migration. *Development* 130, 6375–6385.

811. Liu, X. and Smagghe, G. (2019). Roles of the insulin signaling pathway in insect development and organ growth. *Peptides* 122, 169923.

812. Llinares-Benadero, C. and Borrell, V. (2019). Deconstructing cortical folding: genetic, cellular and mechanical determinants. *Nat. Rev. Neurosci.* 20, 161–176.

813. Lo Celso, C., Prowse, D.M., and Watt, F.M. (2004). Transient activation of β-catenin signalling in adult mouse epidermis is sufficient to induce new hair follicles but continuous activation is required to maintain hair follicle tumours. *Development* 131, 1787–1799.

814. Loison, L. (2013). Georges Teissier (1900–1972) and the Modern Synthesis in France. *Genetics* 195, 295–302.

815. Long, H.K., Prescott, S.L., and Wysocka, J. (2016). Ever-changing landscapes: transcriptional enhancers in development and evolution. *Cell* 167, 1170–1187.

816. Long, J.A. and Cloutier, R. (2020). The unexpected origin of fingers. *Sci. Am.* 322(6), 46–53.

817. Long, K.R. and Huttner, W.B. (2019). How the extracellular matrix shapes neural development. *Open Biol.* 9, 180216.

818. Long, K.R., Newland, B., Florio, M., Kalebic, N., Langen, B., Kolterer, A., Wimberger, P., and Huttner, W.B. (2018). Extracellular matrix components HAPLN1, lumican, and collagen I cause hyaluronic acid-dependent folding of the developing human neocortex. *Neuron* 99, 702–719.

819. Lui, J.C. and Baron, J. (2011). Mechanisms limiting body growth in mammals. *Endocr. Rev.* 32, 422–440.

820. Lunde, K.B. and Johnson, P.T.J. (2012). A practical guide for the study of malformed amphibians and their causes. *J. Herpetol.* 46, 429–441.

821. Luo, S.-J., Liu, Y.-C., and Xu, X. (2019). Tigers of the world: genomics and conservation. *Annu. Rev. Anim. Biosci.* 7, 521–548.

822. Lyon, M.F. (1999). X-chromosome inactivation. *Curr. Biol.* 9, R235–R237.

823. Lyons, L.A., Imes, D.L., Rah, H.C., and Grahn, R.A. (2005). Tyrosinase mutations associated with Siamese and Burmese patterns in the domestic cat (*Felis catus*). *Anim. Genet.* 36, 119–126.

824. Maartens, A. (2017). *On Growth and Form* in context: an interview with Matthew Jarron. *Development* 144, 4199–4202.

825. Maartens, A.P. and Brown, N.H. (2015). Anchors and signals: the diverse roles of integrins in development. *Curr. Top. Dev. Biol.* 112, 233–272.

826. Maas, A.-H. (1948). Über die Auslösbarkeit von Temperatur-Modifikationen während der Embryonal-Entwicklung von *Drosophila melanogaster* Meigen. *W. Roux Arch. Entw.-Mech. Org.* 143, 515–572.

827. Macabenta, F. and Stathopoulos, A. (2019). Sticking to a plan: adhesion and signaling control spatial organization of cells within migrating collectives. *Curr. Opin. Genet. Dev.* 57, 39–46.

828. MacArthur, J.W. and Ford, N. (1937). *A Biological Study of the Dionne Quintuplets: An Identical Set*. University of Toronto Press, Toronto.

829. MacKenzie, T.C., Crombleholme, T.M., Johnson, M.P., Schnaufer, L., Flake, A.W., Hedrick, H.L., Howell, L.J., and Adzick, N.S. (2002). The natural history of prenatally diagnosed conjoined twins. *J. Pediatr. Surg.* 37, 303–309.

830. Madhavan, M.M. and Schneiderman, H.A. (1977). Histological analysis of the dynamics of growth of imaginal discs and histoblast nests during the larval development of *Drosophila melanogaster*. *W. Roux Arch. Dev. Biol.* 183, 269–305.

831. Maeda, R.K. and Karch, F. (2006). The ABC of the BX-C: the bithorax complex explained. *Development* 133, dev02323.

832. Maeda, R.K. and Karch, F. (2015). The *open for business* model of the bithorax complex in *Drosophila*. *Chromosoma* 124, 293–307.

833. Magklara, A. and Lomvardas, S. (2013). Stochastic gene expression in mammals: lessons from olfaction. *Trends Cell Biol.* 23, 449–456.

834. Malacinski, G.M., ed. (1990). *Cytoplasmic Organization Systems*. McGraw-Hill, New York.

835. Malagón, J.N., Ahuja, A., Sivapatham, G., Hung, J., Lee, J., Muñoz-Gómez, S.A., Atallah, J., Singh, R.S., and Larsen, E. (2014). Evolution of *Drosophila* sex comb length illustrates the inextricable interplay between selection and variation. *PNAS* 111, E4103–E4109.

836. Mallarino, R., Henegar, C., Mirasierra, M., Manceau, M., Schradin, C., Vallejo, M., Beronja, S., Barsh, G.S., and Hoekstra, H.E. (2016). Developmental mechanisms of stripe patterns in rodents. *Nature* 539, 518–523.

837. Mallo, M. (2018). Reassessing the role of *Hox* genes during vertebrate development and evolution. *Trends Genet.* 34, 209–217.

838. Malmström, T. and Kröger, R.H.H. (2006). Pupil shapes and lens optics in the eyes of terrestrial vertebrates. *J. Exp. Biol.* 209, 18–25.

839. Manmadhan, S. and Ehmer, U. (2019). Hippo signaling in the liver: a long and ever-expanding story. *Front. Cell Dev. Biol.* 7, 33.

840. Mann, R.S. and Morata, G. (2000). The developmental and molecular biology of genes that subdivide the body of *Drosophila*. *Annu. Rev. Cell Dev. Biol.* 16, 243–271.

841. Manocha, S., Farokhnia, N., Khosropanah, S., Bertol, J.W., Junior, J.S., and Fakhouri, W.D. (2019). Systematic review of hormonal and genetic factors involved in the nonsyndromic disorders of the lower jaw. *Dev. Dynamics* 248, 162–172.

842. Marcellini, S., Gibert, J.-M., and Simpson, P. (2005). *achaete*, but not *scute*, is dispensable for the peripheral nervous system of *Drosophila*. *Dev. Biol.* 285, 545–553.

843. Marcon, L. and Sharpe, J. (2012). Turing patterns in development: what about the horse part? *Curr. Opin. Genet. Dev.* 22, 578–584.

844. Marmor, M.F., Choi, S.S., Zawadzki, R.J., and Werner, J.S. (2008). Visual insignificance of the foveal pit. *Arch. Ophthalmol.* 126, 907–913.

845. Marsh, J.L. and Theisen, H. (1999). Regeneration in insects. *Semin. Cell Dev. Biol.* 10, 365–375.

846. Martin, A.C. and Goldstein, B. (2014). Apical constriction: themes and variations on a cellular mechanism driving morphogenesis. *Development* 141, 1987–1998.

847. Martín, M., Organista, M.F., and de Celis, J.F. (2016). Structure of developmental gene regulatory networks from the perspective of cell fate-determining genes. *Transcription* 7, 32–37.

848. Martinez Arias, A. and Steventon, B. (2018). On the nature and function of organizers. *Development* 145, dev159525.

849. Martinez Arias, A., Nichols, J., and Schröter, C. (2013). A molecular basis for developmental plasticity in early mammalian embryos. *Development* 140, 3499–3510.

850. Martini, F.H., Ober, W.C., Garrison, C.W., Welch, K., Hutchings, R.T., and Ireland, K. (2004). *Fundamentals of Anatomy and Physiology*, 6th ed. Benjamin Cummings, San Francisco, CA.

851. Mason, C. and Guillery, R. (2019). Conversations with Ray Guillery on albinism: linking Siamese cat visual pathway connectivity to mouse retinal development. *Eur. J. Neurosci.* 49, 913–927.

852. Massey, J.H., Chung, D., Siwanowicz, I., Stern, D.L., and Wittkopp, P.J. (2019). The *yellow* gene influences *Drosophila* male mating success through sex comb melanization. *eLife* 8, e49388.

853. Matis, M. (2020). The mechanical role of microtubules in tissue remodeling. *BioEssays* 42, 1900244.

854. Matis, M. and Axelrod, J. (2013). Regulation of PCP by the Fat signaling pathway. *Genes Dev.* 27, 2207–2220.

855. Matsuda, K., Gotoh, H., Tajika, Y., Sushida, T., Aonuma, H., Niimi, T., Akiyama, M., Inoue, Y., and Kondo, S. (2017). Complex furrows in a 2D epithelial sheet code the 3D structure of a beetle horn. *Sci. Rep.* 7, 13939.

856. Matt, G. and Umen, J. (2016). Volvox: a simple algal model for embryogenesis, morphogenesis and cellular differentiation. *Dev. Biol.* 419, 99–113.

857. May-Simera, H. and Kelley, M.W. (2012). Planar cell polarity in the inner ear. *Curr. Top. Dev. Biol.* 101, 111–140.

858. Maynard Smith, J., Burian, R., Kauffman, S., Alberch, P., Campbell, J., Goodwin, B., Lande, R., Raup, D., and Wolpert, L. (1985). Developmental constraints and evolution. *Q. Rev. Biol.* 60, 265–287.

859. Mayor, R. and Theveneau, E. (2013). The neural crest. *Development* 140, 2247–2251.

860. McAvoy, J.W., Dawes, L.J., Sugiyama, Y., and Lovicu, F.J. (2017). Intrinsic and extrinsic regulatory mechanisms are required to form and maintain a lens of the correct size and shape. *Exp. Eye Res.* 156, 34–40.

861. McClure, K.D. and Schubiger, G. (2005). Developmental analysis and squamous morphogenesis of the peripodial epithelium in *Drosophila* imaginal discs. *Development* 132, 5033–5042.

862. McElreath, R. (2018). Sizing up human brain evolution. *Nature* 557, 496–497.

863. McGhee, G.R., Jr. (2011). *Convergent Evolution: Limited Forms Most Beautiful*. Vienna Series in Theoretical Biology, Müller, G.B., Wagner, G.P., and Callebaut, W., editors. MIT Press, Cambridge, MA.

864. McHugo, G.P., Dover, M.J., and MacHugh, D.E. (2019). Unlocking the origins and biology of domestic animals using ancient DNA and paleogenomics. *BMC Biol.* 17, 98.

865. McKay, D.J., Estella, C., and Mann, R.S. (2009). The origins of the *Drosophila* leg revealed by the *cis*-regulatory architecture of the *Distalless* gene. *Development* 136, 61–71.

866. McKenna, D.D. and Farrell, B.D. (2010). 9-Genes reinforce the phylogeny of holometabola and yield alternate views on the phylogenetic placement of Strepsiptera. *PLoS ONE* 5(7), e11887.

867. Mcketton, L., Kelly, K.R., and Schneider, K.A. (2014). Abnormal lateral geniculate nucleus and optic chiasm in human albinism. *J. Comp. Neurol.* 522, 2680–2687.

868. McLean, W.H.I. (2008). Combing the genome for the root cause of baldness. *Nat. Genet.* 11, 1270–1271.

869. McNeill, H. (2010). Planar cell polarity: keeping hairs straight is not so simple. *Cold Spring Harb. Perspect. Biol.* 2, a003376.

870. McPherron, A.C. and Lee, S.-J. (1997). Double muscling in cattle due to mutations in the myostatin gene. *PNAS* 94, 12457–12461.

871. McPherron, A.C. and Lee, S.-J. (2002). Suppression of body fat accumulation in myostatin-deficient mice. *J. Clin. Invest.* 109, 595–601.

872. McPherron, A.C., Lawler, A.M., and Lee, S.-J. (1997). Regulation of skeletal muscle mass in mice by a new TGF-β superfamily member. *Nature* 387, 83–90.

873. Meinhardt, H. (2009). Models for the generation and interpretation of gradients. *Cold Spring Harb. Perspect. Biol.* 1, a001362.

874. Meinhardt, H. and Gierer, A. (1974). Applications of a theory of biological pattern formation based on lateral inhibition. *J. Cell Sci.* 15, 321–346.

875. Meinhardt, H. and Gierer, A. (2000). Pattern formation by local self-activation and lateral inhibition. *BioEssays* 22, 753–760.

876. Men, W., Falk, D., Sun, T., Chen, W., Li, J., Yin, D., Zang, L., and Fan, M. (2014). The corpus callosum of Albert Einstein's brain: another clue to his high intelligence? *Brain* 137, 1–8.

877. Mermoud, J.E., Rowbotham, S.P., and Varga-Weisz, P.D. (2011). Keeping chromatin quiet: how nucleosome remodeling restores heterochromatin after replication. *Cell Cycle* 10, 4017–4025.

878. Merrell, A.J. and Stanger, B.Z. (2019). A feedback loop controlling organ size. *Dev. Cell* 48, 425–426.

879. Mian, A., Gabra, N.I., Sharma, T., Topale, N., Gielecki, J., Tubbs, R.S., and Loukas, M. (2017). Conjoined twins: from conception to separation, a review. *Clin. Anat.* 30, 385–396.

880. Michalopoulos, G.K. and DeFrances, M.C. (1997). Liver regeneration. *Science* 276, 60–66.

881. Mickelson, J.R. and Valberg, S.J. (2015). The genetics of skeletal muscle disorders in horses. *Annu. Rev. Anim. Biosci.* 3, 197–217.

882. Miesfeld, J.B. and Brown, N.L. (2019). Eye organogenesis: a hierarchical view of ocular development. *Curr. Top. Dev. Biol.* 132, 351–393.

883. Migeon, B.R. (2016). An overview of X inactivation based on species differences. *Semin. Cell Dev. Biol.* 56, 111–116.

884. Mikeladze-Dvali, T., Desplan, C., and Pistillo, D. (2005). Flipping coins in the fly retina. *Curr. Top. Dev. Biol.* 69, 1–15 (+ color plates).

885. Millar, S.E., Willert, K., Salinas, P.C., Roelink, H., Nusse, R., Sussman, D.J., and Barsh, G.S. (1999). WNT signaling in the control of hair growth and structure. *Dev. Biol.* 207, 133–149.

886. Miller, G.S., Jr. (1931). *Human Hair and Primate Patterning*. Smithsonian Miscellaneous Collections Vol. 85 No. 10. Smithsonian Institution, Washington, DC.

887. Miller, S.W. and Posakony, J.W. (2018). Lateral inhibition: two modes of nonautonomous negative autoregulation by *neuralized*. *PLoS Genet.* 14(7), e1007528.

888. Miller, S.W., Rebeiz, M., Atanasov, J.E., and Posakony, J.W. (2014). Neural precursor-specific expression of multiple *Drosophila* genes is driven by dual enhancer modules with overlapping function. *PNAS* 111, 17194–17199.

889. Mills, L.S., Zimova, M., Oyler, J., Running, S., Abatzoglou, J.T., and Lukacs, P.M. (2013). Camouflage mismatch in seasonal coat color due to decreased snow duration. *PNAS* 110, 7360–7365.

890. Milner, M.J., Bleasby, A.J., and Kelly, S.L. (1984). The role of the peripodial membrane of leg and wing imaginal discs of *Drosophila melanogaster* during evagination and differentiation in vitro. *W. Roux Arch. Dev. Biol.* 193, 180–186.

891. Minor, K.M., Patterson, E.E., Keating, M.K., Gross, S.D., Ekenstedt, K.J., Taylor, S.M., and Mickelson, J.R. (2011). Presence and impact of the exercise-induced collapse associated DNM1 mutation in Labrador retrievers and other breeds. *Vet. J.* 189, 214–219.

892. Mirth, C. and Akam, M. (2002). Joint development in the *Drosophila* leg: cell movements and cell populations. *Dev. Biol.* 246, 391–406.

893. Mirth, C.K. and Shingleton, A.W. (2019). Coordinating development: how do animals integrate plastic and robust developmental processes? *Front. Cell Dev. Biol.* 7, 8.

894. Mito, T., Shinmyo, Y., Kurita, K., Nakamura, T., Ohuchi, H., and Noji, S. (2011). Ancestral functions of Delta/Notch signaling in the formation of body and leg segments in the cricket *Gryllus bimaculatus*. *Development* 138, 3823–3833.

895. Mitogawa, K., Makanae, A., and Satoh, A. (2018). Hyperinnervation improves *Xenopus* laevis limb regeneration. *Dev. Biol.* 433, 276–286.

896. Mittwoch, U. (2000). Genetics of sex determination: exceptions that prove the rule. *Mol. Genet. Metab.* 71, 405–410.

897. Miyashita, T. and Diogo, R. (2016). Evolution of serial patterns in vertebrate pharyngeal apparatus and paired appendages via assimilation of dissimilar units. *Front. Ecol. Evol.* 4, 71.

898. Moczek, A.P. (2010). Phenotypic plasticity and diversity in insects. *Philos. Trans. R. Soc. Lond. B* 365, 593–603.

899. Modolell, J. and Campuzano, S. (1998). The *achaete-scute* complex as an integrating device. *Int. J. Dev. Biol.* 42, 275–282.

900. Mohit, P., Makhijani, K., Madhavi, M.B., Bharathi, V., Lal, A., Sirdesai, G., Reddy, V.R., Ramesh, P., Kannan, R., Dhawan, J., and Shashidhara, L.S. (2006). Modulation of AP and DV signaling pathways by the homeotic gene *Ultrabithorax* during haltere development in *Drosophila*. *Dev. Biol.* 291, 356–367.

901. Moiseff, A. (1989). Binaural disparity cues available to the barn owl for sound localization. *J. Comp. Physiol. A* 164, 629–636.

902. Monier, B. and Suzanne, M. (2015). The morphogenetic role of apoptosis. *Curr. Top. Dev. Biol.* 114, 335–362.

903. Monk, P.B. and Othmer, H.G. (1989). Relay, oscillations and wave propagation in a model of *Dictyostelium discoideum*. In *Lectures on Mathematics in the Life Sciences*, Vol. 21. American Mathematical Society, Providence, RI, pp. 87–122.

904. Monsoro-Burq, A.H. and Levin, M. (2018). Avian models and the study of invariant asymmetry: how the chicken and the egg taught us to tell right from left. *Int. J. Dev. Biol.* 62, 63–77.

905. Montagu, M.F.A. (1962). Time, morphology, and neoteny in the evolution of man. In *Culture and the Evolution of Man*, Montagu, M.F.A., editor. Oxford University Press, New York, pp. 324–342.

906. Montavon, T. and Soshnikova, N. (2014). *Hox* gene regulation and timing in embryogenesis. *Semin. Cell Dev. Biol.* 34, 76–84.

907. Montavon, T., Le Garrec, J.-F., Kerszberg, M., and Duboule, D. (2008). Modeling *Hox* gene regulation in digits: reverse collinearity and the molecular origin of thumbness. *Genes Dev.* 22, 346–359.

908. Montgomery, S.H., Mundy, N.I., and Barton, R.A. (2016). Brain evolution and development: adaptation, allometry and constraint. *Proc. R. Soc. B* 283, 20160433.

909. Moog, U., Felbor, U., Has, C., and Zim, B. (2020). Disorders caused by genetic mosaicism. *Dtsch. Arztebl. Int.* 117, 119–125.

910. Moore, R. and Alexandre, P. (2020). Delta–Notch signaling: the long and the short of a neuron's influence on progenitor fates. *J. Dev. Biol.* 8, 8.

911. Moran, C., Gillies, C.B., and Nicholas, F.W. (1984). Fertile male tortoiseshell cats: mosaicism due to gene instability? *J. Hered.* 75, 397–402.

912. Morata, G., Shlevkov, E., and Pérez-Garijo, A. (2011). Mitogenic signaling from apoptotic cells in *Drosophila*. *Dev. Growth Differ.* 53, 168–176.

913. Moreno-Marmol, T., Cavodeassi, F., and Bovolenta, P. (2018). Setting eyes on the retinal pigment epithelium. *Front. Cell Dev. Biol.* 6, 145.

914. Morgan, T.H. and Bridges, C.B. (1919). The origin of gynandromorphs. In *Contributions to the Genetics of* Drosophila melanogaster. Carnegie Institution of Washington, Washington, DC, pp. 1–122.

915. Moriyama, Y. and De Robertis, E.M. (2018). Embryonic regeneration by relocalization of the Spemann organizer during twinning in *Xenopus*. *PNAS* 115, E4815–E4822.

916. Morris, D. (1967). *The Naked Ape: A Zoologist's Study of the Human Animal*. Random House, New York.

917. Mosher, D.S., Quignon, P., Bustamante, C.D., Sutter, N.B., Mellersh, C.S., Parker, H.G., and Ostrander, E.A. (2007). A mutation in the myostatin gene increases muscle mass and enhances racing performance in heterozygote dogs. *PLoS Genet.* 3(5), e79.

918. Moulton, D.E., Goriely, A., and Chirat, R. (2018). How seashells take shape. *Sci. Am.* 318(4), 68–75.

919. Moya, I.M. and Halder, G. (2019). Hippo–YAP/TAZ signalling in organ regeneration and regenerative medicine. *Nat. Rev. Mol. Cell Biol.* 20, 211–226.

920. Muckli, L., Naumer, M.J., and Singer, W. (2009). Bilateral visual field maps in a patient with only one hemisphere. *PNAS* 106, 13034–13039.

921. Müller, G.B. and Newman, S.A., eds. (2003). *Origination of Organismal Form: Beyond the Gene in Developmental and Evolutionary Biology*. MIT Press, Cambridge, MA.

922. Muller, G.H. (1990). Skin diseases of the Chinese Shar-Pei. *Adv. Clin. Dermatol.* 20, 1655–1670.

923. Murisier, F., Guichard, S., and Beermann, F. (2007). Distinct distal regulatory elements control tyrosinase expression in melanocytes and the retinal pigment epithelium. *Dev. Biol.* 303, 838–847.

924. Murray, J.D. (1981). On pattern formation mechanisms for lepidopteran wing patterns and mammalian coat markings. *Philos. Trans. Roy. Soc. Lond. B* 295, 473–496.

925. Murray, J.D. (1981). A pre-pattern formation mechanism for animal coat markings. *J. Theor. Biol.* 88, 161–199.

926. Murray, J.D. (1988). How the leopard gets its spots. *Sci. Am.* 258(3), 80–87.

927. Murray, J.D. (1989). *Mathematical Biology*. Springer-Verlag, Berlin.

928. Murray, J.D. (1990). Turing's theory of morphogenesis: its influence on modelling biological pattern and form. *Bull. Math. Biol.* 52, 119–152.

929. Murray, J.D. (2012). Vignettes from the field of mathematical biology: the application of mathematics to biology and medicine. *Interface Focus* 2, 397–406.

930. Murray, J.D., Deeming, D.C., and Ferguson, M.W.J. (1990). Size-dependent pigmentation-pattern formation in embryos of *Alligator mississippiensis*: time of initiation of pattern generation mechanism. *Proc. R. Soc. B* 239, 279–293.

931. Mutzel, V., Okamoto, I., Dunkel, I., Saitou, M., Giorgetti, L., Heard, E., and Schulz, E.G. (2019). A symmetric toggle switch explains the onset of random X inactivation in different mammals. *Nat. Struct. Mol. Biol.* 26, 350–360.

932. Nacu, E., Gromberg, E., Oliveira, C.R., Drechsel, D., and Tanaka, E.M. (2016). FGF8 and SHH substitute for anterior–posterior tissue interactions to induce limb regeneration. *Nature* 533, 407–410.

933. Naef, A. (1926). Über die Urformen der Anthropomorphen und die Stammesgeschichte des Menschenschädels. *Naturwiss.* 14, 445–452.

934. Närhi, K., Järvinen, E., Birchmeier, W., Taketo, M.M., Mikkola, M.L., and Thesleff, I. (2008). Sustained epithelial β-catenin activity induces precocious hair development

but disrupts hair follicle down-growth and hair shaft formation. *Development* 135, 1019–1028.

935. Nascone, N. and Mercola, M. (1997). Organizer induction determines left–right asymmetry in *Xenopus*. *Dev. Biol.* 189, 68–78.

936. Nasoori, A. (2020). Formation, structure, and function of extra-skeletal bones in mammals. *Biol. Rev. Camb. Philos. Soc.* 95, 986–1019.

937. Natori, K., Tajiri, R., Furukawa, S., and Kojima, T. (2012). Progressive tarsal patterning in the *Drosophila* by temporally dynamic regulation of transcription factor genes. *Dev. Biol.* 361, 450–462.

938. Negre, B. and Ruiz, A. (2007). HOM-C evolution in *Drosophila*: is there a need for *Hox* gene clustering? *Trends Genet.* 23, 55–59.

939. Negre, B. and Simpson, P. (2009). Evolution of the achaete-scute complex in insects: convergent duplication of proneural genes. *Trends Genet.* 25, 147–152.

940. Negre, B., Casillas, S., Suzanne, M., Sánchez-Herrero, E., Akam, M., Nefedov, M., Barbadilla, A., de Jong, P., and Ruiz, A. (2005). Conservation of regulatory sequences and gene expression patterns in the disintegrating *Drosophila Hox* complex. *Genome Res.* 15, 692–700.

941. Nelson, C.E., Morgan, B.A., Burke, A.C., Laufer, E., DiMambro, E., Murtaugh, L.C., Gonzales, E., Tessarollo, L., Parada, L.F., and Tabin, C. (1996). Analysis of *Hox* gene expression in the chick limb bud. *Development* 122, 1449–1466.

942. Nelson, C.M. (2016). On buckling morphogenesis. *J. Biomech. Eng.* 138, 021005.

943. Nemec, S., Luxey, M., Jain, D., Sung, A.H., Pastinen, T., and Drouin, J. (2017). *Pitx1* directly modulates the core limb development program to implement hindlimb identity. *Development* 144, 3325–3335.

944. Nerurkar, N.L., Mahadevan, L., and Tabin, C.J. (2017). BMP signaling controls buckling forces to modulate looping morphogenesis of the gut. *PNAS* 114, 2277–2282.

945. Neto-Silva, R.M., Wells, B.S., and Johnston, L.A. (2009). Mechanisms of growth and homeostasis in the *Drosophila* wing. *Annu. Rev. Cell Dev. Biol.* 25, 197–220.

946. Neubauer, S., Gunz, P., Scott, N.A., Hublin, J.-J., and Mitteroecker, P. (2020). Evolution of brain lateralization: a shared hominid pattern of endocranial asymmetry is much more variable in humans than in great apes. *Sci. Adv.* 6, eaax9935.

947. Neumann, C.J. and Cohen, S.M. (1998). Boundary formation in *Drosophila* wing: Notch activity attenuated by the POU protein Nubbin. *Science* 281, 409–413.

948. Neveu, M.M., Holder, G.E., Ragge, N.K., Sloper, J.J., Collin, J.R.O., and Jeffery, G. (2006). Early midline interactions are important in mouse optic chiasm formation but are not critical in man: a significant distinction between man and mouse. *Eur. J. Neurosci.* 23, 3034–3042.

949. Ng, C.S. and Kopp, A. (2008). Sex combs are important for male mating success in *Drosophila melanogaster*. *Behav. Genet.* 38, 195–201.

950. Nguyen, H.Q., Lee, S.D., and Wu, C.-T. (2019). Paircounting. *Trends Genet.* 36, 787–789.

951. Nicholas, F.W. and Hobbs, M. (2013). Mutation discovery for Mendelian traits in non-laboratory animals: a review of achievements up to 2012. *Anim. Genet.* 45, 157–170.

952. Niehuis, O., Hartig, G., Grath, S., Pohl, H., Lehmann, J., Tafer, H., Donath, A., Krauss, V., Eisenhardt, C., Hertel, J., Petersen, M., Mayer, C., Meusemann, K., Peters, R.S., Stadler, P.F., Beutel, R.G., Bornberg-Bauer, E., McKenna, D.D., and

Misof, B. (2012). Genomic and morphological evidence converge to resolve the enigma of Strepsiptera. *Curr. Biol.* 22, 1309–1313.

953. Nijhout, H.F. (1999). Control mechanisms of polyphenic development in insects. *BioScience* 49, 181–192.

954. Nijhout, H.F. (2001). Elements of butterfly wing patterns. *J. Exp. Zool.* 291, 213–225.

955. Nijhout, H.F. (2003). The control of growth. *Development* 130, 5863–5867.

956. Nijhout, H.F. (2003). Development and evolution of adaptive polyphenisms. *Evol. Dev.* 5, 9–18.

957. Nijhout, H.F. (2010). Molecular and physiological basis of colour pattern formation. *Adv. Insect Physiol.* 38, 219–265.

958. Nijhout, H.F. and German, R.Z. (2012). Developmental causes of allometry: new models and implications for phenotypic plasticity and evolution. *Integr. Comp. Biol.* 52, 43–52.

959. Nijhout, H.F. and McKenna, K.Z. (2019). Allometry, scaling, and ontogeny of form: an introduction to the symposium. *Integr. Comp. Biol.* 59, 1275–1280.

960. Nissen, S.B., Perera, M., Gonzalez, J.M., Morgani, S.M., Jensen, M.H., Sneppen, K., Brickman, J.M., and Trusina, A. (2017). Four simple rules that are sufficient to generate the mammalian blastocyst. *PLoS Biol.* 15, e2000737.

961. Nitzan, E., Krispin, S., Pfaltzgraff, E.R., Klar, A., Labosky, P.A., and Kalcheim, C. (2013). A dynamic code of dorsal neural tube genes regulates the segregation between neurogenic and melanogenic neural crest cells. *Development* 140, 2269–2279.

962. Nogare, D.D. and Chitnis, A.B. (2017). Self-organizing spots get under your skin. *PLoS Biol.* 15(12), e2004412.

963. Nonaka, S., Yoshiba, S., Watanabe, D., Ikeuchi, S., Goto, T., Marshall, W.F., and Hamada, H. (2005). De novo formation of left–right asymmetry by posterior tilt of nodal cilia. *PLoS Biol.* 3(8), e268.

964. Noonan, J.P. (2009). Regulatory DNAs and the evolution of human development. *Curr. Opin. Genet. Dev.* 19, 557–564.

965. Nousbeck, J., Burger, B., Fuchs-Telem, D., Pavlovsky, M., Fenig, S., Sarig, O., Itin, P., and Sprecher, E. (2011). A mutation in a skin-specific isoform of *SMARCAD1* causes autosomal-dominant adermatoglyphia. *Am. J. Hum. Genet.* 89, 302–307.

966. Nüsslein-Volhard, C. (2019). *Animal Beauty: On the Evolution of Biological Aesthetics.* MIT Press, Cambridge, MA.

967. Nweeia, M.T., Eichmiller, F.C., Hauschka, P.V., Tyler, E., Mead, J.G., Potter, C.W., Angnatsiak, D.P., Richard, P.R., Orr, J.R., and Black, S.R. (2012). Vestigial tooth anatomy and tusk nomenclature for *Monodon monoceros. Anat. Rec.* 295, 1006–1016.

968. Nyholt, D.R., Gillespie, N.A., Heath, A.C., and Martin, N.G. (2003). Genetic basis of male pattern baldness. *J. Invest. Dermatol.* 121, 1561–1564.

969. O'Brien, S.J. (2004). Cats. *Curr. Biol.* 14, R988–R989.

970. O'Connell, J.E.A. (1976). Craniopagus twins: surgical anatomy and embryology and their implications. *J. Neurol., Neurosurg., and Psychiatry* 39, 1–22.

971. O'Grady, P. and DeSalle, R. (2018). Hawaiian *Drosophila* as an evolutionary model clade: days of future past. *BioEssays* 40, 1700246.

972. Oda, H., Iwasaki-Yokozawa, S., Usui, T., and Akiyama-Oda, Y. (2020). Experimental duplication of bilaterian body axes in spider embryos: Holm's organizer and self-regulation of embryonic fields. *Dev. Genes Evol.* 230, 49–63.

973. Oetting, W.S. (2000). The tyrosinase gene and oculocutaneous albinism Type 1 (OCA1): a model for understanding the molecular biology of melanin formation. *Pigment Cell Res.* 13, 320–325.

974. Okada, T.S. (2004). From embryonic induction to cell lineages: revisiting old problems for modern study. *Int. J. Dev. Biol.* 48, 739–742.

975. Olman, C.A., Bao, P., Engel, S.A., Grant, A.N., Purington, C., Qiu, C., Schallmo, M.-P., and Tjan, B.S. (2018). Hemifield columns co-opt ocular dominance column structure in human achiasma. *NeuroImage* 164, 59–66.

976. Olson, D.J., Oh, D., and Houston, D.W. (2015). The dynamics of plus end polarization and microtubule assembly during *Xenopus* cortical rotation. *Dev. Biol.* 401, 249–263.

977. Olson, M.E. (2012). The developmental renaissance in adaptationism. *Trends Ecol. Evol.* 27, 278–287.

978. Olsson, M., Meadows, J.R.S., Truvé, K., Pielberg, G.R., Puppo, F., Mauceli, E., Quilez, J., Tonomura, N., Zanna, G., Docampo, M.J., Bassols, A., Avery, A.C., Karlsson, E.K., Thomas, A., Kastner, D.L., Bongcam-Rudloff, E., Webster, M.T., Sanchez, A., Hedhammar, A., Remmers, E.F., Andersson, L., Ferrer, L., Tintle, L., and Lindblad-Toh, K. (2011). A novel unstable duplication upstream of *HAS2* predisposes to a breed-defining skin phenotype and a periodic fever syndrome in Chinese Shar-Pei dogs. *PLoS Genet.* 7(3), e1001332.

979. Oppenheim, R.W. (1991). Cell death during development of the nervous system. *Annu. Rev. Neurosci.* 14, 453–501.

980. Orenic, T.V., Held, L.I., Jr., Paddock, S.W., and Carroll, S.B. (1993). The spatial organization of epidermal structures: *hairy* establishes the geometrical pattern of *Drosophila* leg bristles by delimiting the domains of *achaete* expression. *Development* 118, 9–20.

981. Ortolani, A. (1999). Spots, stripes, tail tips and dark eyes: predicting the function of carnivore colour patterns using the comparative method. *Biol. J. Linnean Soc.* 67, 433–476.

982. Ostrander, E.A., Wayne, R.K., Freedman, A.H., and Davis, B.W. (2017). Demographic history, selection and functional diversity of the canine genome. *Nat. Rev. Genet.* 18, 705–720.

983. Othmer, H.G., Painter, K., Umulis, D., and Xue, C. (2009). The intersection of theory and application in elucidating pattern formation in developmental biology. *Math. Model. Nat. Phenom.* 4(4), 3–82.

984. Outters, P., Jaeger, S., Zaarour, N., and Ferrier, P. (2015). Long-range control of V(D) J recombination & allelic exclusion: modeling views. *Adv. Immunol.* 128, 363–413.

985. Pagliara, V., Nasso, R., Ascione, A., Masullo, M., and Arcone, R. (2019). Myostatin and plasticity of skeletal muscle tissue. *J. Hum. Sport Exercise* 14(5proc), S1931–S1937.

986. Palmer, A.R. (2005). Antisymmetry. In *Variation: A Central Concept in Biology*, Hallgrímsson, B. and Hall, B.K., editors. Elsevier Academic Press, New York, pp. 359–397.

987. Palmer, A.R. and Strobeck, C. (1986). Fluctuating asymmetry: measurement, analysis, patterns. *Annu. Rev. Ecol. Syst.* 17, 391–421.

988. Pan, Y., Liu, Z., Shen, J., and Kopan, R. (2005). Notch1 and 2 cooperate in limb ectoderm to receive an early Jagged2 signal regulating interdigital apoptosis. *Dev. Biol.* 286, 472–482.

989. Pan, Y., Tsai, C.-J., Ma, B., and Nussinov, R. (2010). Mechanisms of transcription factor selectivity. *Trends Genet.* 26, 75–83.

990. Pandya, P., Orgaz, J.L., and Sanz-Moreno, V. (2017). Actomyosin contractility and collective migration: may the force be with you. *Curr. Opin. Cell Biol.* 48, 87–96.

991. Papert, S. (1980). *Mindstorms: Children, Computers, and Powerful Ideas.* Basic Books, New York.

992. Parchure, A., Vyas, N., and Mayor, S. (2018). Wnt and Hedgehog: secretion of lipid-modified morphogens. *Trends Cell Biol.* 28, 157–170.

993. Park, K., Kang, J., Subedi, K.P., Ha, J.-H., and Park, C.S. (2008). Canine polydactyl mutations with heterogeneous origin in the conserved intronic sequence of LMBR1. *Genetics* 179, 2163–2172.

994. Parker, H.G., Harris, A., Dreger, D.L., Davis, B.W., and Ostrander, E.A. (2017). The bald and the beautiful: hairlessness in domestic dog breeds. *Philos. Trans. R. Soc. Lond. B* 372, 20150488.

995. Parker, H.G., Shearin, A.L., and Ostrander, E.A. (2010). Man's best friend becomes biology's Best in Show: genome analyses in the domestic dog. *Annu. Rev. Genet.* 44, 309–336.

996. Parker, H.G., VonHoldt, B.M., Quignon, P., Margulies, E.H., Shao, S., Mosher, D.S., Spady, T.C., Elkahloun, A., Cargill, M., Jones, P.G., Maslen, C.L., Acland, G.M., Sutter, N.B., Kuroki, K., Bustamante, C.D., Wayne, R.K., and Ostrander, E.A. (2009). An expressed Fgf4 retrogene is associated with breed-defining chondro-dysplasia in domestic dogs. *Science* 325, 995–998.

997. Parks, A.L., Klueg, K.M., Stout, J.R., and Muskavitch, M.A.T. (2000). Ligand endocytosis drives receptor dissociation and activation in the Notch pathway. *Development* 127, 1373–1385.

998. Parks, H.B. (1936). Cleavage patterns in *Drosophila* and mosaic formation. *Ann. Ent. Soc. Am.* 29, 350–392.

999. Pass, G. (2018). Beyond aerodynamics: the critical roles of the circulatory and tracheal systems in maintaining insect wing functionality. *Arthropod Struct. Dev.* 47, 391–407.

1000. Pastor-Pareja, J.C., Grawe, F., Martín-Blanco, E., and García-Bellido, A. (2004). Invasive cell behavior during *Drosophila* imaginal disc eversion is mediated by the JNK signaling cascade. *Dev. Cell* 7, 387–399.

1001. Patel, S.H., Camargo, F.D., and Yimlamai, D. (2017). Hippo signaling in the liver regulates organ size, cell fate, and carcinogenesis. *Gastroenterology* 152, 533–545.

1002. Patwari, P. and Lee, R.T. (2008). Mechanical control of tissue morphogenesis. *Circ. Res.* 103, 234–243.

1003. Pauciullo, A., Knorr, C., Perucatti, A., Iannuzzi, A., Iannuzzi, L., and Erhardt, G. (2016). Characterization of a very rare case of living ewe–buck hybrid using classical and molecular cytogenetics. *Sci. Rep.* 6, 34781.

1004. Pavan, W.J. and Sturm, R.A. (2019). The genetics of human and hair pigmentation. *Annu. Rev. Genom. Hum. Genet.* 20, 41–72.

1005. Pavlopoulos, A. and Akam, M. (2011). Hox gene *Ultrabithorax* regulates distinct sets of target genes at successive stages of *Drosophila* haltere morphogenesis. *PNAS* 108, 2855–2860.

1006. Pearl, E.J., Li, J., and Green, J.B.A. (2017). Cellular systems for epithelial invagin-ation. *Philos. Trans. R. Soc. Lond. B* 372, 20150526.

1007. Pearson, H. (2007). The roots of accomplishment. *Nature* 446, 20–21.

1008. Pecze, L. (2018). A solution to the problem of proper segment positioning in the course of digit formation. *BioSystems* 173, 266–272.

1009. Pedersen, A.S., Berg, L.C., Almstrup, K., and Thomsen, P.D. (2014). A tortoiseshell male cat: chromosome analysis and histologic examination of the testis. *Cytogenet. Genome Res.* 142, 107–111.

1010. Pener, M.P. and Simpson, S.J. (2009). Locust phase polyphenism: an update. *Adv. Insect Physiol.* 36, 1–272.

1011. Peng, Y. and Axelrod, J.D. (2012). Asymmetric protein localization in planar cell polarity: mechanisms, puzzles, and challenges. *Curr. Top. Dev. Biol.* 101, 33–53.

1012. Pennisi, E. (2018). Buying time. *Science* 362, 988–991.

1013. Penzo-Méndez, A.I. and Stanger, B.Z. (2015). Organ-size regulation in mammals. *Cold Spring Harb. Perspect. Biol.* 7, a019240.

1014. Pereira, G.L., de Matteis, R., Regitano, L.C.A., Chardulo, L.A.L., and Curi, R.A. (2016). MSTN, CKM, and DMRT3 gene variants in different lines of Quarter Horses. *J. Equine Vet. Sci.* 39, 33–37.

1015. Pérez-Gómez, R., Haro, E., Fernández-Guerrero, M., Bastida, M.F., and Ros, M.A. (2018). Role of Hox genes in regulating digit patterning. *Int. J. Dev. Biol.* 62, 797–805.

1016. Perrimon, N., Pitsouli, C., and Shilo, B.-Z. (2012). Signaling mechanisms controlling cell fate and embryonic patterning. *Cold Spring Harb. Perspect. Biol.* 4, a005975.

1017. Peterson, T. and Müller, G.B. (2016). Phenotypic novelty in EvoDevo: the distinction between continuous and discontinuous variation and its importance in evolutionary theory. *Evol. Biol.* 43, 314–335.

1018. Petit, F., Sears, K.E., and Ahituv, N. (2017). Limb development: a paradigm of gene regulation. *Nat. Rev. Genet.* 18, 245–258.

1019. Petrij, F., van Veen, K., Mettler, M., and Brückmann, V. (2001). A second acromelanistic allelomorph at the albino locus of the Mongolian gerbil (*Meriones unguiculatus*). *J. Hered.* 92, 74–78.

1020. Pfennig, D.W. (1992). Polyphenism in spadefoot toad tadpoles as a locally adjusted evolutionary stable strategy. *Evolution* 46, 1408–1420.

1021. Pfennig, D.W. (1999). Cannibalistic tadpoles that pose the greatest threat to kin are most likely to discriminate kin. *Proc. R. Soc. Lond. B* 266, 57–62.

1022. Pfennig, D.W. and Collins, J.P. (1993). Kinship affects morphogenesis in cannibalistic salamanders. *Nature* 362, 836–838.

1023. Pfister, K., Shook, D.R., Chang, C., Keller, R., and Skoglund, P. (2016). Molecular model for force production and transmission during vertebrate gastrulation. *Development* 143, 715–727.

1024. Phillips, R.G., Warner, N.L., and Whittle, J.R.S. (1999). Wingless signaling leads to an asymmetric response to Decapentaplegic-dependent signaling during sense organ patterning on the notum of *Drosophila melanogaster*. *Dev. Biol.* 207, 150–162.

1025. Plonka, P.M., Passeron, T., Brenner, M., Tobin, D.J., Shibahara, S., Thomas, A., Slominski, A., Kadekaro, A.L., Hershkovitz, D., Peters, E., Nordlund, J.J., Abdel-Malek, Z., Takeda, K., Paus, R., Ortonne, J.P., Hearing, V.J., and Schalleruter, K.U. (2009). What are melanocytes really doing all day long...? *Exp. Dermatol.* 18, 799–819.

1026. Plouffe, S.W., Hong, A.W., and Guan, K.-L. (2015). Disease implications of the Hippo/YAP pathway. *Trends Mol. Med.* 21, 212–222.

1027. Poodry, C.A. (1975). A temporal pattern in the development of sensory bristles in *Drosophila*. *W. Roux Arch. Dev. Biol.* 178, 203–213.

1028. Poodry, C.A. and Schneiderman, H.A. (1970). The ultrastructure of the developing leg of *Drosophila melanogaster*. *W. Roux Arch. Dev. Biol.* 166, 1–44.

1029. Poodry, C.A., Hall, L., and Suzuki, D.T. (1973). Developmental properties of *shibir-e^{ts}*: a pleiotropic mutation affecting larval and adult locomotion and development. *Dev. Biol.* 32, 373–386.

1030. Portmann, A. (1967). *Animal Forms and Patterns: A Study of the Appearance of Animals*. Schocken Books, New York.

1031. Posakony, J.W. (1994). Nature versus nurture: asymmetric cell divisions in *Drosophila* bristle development. *Cell* 76, 415–418.

1032. Price, J. and Allen, S. (2004). Exploring the mechanisms regulating regeneration of deer antlers. *Philos. Trans. R. Soc. Lond. B* 359, 809–822.

1033. Price, J.S., Oyajobi, B.O., Oreffo, R.O., and Russell, R.G. (1994). Cells cultured from the growing tip of red deer antler express alkaline phosphatase and proliferate in response to insulin-like growth factor-I. *J. Endocrinol.* 143, R9–R16.

1034. Prieur, D.S. and Rebsam, A. (2017). Retinal axon guidance at the midline: chiasmatic misrouting and consequences. *Dev. Neurobiol.* 77, 844–860.

1035. Pringle, J.W.S. (1948). The gyroscopic mechanism of the halteres of diptera. *Philos. Trans. Roy. Soc. Lond. B* 233, 347–384.

1036. Projecto-Garcia, J., Biddle, J.F., and Ragsdale, E.J. (2017). Decoding the architecture and origins of mechanisms for developmental polyphenism. *Curr. Opin. Genet. Dev.* 47, 1–8.

1037. Protas, M.E. and Patel, N.H. (2008). Evolution of color patterns. *Annu. Rev. Cell Dev. Biol.* 24, 425–446.

1038. Pruvost, M., Bellone, R., Benecke, N., Sandoval-Castellanos, E., Cieslak, M., Kuznetsova, T., Morales-Muñiz, A., O'Connor, T., Reissmann, M., Hofreiter, M., and Ludwig, A. (2011). Genotypes of predomestic horses match phenotypes painted in Paleolithic works of cave art. *PNAS* 108, 18626–18630.

1039. Purcell, R. (1997). *Special Cases: Natural Anomalies and Historical Monsters*. Chronicle Books, San Francisco, CA.

1040. Purves, D., Riddle, D.R., and LaMantia, A.-S. (1992). Iterated patterns of brain circuitry (or how the cortex gets its spots). *Trends Neurosci.* 15, 362–368.

1041. Quammen, D. (2014). People of the horse. *Nat. Geogr.* 225(3), 104–126.

1042. Rabah, S., Salati, S., and Wani, S. (2008). Mirror hand deformity: a rare congenital anomaly of the upper limb. *Internet J. Surg.* 21(1), 1–5.

1043. Ramain, P., Khechumian, R., Khechumian, K., Arbogast, N., Ackermann, C., and Heitzler, P. (2000). Interactions between Chip and the Achaete/Scute-Daughterless heterodimers are required for Pannier-driven proneural patterning. *Mol. Cell* 6, 781–790.

1044. Ramsden, C.A., Bankier, A., Brown, T.J., Cowen, P.S.J., Frost, G.I., McCallum, D.D., Studdert, V.P., and Fraser, J.R.E. (2000). A new disorder of hyaluronan metabolism associated with generalized folding and thickening of the skin. *J. Pediatr.* 136, 62–68.

1045. Randall, V.A. (2007). Hormonal regulation of hair follicles exhibits a biological paradox. *Semin. Cell Dev. Biol.* 18, 274–285.

1046. Raser, J.M. and O'Shea, E.K. (2005). Noise in gene expression: origins, consequences, and control. *Science* 309, 2010–2013.

1047. Raspopovic, J., Marcon, L., Russo, L., and Sharpe, J. (2014). Digit patterning is controlled by a Bmp-Sox9-Wnt Turing network modulated by morphogen gradients. *Science* 345, 566–570.

1048. Rasskin-Gutman, D. and De Renzi, M., eds. (2009). *Pere Alberch: The Creative Trajectory of an Evo-Devo Biologist*. Universitat de València, València.

1049. Rauskolb, C. (2001). The establishment of segmentation in the *Drosophila* leg. *Development* 128, 4511–4521.

1050. Rauskolb, C. and Irvine, K.D. (1999). Notch-mediated segmentation and growth control of the *Drosophila* leg. *Dev. Biol.* 210, 339–350.

1051. Rebeiz, M., Miller, S.W., and Posakony, J.W. (2011). Notch regulates *numb*: integration of conditional and autonomous cell fate specification. *Development* 138, 215–225.

1052. Red-Horse, K. and Siekmann, A.F. (2019). Veins and arteries build hierarchical branching patterns differently: bottom-up versus top-down. *BioEssays* 41, e1800198.

1053. Reddy, G.V. and Rodrigues, V. (1999). A glial cell arises from an additional division within the mechanosensory lineage during development of the microchaete on the *Drosophila* notum. *Development* 126, 4617–4622.

1054. Reh, T.A. and Constantine-Paton, M. (1985). Eye-specific segregation requires neural activity in three-eyes *Rana pipiens*. *J. Neurosci.* 5, 1132–1143.

1055. Reik, E.F. (1976). Four-winged diptera from the upper Permian of Australia. *Proc. Linn. Soc. New South Wales* 101(4), 250–255.

1056. Reilly, P.R. (2008). *The Strongest Boy in the World: How Genetic Information Is Reshaping Our Lives*. Cold Spring Harbor Laboratory Press, Cold Spring Harbor, NY.

1057. Reiss, J.O., Burke, A.C., Archer, C., De Renzi, M., Dopazo, H., Etxeberría, A., Gale, E.A., Hinchliffe, J.R., de la Rosa Garcia, L.N., Rose, C.S., Rasskin-Gutman, D., and Müller, G.B. (2009). Pere Alberch: originator of EvoDevo. *Biol. Theory* 3, 351–356.

1058. Reiter, F., Wienerroither, S., and Stark, A. (2017). Combinatorial function of transcription factors and cofactors. *Curr. Opin. Genet. Dev.* 43, 73–81.

1059. Reno, P.L., Kjosness, K.M., and Hines, J.E. (2016). The role of Hox in pisiform and calcaneus growth plate formation and the nature of the zeugopod/autopod boundary. *J. Exp. Zool. B Mol. Dev. Evol.* 326, 303–321.

1060. Rensberger, B. (1998). *Life Itself: Exploring the Realm of the Living Cell*. Oxford University Press, New York.

1061. Ressurreição, M., Warrington, S., and Strutt, D. (2018). Rapid disruption of Dishevelled activity uncovers an intercellular role in maintenance of Prickle in core planar polarity protein complexes. *Cell Rep.* 25, 1415–1424.

1062. Restrepo, S., Zartman, J.J., and Basler, K. (2014). Coordination of patterning and growth by the morphogen DPP. *Curr. Biol.* 24, R245–R255.

1063. Rhyu, M.S., Jan, L.Y., and Jan, Y.N. (1994). Asymmetric distribution of numb protein during division of the sensory organ precursor cell confers distinct fates to daughter cells. *Cell* 76, 477–491.

1064. Rice, D.S., Goldowitz, D., Williams, R.W., Hamre, K., Johnson, P.T., Tan, S.-S., and Reese, B.E. (1999). Extrinsic modulation of reginal ganglion cell projections: analysis of the albino mutation in pigmentation mosaic mice. *Dev. Biol.* 216, 41–56.

1065. Rice, G.R., Barmina, O., Luecke, D., Hu, K., Arbeitman, M., and Kopp, A. (2019). Modular tissue-specific regulation of *doublesex* underpins sexually dimorphic development in *Drosophila*. *Development* 146, dev178285.

1066. Rice, S.H. (2002). The role of heterochrony in primate brain evolution. In *Human Evolution Through Developmental Change*, Minugh-Purvis, N. and McNamara, K.J., editors. Johns Hopkins University Press, Baltimore, MD, pp. 154–170.

1067. Richardson, J. and Simpson, P. (2006). A conserved *trans*-regulatory landscape for *scute* expression on the notum of cyclorrhaphous Diptera. *Dev. Genes Evol.* 216, 29–38.

1068. Richardson, M.K., Gobes, S.M.H., van Leeuwen, A.C., Polman, J.A.E., Pieau, C., and Sánchez-Villagra, M.R. (2009). Heterochrony in limb evolution: developmental mechanisms and natural selection. *J. Exp. Zool. B Mol. Dev. Evol.* 312, 639–664.

1069. Richelle, J. and Ghysen, A. (1979). Determination of sensory bristles and pattern formation in *Drosophila*. I. A model. *Dev. Biol.* 70, 418–437.

1070. Rieppel, O. (2001). Turtles as hopeful monsters. *BioEssays* 23, 987–991.

1071. Rimbault, M., Beale, H.C., Schoenebeck, J.J., Hoopes, B.C., Allen, J.J., Kilroy-Glynn, P., Wayne, R.K., Sutter, N.B., and Ostrander, E.A. (2013). Derived variants at six genes explain nearly half of size reduction in dog breeds. *Genome Res.* 23, 1985–1995.

1072. Robinson, R. (1959). Genetic studies of the Syrian hamster. II. Partial albinism. *Heredity* 13, 165–177.

1073. Rodríguez-Carballo, E., Lopez-Delisle, L., Zhan, Y., Fabre, P.J., Beccari, L., El-Idrissi, I., Huynh, T.H.N., Ozadam, H., Dekker, J., and Duboule, D. (2017). The HoxD cluster is a dynamic and resilient TAD boundary controlling the segregation of antagonistic regulatory landscapes. *Genes Dev.* 31, 2264–2281.

1074. Rodriguez-Estaban, C., Schwabe, J.W.R., de la Peña, J., Foys, B., Eshelman, B., and Izpisua Belmonte, J.C. (1997). *Radical fringe* positions the apical ectodermal ridge at the dorsoventral boundary of the vertebrate limb. *Nature* 386, 360–366.

1075. Rogers, G.E. (2004). Hair follicle differentiation and regulation. *Int. J. Dev. Biol.* 48, 163–170.

1076. Rolian, C. (2014). Genes, development, and evolvability in primate evolution. *Evol. Anthrop.* 23, 93–104.

1077. Rolian, C. (2020). Endochondral ossification and the evolution of limb proportions. *Wiley Interdiscip. Rev. Dev. Biol.* 2020, e373.

1078. Rollo, C.D. (1995). *Phenotypes: Their Epigenetics, Ecology and Evolution*. Chapman & Hall, New York.

1079. Romani, S., Campuzano, S., Macagno, E.R., and Modolell, J. (1989). Expression of *achaete* and *scute* genes in *Drosophila* imaginal discs and their function in sensory organ development. *Genes Dev.* 3, 997–1007.

1080. Romero, D.M., Bahi-Buisson, N., and Francis, F. (2018). Genetics and mechanisms leading to human cortical malformations. *Semin. Cell Dev. Biol.* 76, 33–75.

1081. Rongioletti, F., Merlo, G.R., Cinotti, E., Fausti, V., Cozzani, E., Cribier, B., Metze, D., Calonje, E., Kanitakis, J., Kempf, W., Stefanato, C.M., Marinho, E., and Parodi, A. (2013). Scleromyxedema: a multicenter study of characteristics, comorbidities, course, and therapy in 30 patients. *J. Am. Acad. Dermatol.* 69, 66–72.

1082. Rooney, M.F., Hill, E.W., Kelly, V.P., and Porter, R.K. (2018). The "speed gene" effect of myostatin arises in Thoroughbred horses due to a promoter proximal SINE insertion. *PLoS ONE* 13(10), e0205664.

1083. Rørth, P. (2012). Fellow travellers: emergent properties of collective cell migration. *EMBO Rep.* 13(11), 984–991.

1084. Roselló-Díez, A., Arques, C.G., Delgado, I., Giovinazzo, G., and Torres, M. (2014). Diffusible signals and epigenetic timing cooperate in late proximo-distal limb patterning. *Development* 141, 1534–1543.

1085. Rosenberger, A.L. and Preuschoft, H. (2012). Evolutionary morphology, cranial biomechanics and the origins of tarsiers and anthropoids. *Palaeobiodivers. Palaeoenviron.* 92, 507–525.

1086. Ross, C.M. (1969). Generalized folded skin with an underlying lipomatous nevus: "the Michelin Tire Baby". *Arch. Derm.* 100, 320–323.

1087. Ross, C.M. (1972). Generalized folded skin with an underlying lipomatous nevus: the Michelin Tire Baby. *Arch. Derm.* 106, 766.

1088. Roth, G. and Dicke, U. (2019). Origin and evolution of human cognition. *Progr. Brain Res.* 250, 285–316.

1089. Roy, S., Shashidhara, L.S., and VijayRaghavan, K. (1997). Muscles in the *Drosophila* second thoracic segment are patterned independently of autonomous homeotic gene function. *Curr. Biol.* 7, 222–227.

1090. Rozowski, M. (2002). Establishing character correspondence for sensory organ traits in flies: sensory organ development provides insights for reconstructing character evolution. *Mol. Phylogenet. Evol.* 24, 400–411.

1091. Rozowski, M. and Akam, M. (2002). Hox gene control of segment-specific bristle patterns in *Drosophila*. *Genes Dev.* 16, 1150–1162.

1092. Rudel, D. and Sommer, R.J. (2003). The evolution of developmental mechanisms. *Dev. Biol.* 264, 15–37.

1093. Ruiz-Losada, M., Blom-Dahl, D., Córdoba, S., and Estella, C. (2018). Specification and patterning of *Drosophila* appendages. *J. Dev. Biol.* 6, jdb6030017.

1094. Russell, L.B., ed. (1978). *Genetic Mosaics and Chimeras in Mammals*. Plenum, New York.

1095. Russell, M.A. (1974). Pattern formation in the imaginal discs of a temperature-sensitive cell-lethal mutant of *Drosophila melanogaster*. *Dev. Biol.* 40, 24–39.

1096. Russell, M.A., Girton, J.R., and Morgan, K. (1977). Pattern formation in a ts-cell-lethal mutant of *Drosophila*: the range of phenotypes induced by larval heat treatments. *W. Roux Arch. Dev. Biol.* 183, 41–59.

1097. Rusting, R.L. (2001). Hair: why it grows, why it stops. *Sci. Am.* 284(6), 70–79.

1098. Ruvinsky, I. and Gibson-Brown, J.J. (2000). Genetic and developmental bases of serial homology in vertebrate limb evolution. *Development* 127, 5233–5244.

1099. Sabarís, G., Laiker, I., Noon, E.P.-B., and Frankel, N. (2019). Actors with multiple roles: pleiotropic enhancers and the paradigm of enhancer modularity. *Trends Genet.* 35, 423–433.

1100. Saha, M. (1991). Spemann seen through a lens. In *A Conceptual History of Modern Embryology*, Gilbert, S.F., editor. Plenum, New York, pp. 91–108.

1101. Saito, K., Nomura, S., Yamamoto, S., Niiyama, R., and Okabe, Y. (2017). Investigation of hindwing folding in ladybird beetles by artificial elytron transplantation and microcomputed tomography. *PNAS* 114, 5624–5628.

1102. Saito, K., Yamamoto, S., Maruyama, M., and Okabe, Y. (2014). Asymmetric hindwing foldings in rove beetles. *PNAS* 111, 16349–16352.

1103. Saiz-Lopez, P., Chinnaiya, K., Campa, V.M., Delgado, I., Ros, M.A., and Towers, M. (2015). An intrinsic timer specifies distal structures of the vertebrate limb. *Nat. Commun.* 6, 8108.

1104. Sakai, M. (2008). Cell-autonomous and inductive processes among three embryonic domains control dorsal–ventral and anterior–posterior development of *Xenopus laevis*. *Dev. Growth Differ.* 50, 49–62.

1105. Salazar-Ciudad, I., Jernvall, J., and Newman, S.A. (2003). Mechanisms of pattern formation in development and evolution. *Development* 130, 2027–2037.

1106. Salser, S.J. and Kenyon, C. (1996). A *C. elegans* Hox gene switches on, off, on and off again to regulate proliferation, differentiation and morphogenesis. *Development* 122, 1651–1661.

1107. Samlaska, C.P., James, W.D., and Sperling, L.C. (1989). Scalp whorls. *J. Am. Acad. Dermatol.* 21, 553–556.

1108. Sánchez-Villagra, M.R. and Menke, P.R. (2005). The mole's thumb: evolution of the hand skeleton in talpids (Mammalia). *Zoology* 108, 3–12.

1109. Sander, K. and Faessler, P.E. (2001). Introducing the Spemann–Mangold organizer: experiments and insights that generated a key concept in developmental biology. *Int. J. Dev. Biol.* 45, 1–11.

1110. Sanicola, M., Sekelsky, J., Elson, S., and Gelbart, W.M. (1995). Drawing a stripe in *Drosophila* imaginal disks: negative regulation of *decapentaplegic* and *patched* expression by *engrailed*. *Genetics* 139, 745–756.

1111. Santamaría, P. (1979). Heat shock induced phenocopies of dominant mutants of the Bithorax Complex in *Drosophila melanogaster*. *Mol. Gen. Genet.* 172, 161–163.

1112. Santana, S.E., Alfaro, J.L., and Alfaro, M.E. (2012). Adaptive evolution of facial colour patterns in Neotropical primates. *Proc. R. Soc. B* 279, 2204–2211.

1113. Sarin, K.Y. and Artandi, S.E. (2007). Aging, graying and loss of melanocyte stem cells. *Stem Cell Rev.* 3, 212–217.

1114. Sarnat, H.B. and Netsky, M.G. (1981). *Evolution of the Nervous System.* Oxford University Press, New York.

1115. Sawada, R., Aramaki, T., and Kondo, S. (2018). Flexibility of pigment cell behavior permits the robustness of skin pattern formation. *Genes Cells* 23, 537–545.

1116. Saxena, A., Towers, M., and Cooper, K.L. (2017). The origins, scaling and loss of tetrapod digits. *Philos. Trans. R. Soc. Lond. B* 372, 20150482.

1117. Scheibert, J., Leurent, S., Prevost, A., and Debrégeas, G. (2009). The role of fingerprints in the coding of tactile information probed with a biomimetic sensor. *Science* 323, 1503–1506.

1118. Schmidt-Küntzel, A., Eizirik, E., O'Brien, S.J., and Menotti-Raymond, M. (2005). *Tyrosinase* and *Tyrosinase Related Protein 1* alleles specify domestic cat coat color phenotypes of the *albino* and *brown* loci. *J. Hered.* 96, 289–301.

1119. Schmidt-Küntzel, A., Nelson, G., David, V.A., Schäffer, A.A., Eizirik, E., Roelke, M.E., Kehler, J.S., Hannah, S.S., O'Brien, S.J., and Menotti-Raymond, M. (2009). A domestic cat X chromosome linkage map and the sex-linked *orange* locus: mapping of *orange*, multiple origins and epistasis over *nonagouti*. *Genetics* 181, 1415–1425.

1120. Schmidt-Ullrich, R. and Paus, R. (2005). Molecular principles of hair follicle induction and morphogenesis. *BioEssays* 27, 247–261.

1121. Schneider, I., Kreis, J., Schweickert, A., Blum, M., and Vick, P. (2019). A dual function of FGF signaling in *Xenopus* left–right axis formation. *Development* 146, dev173575.

1122. Schneider, M.R., Schmidt-Ullrich, R., and Paus, R. (2009). The hair follicle as a dynamic miniorgan. *Curr. Biol.* 19, R132–R142.

1123. Schöck, F. and Perrimon, N. (2002). Molecular mechanisms of epithelial morphogenesis. *Annu. Rev. Cell Dev. Biol.* 18, 463–493.

1124. Schoenebeck, J.J. and Ostrander, E.A. (2014). Insights into morphology and disease from the dog genome project. *Annu. Rev. Cell Dev. Biol.* 30, 535–560.

1125. Schoenebeck, J.J., Hutchinson, S.A., Byers, A., Beale, H.C., Carrington, B., Faden, D.L., Rimbault, M., Decker, B., Kidd, J.M., Sood, R., Boyko, A.R., Fondon, J.W., III, Wayne, R.K., Bustamante, C.D., Ciruna, B., and Ostrander, E.A. (2012). Variation of *BMP3* contributes to dog breed skull diversity. *PLoS Genet.* 8(8), e1002849.

1126. Schreiber, A.M., Cai, L., and Brown, D.D. (2005). Remodeling of the intestine during metamorphosis of *Xenopus laevis*. *PNAS* 102, 3720–3725.

1127. Schroeder, T.B.H., Houghtaling, J., Wilts, B.D., and Mayer, M. (2018). It's not a bug, it's a feature: functional materials in insects. *Adv. Mater.* 30, 1705322.

1128. Schubiger, G., Schubiger, M., and Sustar, A. (2012). The three leg imaginal discs of *Drosophila*: "vive la différence". *Dev. Biol.* 369, 76–90.

1129. Schuelke, M., Wagner, K.R., Stolz, L.E., Hübner, C., Riebel, T., Kömen, W., Braun, T., Tobin, J.F., and Lee, S.-J. (2004). Myostatin mutation associated with gross muscle hypertrophy in a child. *N. Engl. J. Med.* 350, 2682–2688.

1130. Schwarzer, W. and Spitz, F. (2014). The architecture of gene expression: integrating dispersed *cis*-regulatory modules into coherent regulatory domains. *Curr. Opin. Genet. Dev.* 27, 74–82.

1131. Schweisguth, F. and Corson, F. (2019). Self-organization in pattern formation. *Dev. Cell* 49, 659–677.

1132. Seher, T.D., Ng, C.S., Signor, S.A., Podlaha, O., Barmina, O., and Kopp, A. (2012). Genetic basis of a violation of Dollo's Law: re-evolution of rotating sex combs in *Drosophila bipectinata*. *Genetics* 192, 1465–1475.

1133. Sengpiel, F. (2008). Binocular vision: only half a brain needed. *Curr. Biol.* 18, R1054–R1056.

1134. Sessions, S.K. and Ruth, S.B. (1990). Explanation for naturally occurring supernumerary limbs in amphibians. *J. Exp. Zool.* 254, 38–47.

1135. Sessions, S.K., Franssen, R.A., and Horner, V.L. (1999). Morphological clues from multilegged frogs: are retinoids to blame? *Science* 284, 800–802.

1136. Shao, X., Ding, Z., Zhao, M., Liu, K., Sun, H., Chen, J., Liu, X., Zhang, Y., Hong, Y., Li, H., and Li, H. (2017). Mammalian Numb protein antagonizes Notch by controlling postendocytic trafficking of the Notch ligand Delta-like 4. *J. Biol. Chem.* 292, 20628–20643.

1137. Sharma, M., Castro-Piedras, I., Simmons, G.E., Jr., and Pruitt, K. (2018). Dishevelled: a masterful conductor of complex Wnt signals. *Cell. Signal.* 47, 52–64.

1138. Shashidhara, L.S., Agrawal, N., Bajpai, R., Bharathi, V., and Sinha, P. (1999). Negative regulation of dorsoventral signaling by the homeotic gene *Ultrabithorax* during haltere development in *Drosophila*. *Dev. Biol.* 212, 491–502.

1139. Shellenbarger, D.L. and Mohler, J.D. (1978). Temperature-sensitive periods and autonomy of pleiotropic effects of $l(1)N^{ts1}$, a conditional Notch lethal in *Drosophila*. *Dev. Biol.* 62, 432–446.

1140. Shelton, P.M.J., Truby, P.R., and Shelton, R.G.J. (1981). Naturally occurring abnormalities (Bruchdreifachbildungen) in the chelae of three species of Crustacea (Decapoda) and a possible explanation. *J. Embryol. Exp. Morphol.* 63, 285–304.

1141. Sherwood, C.C. and Gómez-Robles, A. (2017). Brain plasticity and human evolution. *Annu. Rev. Anthrop.* 46, 399–419.

1142. Sherwood, C.C., Bauernfeind, R., Bianchi, S., Raghanti, M.A., and Hof, P.R. (2012). Human brain evolution writ large and small. *Progr. Brain Res.* 195, 237–254.

1143. Sheth, R., Marcon, L., Bastida, M.F., Junco, M., Quintana, L., Dahn, R., Kmita, M., Sharpe, J., and Ros, M.A. (2012). *Hox* genes regulate digit patterning by controlling the wavelength of a Turing-type mechanism. *Science* 338, 1476–1480.

1144. Shi, Y., Barton, K., De Maria, A., Petrash, J.M., Shiels, A., and Bassnett, S. (2009). The stratified syncytium of the vertebrate lens. *J. Cell Sci.* 122, 1607–1615.

1145. Shi, Y., De Maria, A., Lubura, S., Sikic, H., and Bassnett, S. (2015). The penny pusher: a cellular model of lens growth. *Invest. Ophthalmol. Vis. Sci.* 56, 799–809.

1146. Shimizu-Nishikawa, K., Takahashi, J., and Nishikawa, A. (2003). Intercalary and supernumerary regeneration in the limbs of the frog, *Xenopus laevis. Dev. Dynamics* 227, 563–572.

1147. Shimomura, Y., Agalliu, D., Vonica, A., Luria, V., Wajid, M., Baumer, A., Belli, S., Petukhova, L., Schinzel, A., Brivanlou, A.H., Barres, B.A., and Christiano, A.M. (2010). APCDD1 is a novel Wnt inhibitor mutated in hereditary hypotrichosis simplex. *Nature* 464, 1043–1047.

1148. Shindo, A. (2018). Models of convergent extension during morphogenesis. *Wiley Interdiscip. Rev. Dev. Biol.* 7, e293.

1149. Shindo, A., Inoue, Y., Kinoshita, M., and Wallingford, J.B. (2019). PCP-dependent transcellular regulation of actomyosin oscillation facilitates convergent extension of vertebrate tissue. *Dev. Biol.* 446, 159–167.

1150. Shingleton, A.W. and Frankino, W.A. (2013). New perspectives on the evolution of exaggerated traits. *BioEssays* 35, 100–107.

1151. Shingleton, A.W. and Frankino, W.A. (2018). The (ongoing) problem of relative growth. *Curr. Opin. Insect Sci.* 25, 9–19.

1152. Shingleton, A.W., Frankino, W.A., Flatt, T., Nijhout, H.F., and Emlen, D.J. (2007). Size and shape: the developmental regulation of static allometry in insects. *BioEssays* 29, 536–548.

1153. Shiota, K., Yamada, S., Komada, M., and Ishibashi, M. (2007). Embryogenesis of holoprosencephaly. *Am. J. Med. Genet. A* 143A, 3079–3087.

1154. Shirai, T., Yorimitsu, T., Kiritooshi, N., Matsuzaki, F., and Nakagoshi, H. (2007). Notch signaling relieves the joint-suppressive activity of Defective proventriculus in the *Drosophila* leg. *Dev. Biol.* 312, 147–156.

1155. Shoji, H. and Iwasa, Y. (2005). Labyrinthine versus stright-striped patterns generated by two-dimensional Turing systems. *J. Theor. Biol.* 237, 104–116.

1156. Shoji, H., Iwasa, Y., and Kondo, S. (2003). Stripes, spots, or reversed spots in two-dimensional Turing systems. *J. Theor. Biol.* 224, 339–350.

1157. Sholtis, S.J. and Noonan, J.P. (2010). Gene regulation and the origins of human biological uniqueness. *Trends Genet.* 26, 110–118.

1158. Shorrocks, B. and Croft, D.P. (2009). Necks and networks: a preliminary study of population structure in the reticulated giraffe (*Giraffa camelopardalis reticulata* de Winston). *Afr. J. Ecol.* 47, 374–381.

1159. Shroff, S., Joshi, M., and Orenic, T.V. (2007). Differential Delta expression underlies the diversity of sensory organ patterns among the legs of the *Drosophila* adult. *Mech. Dev.* 124, 43–58.

1160. Sick, S., Reinker, S., Timmer, J., and Schlake, T. (2006). WNT and DKK determine hair follicle spacing through a reaction–diffusion mechanism. *Science* 314, 1447–1450.

1161. Siebel, C. and Lendahl, U. (2017). Notch signaling in development, tissue homeostasis, and disease. *Physiol. Rev.* 97, 1235–1294.

1162. Simpson, P. (1997). Notch signalling in development: on equivalence groups and asymmetric developmental potential. *Curr. Op. Genet. Dev.* 7, 537–542.

1163. Simpson, P. (2007). The stars and stripes of animal bodies: evolution of regulatory elements mediating pigment and bristle patterns in *Drosophila. Trends Genet.* 23, 350–358.

1164. Simpson, P. and Marcellini, S. (2006). The origin and evolution of stereotyped patterns of macrochaetes on the nota of cyclorraphous Diptera. *Heredity* 97, 148–156.

1165. Sinclair, R. (1998). Male pattern androgenetic alopecia. *BMJ* 317, 865–869.

1166. Singh, A., Gupta, R., Zaidi, S.H.H., and Singh, A. (2016). Dermatoglyphics: a brief review. *Int. J. Adv. Integr. Med. Sci.* 1, 111–115.

1167. Singh, A., Tare, M., Puli, O.R., and Kango-Singh, M. (2011). A glimpse into dorso-ventral patterning of the *Drosophila* eye. *Dev. Dynamics* 241, 69–84.

1168. Sjöqvist, M. and Andersson, E.R. (2019). Do as I say, Not(ch) as I do: lateral control of cell fate. *Dev. Biol.* 447, 58–70.

1169. Skaer, N. and Simpson, P. (2000). Genetic analysis of bristle loss in hybrids between *Drosophila melanogaster* and *D. simulans* provides evidence for divergence of *cis*-regulatory sequences in the *achaete-scute* gene complex. *Dev. Biol.* 221, 148–167.

1170. Skaer, N., Pistillo, D., and Simpson, P. (2002). Transcriptional heterochrony of *scute* and changes in bristle pattern between two closely related species of blowfly. *Dev. Biol.* 252, 31–45.

1171. Skeath, J.B. and Carroll, S.B. (1991). Regulation of *achaete-scute* gene expression and sensory organ pattern formation in the *Drosophila* wing. *Genes Dev.* 5, 984–995.

1172. Skoglund, P. and Keller, R. (2010). Integration of planar cell polarity and ECM signaling in elongation of the vertebrate body plan. *Curr. Opin. Cell Biol.* 22, 589–596.

1173. Skulachev, V.P., Holtze, S., Vyssokikh, M.Y., Bakeeva, L.E., Skulachev, M.V., Markov, A.V., Hildebrandt, T.B., and Sandovnichii, V.A. (2017). Neoteny, prolongation of youth: from naked mole rats to "naked apes" (humans). *Physiol. Rev.* 97, 699–720.

1174. Smaers, J.B., Gómez-Robles, A., Parks, A.N., and Sherwood, C.C. (2017). Exceptional evolutionary expansion of prefrontal cortex in great apes and humans. *Curr. Biol.* 27, 714–720.

1175. Smaers, J.B., Mongle, C.S., Safi, K., and Dechmann, D.K.N. (2019). Allometry, evolution and development of neocortex size in mammals. *Progr. Brain Res.* 250, 83–107.

1176. Smith, D.J., Montenegro-Johnson, T.D., and Lopes, S.S. (2019). Symmetry-breaking cilia-driven flow in embryogenesis. *Ann. Rev. Fluid Mech.* 51, 105–128.

1177. Smith, K.K. (2003). Time's arrow: heterochrony and the evolution of development. *Int. J. Dev. Biol.* 47, 613–621.

1178. Smith-Bolton, R.K., Worley, M.I., Kanda, H., and Hariharan, I.K. (2009). Regenerative growth in *Drosophila* imaginal discs is regulated by Wingless and Myc. *Dev. Cell* 16, 797–809.

1179. Soder, A.I., Hoare, S.F., Muire, S., Balmain, A., Parkinson, E.K., and Keith, W.N. (1997). Mapping of the gene for the mouse telomerase RNA component, Terc, to chromosome 3 by fluorescence *in situ* hybridization and mouse chromosome painting. *Genomics* 41, 293–294.

1180. Soler, C., Daczewska, M., Da Ponte, J.P., Dastugue, B., and Jagla, K. (2004). Coordinated development of muscles and tendons of the *Drosophila* leg. *Development* 131, 6041–6051.

1181. Solnica-Krezel, L. and Sepich, D.S. (2012). Gastrulation: making and shaping germ layers. *Annu. Rev. Cell Dev. Biol.* 28, 687–717.

1182. Sommer, R.J. (2020). Phenotypic plasticity: from theory and genetics to current and future challenges. *Genetics* 215, 1–13.

1183. Souder, W. (2000). *A Plague of Frogs: Unraveling an Environmental Mystery*. University of Minnesota Press, Minneapolis, MN.

1184. Soukup, V., Horácek, I., and Cerny, R. (2013). Development and evolution of the vertebrate primary mouth. *J. Anat.* 222, 79–99.

1185. Sousa, A.M.M., Meyer, K.A., Santpere, G., Gulden, F.O., and Sestan, N. (2017). Evolution of the human nervous system function, structure, and development. *Cell* 170, 226–247.

1186. Spéder, P. and Noselli, S. (2007). Left–right asymmetry: class I myosins show the direction. *Curr. Opin. Cell Biol.* 19, 82–87.

1187. Spemann, H. (1938). *Embryonic Development and Induction*. Yale University Press, New Haven, CT.

1188. Spemann, H. and Falkenberg, H. (1919). Über asymmetrische Entwicklung und Situs inversus viscerum bei Zwillingen und Doppelbildungen. *Archiv. für Entwicklungsmechanik* 45, 371–422.

1189. Spencer, R. (2000). Craniopagus conjoined twins: typical, parasitic, and intracranial fetus-in-fetu. *Neurosurg. Quart.* 10, 60–79.

1190. Spencer, R. (2000). Theoretical and analytical embryology of conjoined twins. Part I. Embryogenesis. *Clin. Anat.* 13, 36–53.

1191. Spencer, R. (2000). Theoretical and analytical embryology of conjoined twins. Part II. Adjustments to union. *Clin. Anat.* 13, 97–120.

1192. Spencer, R. (2003). *Conjoined Twins: Developmental Malformations and Clinical Implications*. Johns Hopkins University Press, Baltimore, MD.

1193. Sponenberg, D.P., Carr, G., Simak, E., and Schwink, K. (1990). The inheritance of the leopard complex of spotting patterns in horses. *J. Hered.* 81, 323–331.

1194. St. Johnston, D. (2015). The renaissance of developmental biology. *PLoS Biol.* 13, e1002149.

1195. Stahl, A.L., Charlton-Perkins, M., Buschbeck, E.K., and Cook, T.A. (2017). The cuticular nature of corneal lenses in *Drosophila melanogaster*. *Dev. Genes Evol.* 227, 271–278.

1196. Stanford, P.K. (2005). August Weismann's theory of the germ-plasm and the problem of unconceived alternatives. *Hist. Philos. Life Sci.* 27, 163–199.

1197. Steed, L. (2020). Kitty scary: wrinkly sphynx cat has stolen hearts thanks to his unique appearance. *The Sun* (9 March 2020).

1198. Steinberg, M.S. (2007). Differential adhesion in morphogenesis: a modern view. *Curr. Opin. Genet. Dev.* 17, 281–286.

1199. Steiner, E. (1976). Establishment of compartments in the developing leg imaginal discs of *Drosophila melanogaster*. *W. Roux Arch. Dev. Biol.* 180, 9–30.

1200. Stent, G.S. (1978). *Paradoxes of Progress*. W. H. Freeman, San Francisco, CA.

1201. Stern, C. (1954). Genes and developmental patterns. *Caryologia* (suppl: Proc. 9th Int. Congr. Genet. Part I), 355–369.

1202. Stern, C. (1954). Two or three bristles. *Am. Sci.* 42, 213–247.

1203. Stern, C. (1968). *Genetic Mosaics and Other Essays*. Harvard University Press, Cambridge, MA.

1204. Stern, C. (1969). Richard Benedict Goldschmidt: April 12, 1878 – April 24, 1958. *Persp. Biol. Med.* 12, 178–203.

1205. Stern, C.D. (2019). The 'omics revolution: how an obsession with compiling lists is threatening the ancient art of experimental design. *BioEssays* 41, 1900168.

1206. Stern, D. (2006). Morphing into shape. *Science* 313, 50–51.

1207. Sternberg, P.W. (2019). Ablating the fixed lineage conjecture: commentary on Kimble 1981. *Dev. Biol.* 446, 1–16.

1208. Stevens, J.L., Edgerton, V.R., and Mitton, S. (1971). Gross anatomy of the hindlimb skeletal system of the *Galago senegalensis*. *Primates* 12(3–4), 313–321.

1209. Stevenson, R.D., Hill, M.F., and Bryant, P.J. (1995). Organ and cell allometry in Hawaiian *Drosophila*: how to make a big fly. *Proc. Roy. Soc. Lond. B* 259, 105–110.

1210. Stinckens, A., Luyten, T., Bijttebier, J., Van den Maagdenberg, K., Dieltines, D., Janssens, S., De Smet, S., Georges, M., and Buys, N. (2008). Characterization of the complete porcine *MSTN* gene and expression levels in pig breeds differing in muscularity. *Anim. Genet.* 39, 586–596.

1211. Stock, G.B. and Bryant, S.V. (1981). Studies of digit regeneration and their implications for theories of development and evolution of vertebrate limbs. *J. Exp. Zool.* 216, 423–433.

1212. Stockard, C.R. (1941). *The Genetic and Endocrinic Basis for Differences in Form and Behavior*. American Anatomical Memoirs, No. 19. Wistar Institute of Anatomy and Biology, Philadelphia, PA.

1213. Stocum, D.L. (2000). Frog limb deformities: an "eco-devo" riddle wrapped in multiple hypotheses surrounded by insufficient data. *Teratology* 62, 147–150.

1214. Stollewerk, A., Schoppmeier, M., and Damen, W.G.M. (2003). Involvement of *Notch* and *Delta* genes in spider segmentation. *Nature* 423, 863–865.

1215. Stone, J.L. and Goodrich, J.T. (2006). The craniopagus malformation: classification and implications for surgical separation. *Brain* 129, 1084–1095.

1216. Stopper, G.F., Hecker, L., Franssen, R.A., and Sessions, S.K. (2002). How trematodes cause limb deformities in amphibians. *J. Exp. Zool.* 294, 252–263.

1217. Striedter, G., Srinivasan, S., and Monuki, E.S. (2015). Cortical folding: when, where, how, and why? *Annu. Rev. Neurosci.* 38, 291–307.

1218. Struhl, G. (1984). Splitting the bithorax complex of *Drosophila*. *Nature* 308, 454–457.

1219. Struhl, G. and Basler, K. (1993). Organizing activity of wingless protein in *Drosophila*. *Cell* 72, 527–540.

1220. Strutt, D. (2002). The asymmetric subcellular localisation of components of the planar polarity pathway. *Semin. Cell Dev. Biol.* 13, 225–231.

1221. Strutt, D. (2009). Gradients and the specification of planar polarity in the insect cuticle. *Cold Spring Harb. Perspect. Biol.* 1, a000489.

1222. Strutt, D., Johnson, R., Cooper, K., and Bray, S. (2002). Asymmetric localization of Frizzled and the determination of Notch-dependent cell fate in the *Drosophila* eye. *Curr. Biol.* 12, 813–824.

1223. Sturtevant, A.H. (1913). The Himalayan rabbit case, with some considerations on multiple allelomorphs. *Am. Nat.* 47, 234–238.

1224. Sturtevant, A.H. (1929). The claret mutant type of *Drosophila simulans:* a study of chromosome elimination and of cell-lineage. *Z. wiss. Zool.* 135, 323–356.

1225. Sturtevant, A.H. (1970). Studies on the bristle pattern of *Drosophila. Dev. Biol.* 21, 48–61.

1226. Sturtevant, A.H. (2001). Reminiscences of T. H. Morgan. *Genetics* 159, 1–5.

1227. Suchy, F. and Nakauchi, H. (2017). Lessons from interspecies mammalian chimeras. *Annu. Rev. Cell Dev. Biol.* 33, 203–217.

1228. Suchy, F. and Nakauchi, H. (2018). Interspecies chimeras. *Curr. Opin. Genet. Dev.* 52, 36–41.

1229. Sui, L., Alt, S., Weigert, M., Dye, N., Eaton, S., Jug, F., Myers, E.W., Jülicher, F., Salbreux, G., and Dahmann, C. (2018). Differential lateral and basal tension drive folding of *Drosophila* wing discs through two distinct mechanisms. *Nat. Commun.* 9, 4620.

1230. Suijkerbuijk, S.J.E., van Osch, M.H.J., Bos, F.L., Hanks, S., Rahman, N., and Kops, G.J.P.L. (2010). Molecular causes for BUBR1 dysfunction in the human cancer predisposition syndrome mosaic variegated aneuploidy. *Cancer Res.* 70, 4891–4900.

1231. Summerbell, D., Lewis, J.H., and Wolpert, L. (1973). Positional information in chick limb morphogenesis. *Nature* 244, 492–496.

1232. Sun, B., Tu, J., Liang, Q., Cheng, X., Fan, X., Li, Y., Wallbank, R.W.R., and Yang, M. (2019). Expression of mammalian *ASH1* and *ASH4* in *Drosophila* reveals opposing functional roles in neurogenesis. *Gene* 688, 132–139.

1233. Sun, G. and Irvine, K.D. (2014). Control of growth during regeneration. *Curr. Top. Dev. Biol.* 108, 95–120.

1234. Sun, J., Ling, M., Wu, W., Bhushan, B., and Tong, J. (2014). The hydraulic mechanism of the unfolding of hind wings in *Dorcus titanus platymelus* (Order: Coleoptera). *Int. J. Mol. Sci.* 15, 6009–6018.

1235. Sun, M., Li, N., Dong, W., Chen, Z., Liu, Q., Xu, Y., He, G., Shi, Y., Li, X., Hao, J., Luo, Y., Shang, D., Lv, D., Ma, F., Zhang, D., Hua, R., Lu, C., Wen, Y., Cao, L., Irvine, A.D., McLean, W.H.I., Dong, Q., Wang, M.-R., Yu, J., He, L., Lo, W.H.Y., and Zhang, X. (2009). Copy-number mutations on chromosome 17q24.2–q24.3 in congenital generalized hypertrichosis terminalis with or without gingival hyperplasia. *Am. J. Hum. Genet.* 84, 807–813.

1236. Sun, T. and Hevner, R.F. (2014). Growth and folding of the mammalian cerebral cortex: from molecules to malformations. *Nat. Rev. Neurosci.* 15, 217–232.

1237. Sundaram, M.V. and Cohen, J.D. (2017). Time to make the doughnuts: building and shaping seamless tubes. *Semin. Cell Dev. Biol.* 67, 123–131.

1238. Sutter, N.B., Bustamante, C.D., Chase, K., Gray, M.M., Zhao, K., Zhu, L., Padhukasahasram, B., Karlins, E., Davis, S., Jones, P.G., Quignon, P., Johnson, G.S., Parker, H.G., Fretwell, N., Mosher, D.S., Lawler, D.F., Satyaraj, E., Nordborg, M., Lark, K.G., Wayne, R.K., and Ostrander, E.A. (2007). A single *IGF1* allele is a major determinant of small size in dogs. *Science* 316, 112–115.

1239. Suttie, J.M., Gluckman, P.D., Butler, J.H., Fennessy, P.F., Corson, I.D., and Laas, F.J. (1985). Insulin-like growth factor 1 (IGF-1) antler-stimulating hormone? *Endocrinology* 116, 846–848.

1240. Suzanne, M. (2016). Molecular and cellular mechanisms involved in leg joint morphogenesis. *Semin. Cell Dev. Biol.* 55, 131–138.

1241. Suzuki, D.G. (2016). Two-headed mutants of the lamprey, a basal vertebrate. *Zoological Lett.* 2, 22.

1242. Suzuki, D.T. (1970). Temperature-sensitive mutations in *Drosophila melanogaster*. *Science* 170, 695–706.

1243. Swett, F.H. (1926). On the production of double limbs in amphibians. *J. Exp. Zool.* 44, 419–473.

1244. Symmons, O., Pan, L., Remeseiro, S., Aktas, T., Klein, F., Huber, W., and Spitz, F. (2016). The Shh topological domain facilitates the action of remote enhancers by reducing the effects of genomic distances. *Dev. Cell* 39, 529–543.

1245. Szabad, J., Schüpbach, T., and Wieschaus, E. (1979). Cell lineage and development in the larval epidermis of *Drosophila melanogaster*. *Dev. Biol.* 73, 256–271.

1246. Szabo, K.T. (1989). *Congenital Malformations in Laboratory and Farm Animals*. Academic Press, New York.

1247. Szebenyi, A.L. (1969). Cleaning behaviour in *Drosophila melanogaster*. *Anim. Behav.* 17, 641–651.

1248. Szenker-Ravi, E., Altunoglu, U., Leushacke, M., Bosso-Lefèvre, C., Khatoo, M., Tran, H.T., Naert, T., Noelanders, R., Hajamohideen, A., Beneteau, C., de Sousa, S.B., Karaman, B., Latypova, X., Başaran, S., Yücel, E.B., Tan, T.T., Vlaminck, L., Nayak, S.S., Shukla, A., Girisha, K.M., Le Caignec, C., Soshnikova, N., Uyguner, Z.O., Vleminckx, K., Barker, N., Kayserili, H., and Reversade, B. (2018). RSPO2 inhibition of RNF43 and ZNRF3 governs limb development independently of LGR4/5/6. *Nature* 557, 564–569.

1249. Tabata, T., Schwartz, C., Gustavson, E., Ali, Z., and Kornberg, T.B. (1995). Creating a *Drosophila* wing de novo, the role of *engrailed*, and the compartment border hypothesis. *Development* 121, 3359–3369.

1250. Taber, L.A. (2014). Morphometrics: transforming tubes into organs. *Curr. Opin. Genet. Dev.* 27, 7–13.

1251. Tabin, C. and Laufer, E. (1993). *Hox* genes and serial homology. *Nature* 361, 692–693.

1252. Tabin, C.J. (1992). Why we have (only) five fingers per hand: Hox genes and the evolution of paired limbs. *Development* 116, 289–296.

1253. Tadin-Strapps, M., Salas-Alanis, J.C., Moreno, L., Warburton, D., Martinez-Mir, A., and Christiano, A.M. (2003). Congenital universal hypertrichosis with deafness and dental anomalies inherited as an X-linked trait. *Clin. Genet.* 63, 418–422.

1254. Tajiri, R. (2017). Cuticle itself as a central and dynamic player in shaping cuticle. *Curr. Opin. Insect Sci.* 19, 30–35.

1255. Tajiri, R., Misaki, K., Yonemura, S., and Hayashi, S. (2010). Dynamic shape changes of ECM-producing cells drive morphogenesis of ball-and-socket joints in the fly leg. *Development* 137, 2055–2063.

1256. Tajiri, R., Misaki, K., Yonemura, S., and Hayashi, S. (2011). Joint morphology in the insect leg: evolutionary history inferred from *Notch* loss-of-function phenotypes in *Drosophila*. *Development* 138, 4621–4626.

1257. Takahashi, H., Abe, M., and Kuroda, R. (2019). GSK3β controls the timing and pattern of the fifth spiral cleavage at the 2–4 cell stage in *Lymnaea stagnalis*. *Dev. Genes Evol.* 229, 73–81.

1258. Takechi, M., Adachi, N., Hirai, T., Kuratani, S., and Kuraku, S. (2013). The Dlx genes as clues to vertebrate genomics and craniofacial evolution. *Semin. Cell Dev. Biol.* 24, 110–118.

1259. Tallinen, T., Chung, J.Y., Biggins, J.S., and Mahadevan, L. (2014). Gyrification from constrained cortical expansion. *PNAS* 111, 12667–12672.

1260. Tallinen, T., Chung, J.Y., Rousseau, F., Girard, N., Lefèvre, J., and Mahadevan, L. (2016). On the growth and form of cortical convolutions. *Nat. Phys.* 12, 588–593.

1261. Tanaka, E.M. (2016). The molecular and cellular choreography of appendage regeneration. *Cell* 165, 1598–1608.

1262. Tanaka, K., Barmina, O., and Kopp, A. (2009). Distinct developmental mechanisms underlie the evolutionary diversification of *Drosophila* sex combs. *PNAS* 106, 4764–4769.

1263. Tanaka, K., Barmina, O., Sanders, L.E., Arbeitman, M.N., and Kopp, A. (2011). Evolution of sex-specific traits through changes in HOX-dependent *doublesex* expression. *PLoS Biol.* 9(8), e1001131.

1264. Tanaka, Y., Okada, Y., and Hirokawa, N. (2005). FGF-induced vesicular release of Sonic hedgehog and retinoic acid in leftward nodal flow is critical for left–right determination. *Nature* 435, 172–177.

1265. Tao, X., Chen, X., and Tian, J. (2012). Fingerprint recognition with identical twin fingerprints. *PLoS ONE* 7(4), e35704.

1266. Tapaltsyan, V., Charles, C., Hu, J., Mindell, D., Ahituv, N., Wilson, G.M., Black, B.L., Viriot, L., and Klein, O.D. (2016). Identification of novel *Fgf* enhancers and their role in dental evolution. *Evol. Dev.* 18, 31–40.

1267. Tapon, N., Harvey, K.F., Bell, D.W., Wahrer, D.C.R., Schiripo, T.A., Haber, D.A., and Hariharan, I.K. (2002). *salvador* promotes both cell cycle exit and apoptosis in *Drosophila* and is mutated in human cancer cell lines. *Cell* 110, 467–478.

1268. Tate, E. (1992). Professor marvels at mutant toad. Eyes inside its mouth upside down to boot. *The Hamilton Spectator* Metro Section (Sept. 4, 1992), B6.

1269. Taub, R. (2004). Liver regeneration: from myth to mechanism. *Nat. Rev. Mol. Cell Biol.* 5, 836–847.

1270. Tautz, D. (1996). Selector genes, polymorphisms, and evolution. *Science* 271, 160–161.

1271. Taylor, G.K. (2001). Mechanics and aerodynamics of insect flight control. *Biol. Rev.* 76, 449–471.

1272. Taylor, J. and Adler, P.N. (2008). Cell rearrangement and cell division during the tissue level morphogenesis of evaginating *Drosophila* imaginal discs. *Dev. Biol.* 313, 739–751.

1273. te Welscher, P., Zuniga, A., Kuijper, S., Drenth, T., Goedemans, H.J., Meijlink, F., and Zeller, R. (2002). Progression of vertebrate limb development through SHH-mediated counteraction of Gli3. *Science* 298, 827–830.

1274. Theisen, H., Syed, A., Nguyen, B.T., Lukacsovich, T., Purcell, J., Srivastava, G.P., Iron, D., Gaudenz, K., Nie, Q., Wan, F.Y.M., Waterman, M.L., and Marsh, J.L. (2007). Wingless directly represses DPP morphogen expression via an Armadillo/TCF/Brinker complex. *PLoS ONE* 2(1), e142.

1275. Theissen, G. (2006). The proper place of hopeful monsters in evolutionary biology. *Theory Biosci.* 124, 349–369.

1276. Theveneau, E. and Mayor, R. (2012). Neural crest delamination and migration: from epithelium-to-mesenchyme transition to collective cell migration. *Dev. Biol.* 366, 34–54.

1277. Thistle, R., Cameron, P., Ghorayshi, A., Dennison, L., and Scott, K. (2012). Contact chemoreceptors mediate male–male repulsion and male–female attraction during *Drosophila* courtship. *Cell* 149, 1140–1151.

1278. Thomas, C. and Strutt, D. (2012). The roles of the cadherins Fat and Dachsous in planar polarity specification in *Drosophila. Dev. Dynamics* 241, 27–39.

1279. Thompson, D.B. (2019). Diet-induced plasticity of linear static allometry is not so simple for grasshoppers: genotype–environment interaction in ontogeny is masked by convergent growth. *Integr. Comp. Biol.* 59, 1382–1398.

1280. Thompson, D.W. (1917). *On Growth and Form.* Cambridge University Press, Cambridge.

1281. Thompson, D.W. (1942). *On Growth and Form*, 2nd ed. Cambridge University Press, Cambridge.

1282. Tickle, C. and Towers, M. (2017). Sonic Hedgehog signaling in limb development. *Front. Cell Dev. Biol.* 5, 14.

1283. Tingler, M., Kurz, S., Maerker, M., Ott, T., Fuhl, F., Schweickert, A., LeBlanc-Straceski, J.M., Noselli, S., and Blum, M. (2018). A conserved role of the unconventional Myosin 1d in laterality determination. *Curr. Biol.* 28, 810–816.

1284. Tisler, M., Schweickert, A., and Blum, M. (2017). *Xenopus*, an ideal model organism to study laterality in conjoined twins. *Genesis* 55, e22993.

1285. Tisler, M., Thumberger, T., Schneider, I., Schweickert, A., and Blum, M. (2017). Leftward flow determines laterality in conjoined twins. *Curr. Biol.* 27, 543–548.

1286. Tobias, J.A., Montgomerie, R., and Lyon, B.E. (2012). The evolution of female ornaments and weaponry: social selection, sexual selection and ecological competition. *Philos. Trans. R. Soc. Lond. B* 367, 2274–2293.

1287. Toda, S., Blauch, L.R., Tang, S.K.Y., Morsut, L., and Lim, W.A. (2018). Programming self-organizing multicellular structures with synthetic cell-cell signaling. *Science* 361, 156–162.

1288. Tokita, M. (2015). How the pterosaur got its wings. *Biol. Rev.* 90, 1163–1178.

1289. Tokunaga, C. (1962). Cell lineage and differentiation on the male foreleg of *Drosophila melanogaster. Dev. Biol.* 4, 489–516.

1290. Tokunaga, C. (1978). Genetic mosaic studies of pattern formation in *Drosophila melanogaster*, with special reference to the prepattern hypothesis. In *Genetic Mosaics and Cell Differentiation*, Gehring, W.J., editor. Springer-Verlag, Berlin, pp. 157–204.

1291. Tokunaga, C. (1982). Curt Stern, 1902–1981, in memoriam. *Jpn. J. Genet.* 57, 459–466.

1292. Tomaszewski, R. and Bulandra, A. (2015). Ulnar dimelia: diagnosis and management of a rare congenital anomaly of the upper limb. *J. Orthop.* 12, S121–S124.

1293. Tomita, K., Moriyoshi, K., Nakanishi, S., Guillemot, F., and Kageyama, R. (2000). Mammalian *achaete-scute* and *atonal* homologs regulate neuronal versus glial fate determination in the central nervous system. *EMBO J.* 19, 5460–5472.

1294. Tomoyasu, Y. (2017). *Ultrabithorax* and the evolution of insect forewing/hindwing differentiation. *Curr. Opin. Insect Sci.* 19, 8–15.

1295. Tomoyasu, Y., Arakane, Y., Kramer, K.J., and Denell, R.E. (2009). Repeated co-options of exoskeleton formation during wing-to-elytron evolution in beetles. *Curr. Biol.* 19, 2057–2065.

1296. Tomoyasu, Y., Nakamura, M., and Ueno, N. (1998). Role of Dpp signalling in prepattern formation of the dorsocentral mechanosensory organ in *Drosophila melanogaster. Development* 125, 4215–4224.

1297. Tomoyasu, Y., Wheeler, S.R., and Denell, R.E. (2005). *Ultrabithorax* is required for membranous wing identity in the beetle *Tribolium casteneum. Nature* 433, 643–647.

1298. Tornini, V.A. and Poss, K.D. (2014). Keeping at arm's length during regeneration. *Dev. Cell* 29, 139–145.

1299. Toro, R. (2012). On the possible shapes of the human brain. *Evol. Biol.* 39, 600–612.

1300. Toussaint, S., Llamosi, A., Morino, L., and Youlatos, D. (2020). The central role of small vertical substrates for the origin of grasping in early primates. *Curr. Biol.* 30, 1600–1613.

1301. Tozluoglu, M., Duda, M., Kirkland, N.J., Barrientos, R., Burden, J.J., Muñoz, J.J., and Mao, Y. (2019). Planar differential growth rates initiate precise fold positions in complex epithelia. *Dev. Cell* 51, 299–312.

1302. Triggs-Raine, B. and Natowicz, M.R. (2015). Biology of hyaluronan: insights from genetic disorders of hyaluronan metabolism. *World J. Biol. Chem.* 6, 110–120.

1303. Tripathi, A., Swaroop, S., and Varadarajan, R. (2019). Molecular determinants of temperature-sensitive phenotypes. *Biochemistry* 58, 1738–1750.

1304. Troost, T., Schneider, M., and Klein, T. (2015). A re-examination of the selection of the sensory organ precursor of the bristle sensilla of *Drosophila melanogaster*. *PLoS Genet.* 11(1), e1004911.

1305. Trueb, L. (1973). Bones, frogs, and evolution. In *Evolutionary Biology of the Anurans: Contemporary Research on Major Problems*, Vial, J.L., editor. University of Missouri Press, Columbia, MO, pp. 65–132.

1306. Turing, A.M. (1952). The chemical basis of morphogenesis. *Philos. Trans. Roy. Soc. Lond. B* 237, 37–72.

1307. Udan, R.S., Kango-Singh, M., Nolo, R., Tao, C., and Halder, G. (2003). Hippo promotes proliferation arrest and apoptosis in the Salvador/Warts pathway. *Nat. Cell Biol.* 5, 914–920.

1308. Ulshafer, R.J. and Clavert, A. (1979). The use of avian double monsters in studies on induction of the nervous system. *J. Embryol. Exp. Morphol.* 53, 237–243.

1309. Umair, M., Ahmad, F., Bilal, M., Ahmad, W., and Alfadhel, M. (2018). Clinical genetics of polydactyly: an updated review. *Front. Genet.* 9, 447.

1310. Umulis, D.M. and Othmer, H.G. (2013). Mechanisms of scaling in pattern formation. *Development* 140, 4830–4843.

1311. Urdy, S., Goudemand, N., and Pantalacci, S. (2016). Looking beyond the genes: the interplay between signaling pathways and mechanics in the shaping and diversification of epithelial tissues. *Curr. Top. Dev. Biol.* 119, 227–290.

1312. Usui, K. and Tokita, M. (2018). Creating diversity in mammalian facial morphology: a review of potential developmental mechanisms. *EvoDevo* 9, 15.

1313. Usui, K., Goldstone, C., Gibert, J.-M., and Simpson, P. (2008). Redundant mechanisms mediate bristle patterning on the *Drosophila* thorax. *PNAS* 105, 20112–20117.

1314. Valentin, M.N., Solomon, B.D., Richard, G., Ferreira, C.R., and Kirkorian, A.Y. (2018). Basan gets a new fingerprint: mutations in the skin-specific isoform of *SMARCAD1* cause ectodermal dysplasia syndromes with adermatoglyphia. *Am. J. Med. Genet. A* 176A, 2451–2455.

1315. Valizadeh, A., Majidinia, M., Samadi-Kafil, H., Yousefi, M., and Yousefi, B. (2019). The roles of signaling pathways in liver repair and regeneration. *J. Cell Physiol.* 234, 14966–14974.

1316. van Amerongen, R. and Nusse, R. (2009). Towards an integrated view of Wnt signaling in development. *Development* 136, 3205–3214.

1317. van Arensbergen, J., van Steensel, B., and Bussemaker, H.J. (2014). In search of the determinants of enhancer–promoter interaction specificity. *Trends Cell Biol.* 24, 695–702.

1318. Van Raamsdonk, C.D., Barsh, G.S., Wakamatsu, K., and Ito, S. (2009). Independent regulation of hair and skin color by two G protein-coupled pathways. *Pigment Cell Melanoma Res.* 22, 819–826.

1319. Van Raamsdonk, C.D., Fitch, K.R., Fuchs, H., de Angelis, M.H., and Barsh, G.S. (2004). Effects of G-protein mutations on skin color. *Nat. Genet.* 36, 961–968.

1320. Van Valen, L. (1974). A natural model for the origin of some higher taxa. *J. Herpetol.* 8, 109–121.

1321. Vandamme, N. and Berx, G. (2019). From neural crest cells to melanocytes: cellular plasticity during development and beyond. *Cell. Mol. Life Sci.* 76, 1919–1934.

1322. Vandenberg, L.N., Adams, D.S., and Levin, M. (2012). Normalized shape and location of perturbed craniofacial structures in the *Xenopus* tadpole reveal an innate ability to achieve correct morphology. *Dev. Dynamics* 241, 863–878.

1323. Vandervorst, P. and Ghysen, A. (1980). Genetic control of sensory connections in *Drosophila. Nature* 286, 65–67.

1324. Vargesson, N. (2020). Positional information: a concept underpinning our understanding of developmental biology. *Dev. Dynamics* 249, 298–312.

1325. Venkatasubban, R. (2013). West Texas one-eyed goat appears in "Ripley's Believe It or Not!". *Midland Reporter-Telegram* (Friday, Oct. 18, 2013).

1326. Venot, Q., Blanc, T., Rabia, S.H., Berteloot, L., Ladraa, S., Duong, J.-P., Blanc, E., Johnson, S.C., Hoguin, C., Boccara, O., Sarnacki, S., Boddaert, N., Pannier, S., Martinez, F., Magassa, S., Yamaguchi, J., Knebelmann, B., Merville, P., Grenier, N., Joly, D., Cormier-Daire, V., Michot, C., Bole-Feysot, C., Picard, A., Soupre, V., Lyonnet, S., Sadoine, J., Slimani, L., Chaussain, C., Laroche-Raynaud, C., Guibaud, L., Broissand, C., Amiel, J., Legendre, C., Terzi, F., and Canaud, G. (2018). Targeted therapy in patients with PIK3CA-related overgrowth syndrome. *Nature* 558, 540–546.

1327. Verbrugge, S.A.J., Schönfelder, M., Becker, L., Nezhad, F.Y., de Angelis, M.H., and Wackerhage, H. (2018). Genes whose gain or loss-of-function increases skeletal muscle mass in mice: a systematic literature review. *Front. Physiol.* 9, 553.

1328. Vergara, M.N., Tsissios, G., and Rio-Tsonis, K.D. (2018). Lens regeneration: a historical perspective. *Int. J. Dev. Biol.* 62, 351–361.

1329. Verheyden, J.M. and Sun, X. (2008). An Fgf/*Gremlin* inhibitory feedback loop triggers termination of limb bud outgrowth. *Nature* 454, 638–641.

1330. Verma, P.K. and El-Harouni, A.A. (2015). Review of literature: genes related to postaxial polydactyly. *Front. Pediatr.* 3, 8.

1331. Vicente, C. (2016). An interview with Peter Lawrence. *Development* 143, 183–185.

1332. Vidal, V.P.I., Chaboissier, M.-C., Lützkendorf, S., Cotsarelis, G., Mill, P., Hui, C.-C., Ortone, N., Ortone, J.-P., and Schedl, A. (2005). Sox9 is essential for outer root sheath differentiation and the formation of the hair stem cell compartment. *Curr. Biol.* 15, 1340–1351.

1333. Villares, R. and Cabrera, C.V. (1987). The *achaete-scute* gene complex of *D. melanogaster:* conserved domains in a subset of genes required for neurogenesis and their homology to *myc. Cell* 50, 415–424.

1334. Vinicius, L. (2005). Human encephalization and developmental timing. *J. Human Evol.* 49, 762–776.

1335. Vogel, G. (2012). Turing pattern fingered for digit formation. *Science* 338, 1406.

1336. Vogel, G. (2013). How do organs know when they have reached the right size? *Science* 340, 1156–1157.

1337. Vogt, T.F. and Duboule, D. (1999). Antagonists go out on a limb. *Cell* 99, 563–566.

1338. Vollmer, J., Casares, F., and Iber, D. (2017). Growth and size control during development. *Open Biol.* 7, 170190.

1339. von Kalm, L., Fristrom, D., and Fristrom, J. (1995). The making of a fly leg: a model for epithelial morphogenesis. *BioEssays* 17, 693–702.

1340. von Luschan, F. (1907). Ein Haarmensch. *Zeitschrift für Ethnologie* 39, 425–429.

1341. vonHoldt, B.M., Shuldiner, E., Koch, I.J., Kartzinel, R.Y., Hogan, A., Brubaker, L., Wanser, S., Stahler, D., Wynne, C.D.L., Ostrander, E.A., Sinsheimer, J.S., and Udell, M.A.R. (2017). Structural variants in genes associated with human Williams-Beuren syndrome underlie stereotypical hypersociability in domestic dogs. *Sci. Adv.* 3, e1700398.

1342. Waddington, C.H. (1940). *Organizers and Genes*. Cambridge University Press, Cambridge.

1343. Waddington, C.H. (1956). Genetic assimilation of the *bithorax* phenotype. *Evolution* 10, 1–13.

1344. Waddington, C.H. (1957). *The Strategy of the Genes: A Discussion of Some Aspects of Theoretical Biology*. George Allen & Unwin, London.

1345. Walentek, P., Beyer, T., Thumberger, T., Schweickert, A., and Blum, M. (2012). *ATP4a* Is required for Wnt-dependent *Foxj1* expression and leftward flow in *Xenopus* left–right development. *Cell Rep.* 1, 516–527.

1346. Wallingford, J.B. (2019). We are all developmental biologists. *Dev. Cell* 50, 132–137.

1347. Walsh, C.A. (2019). Rainer W. Guillery and the genetic analysis of brain development. *Eur. J. Neurosci.* 49, 900–908.

1348. Wang, C., Rüther, U., and Wang, B. (2007). The Shh-independent activator function of the full-length Gli3 protein and its role in vertebrate limb digit patterning. *Dev. Biol.* 305, 460–469.

1349. Wang, D., Berg, D., Ba, H., Sun, H., Wang, Z., and Li, C. (2019). Deer antler stem cells are a novel type of cells that sustain full regeneration of a mammalian organ – deer antler. *Cell Death Dis.* 10, 443.

1350. Wang, Y., Dong, L., and Evans, S.E. (2016). Polydactyly and other limb abnormalities in the Jurassic salamander *Chunerpeton* from China. *Palaeobio. Palaeoenv.* 96, 49–59.

1351. Wang, Y., Wang, G.-D., He, Q.-L., Luo, Z.-P., Yang, L., Yao, Q., and Chen, K.-P. (2020). Phylogenetic analysis of achaete-scute complex genes in metazoans. *Mol. Genet. Genomics* 295, 591–606.

1352. Wanninger, A. and Wollesen, T. (2019). The evolution of molluscs. *Biol. Rev.* 94, 102–115.

1353. Warman, P.H. and Ennos, A.R. (2009). Fingerprints are unlikely to increase the friction of primate fingerpads. *J. Exp. Biol.* 212, 2016–2022.

1354. Warren, I. and Smith, H. (2007). Stalk-eyed flies (Diopsidae): modelling the evolution and development of an exaggerated sexual trait. *BioEssays* 29, 300–307.

1355. Warren, I.A., Gotoh, H., Dworkin, I.M., Emlen, D.J., and Lavine, L.C. (2013). A general mechanism for conditional expression of exaggerated sexually-selected traits. *BioEssays* 35, 889–899.

1356. Warrington, S.J., Strutt, H., Fisher, K.H., and Strutt, D. (2017). A dual function for Prickle in regulating frizzled stability during feedback-dependent amplification of planar polarity. *Curr. Biol.* 27, 2784–2797.

1357. Wartlick, O., Mumcu, P., Jülicher, F., and Gonzalez-Gaitan, M. (2011). Understanding morphogenetic growth control: lessons from flies. *Nat. Rev. Mol. Cell Biol.* 12, 594–604.

1358. Watanabe, M. and Kondo, S. (2015). Is pigment patterning in fish skin determined by the Turing mechanism? *Trends Genet.* 31, 88–96.

1359. Watt, K.I., Harvey, K.F., and Gregorevic, P. (2017). Regulation of tissue growth by the mammalian Hippo signaling pathway. *Front. Physiol.* 8, 942.

1360. Weatherbee, S.D., Halder, G., Kim, J., Hudson, A., and Carroll, S. (1998). Ultrabithorax regulates genes at several levels of the wing-patterning hierarchy to shape the development of the *Drosophila* haltere. *Genes Dev.* 12, 1474–1482.

1361. Weaver, C. and Kimelman, D. (2004). Move it or lose it: axis specification in *Xenopus. Development* 131, 3491–3499.

1362. Webb, A.A. and Cullen, C.L. (2010). Coat color and coat color pattern-related neurologic and neuro-ophthalmic diseases. *CVJ* 51, 653–657.

1363. Weber, C., Zhou, Y., Lee, J.G., Looger, L.L., Qian, G., Ge, C., and Capel, B. (2020). Temperature-dependent sex determination is mediated by pSTAT3 repression of *Kdm6b. Science* 368, 303–306.

1364. Weber, M.A. and Sebire, N.J. (2010). Genetics and developmental pathology of twinning. *Semin. Fetal Neonatal Med.* 15, 313–318.

1365. Weiner, J. (1999). *Time, Love, Memory.* Random House, New York.

1366. Weirauch, M.T. and Hughes, T.R. (2010). Conserved expression without conserved regulatory sequence: the more things change, the more they stay the same. *Trends Genet.* 26, 66–74.

1367. Weiss, P. (1939). *Principles of Development.* Henry Holt & Co., New York.

1368. Wellik, D.M. and Capecchi, M.R. (2003). *Hox10* and *Hox11* genes are required to globally pattern the mammalian skeleton. *Science* 301, 363–367.

1369. Welte, M.A., Duncan, I., and Lindquist, S. (1995). The basis for a heat-induced developmental defect: defining crucial lesions. *Genes Dev.* 9, 2240–2250.

1370. Welton, T., Ather, S., Proudlock, F.A., Gottlob, I., and Dineen, R.A. (2017). Altered whole-brain connectivity in albinism. *Hum. Brain Mapp.* 38, 740–752.

1371. Wendelin, D.S., Pope, D.N., and Mallory, S.B. (2003). Hypertrichosis. *J. Am. Acad. Dermatol.* 48, 161–179.

1372. Werdelin, L. and Olsson, L. (1997). How the leopard got its spots: a phylogenetic view of the evolution of felid coat patterns. *Biol. J. Linnean Soc.* 62, 383–400.

1373. Wernet, M.F. and Desplan, C. (2015). Brain wiring in the fourth dimension. *Cell* 162, 20–22.

1374. Wernet, M.F., Mazzoni, E.O., Celik, A., Duncan, D.M., Duncan, I., and Desplan, C. (2006). Stochastic *spineless* expression creates the retinal mosaic for color vision. *Nature* 440, 174–180.

1375. Werts, A.D. and Goldstein, B. (2011). How signaling between cells can orient a mitotic spindle. *Semin. Cell Dev. Biol.* 22, 842–849.

1376. West-Eberhard, M.J. (2003). *Developmental Plasticity and Evolution.* Oxford University Press, New York.

1377. Westgate, G.E., Ginger, R.S., and Green, M.R. (2017). The biology and genetics of curly hair. *Exp. Dermatol.* 26, 483–490.

1378. Westhusin, M. (1997). From mighty mice to mighty cows. *Nat. Genet.* 17, 4–5.

1379. Wexler, J.R., Plachetzki, D.C., and Kopp, A. (2014). Pan-metazoan phylogeny of the DMRT gene family: a framework for functional studies. *Dev. Genes Evol.* 224, 175–181.

1380. White, R.A.H. and Akam, M.E. (1985). *Contrabithorax* mutations cause inappropriate expression of *Ultrabithorax* products in *Drosophila*. *Nature* 318, 567–569.

1381. Whiting, M.F. and Wheeler, W.C. (1994). Insect homeotic transformation. *Nature* 368, 696.

1382. Wianny, F., Kennedy, H., and Dehay, C. (2018). Bridging the gap between mechanics and genetics in cortical folding: ECM as a major driving force. *Neuron* 99, 625–627.

1383. Widelitz, R.B., Baker, R.E., Plikus, M., Lin, C.-M., Maini, P.K., Paus, R., and Chuong, C.M. (2006). Distinct mechanisms underlie pattern formation in the skin and skin appendages. *Birth Defects Res. C Embryo Today* 78, 280–291.

1384. Widmann, T.J. and Dahmann, C. (2009). Dpp signaling promotes the cuboidal-to-columnar shape transition of *Drosophila* wing disc epithelia by regulating Rho1. *J. Cell Sci.* 122, 1362–1373.

1385. Widmann, T.J. and Dahmann, C. (2009). Wingless signaling and the control of cell shape in wing imaginal discs. *Dev. Biol.* 334, 161–173.

1386. Wieschaus, E. (1978). Cell lineage relationships in the *Drosophila* embryo. In *Genetic Mosaics and Cell Differentiation*, Gehring, W.J., editor. Springer-Verlag, Berlin, pp. 97–118.

1387. Wieschaus, E. and Gehring, W. (1976). Gynandromorph analysis of the thoracic disc primordia in *Drosophila melanogaster*. *W. Roux Arch. Dev. Biol.* 180, 31–46.

1388. Wieschaus, E. and Nüsslein-Volhard, C. (2014). Walter Gehring (1939–2014). *Curr. Biol.* 24, R632–R634.

1389. Wieschaus, E. and Nüsslein-Volhard, C. (2016). The Heidelberg screen for pattern mutants of *Drosophila*: a personal account. *Annu. Rev. Cell Dev. Biol.* 32, 1–46.

1390. Wigglesworth, V.B. (1940). Local and general factors in the development of "pattern" in *Rhodnius prolixus* (Hemiptera). *J. Exp. Biol.* 17, 180–200.

1391. Wigglesworth, V.B. (1988). The control of pattern as seen in the integument of an insect. *BioEssays* 9, 23–27.

1392. Wilkins, A.S. (2020). A striking example of developmental bias in an evolutionary process: the "domestication syndrome". *Evol. Dev.* 22, 143–153.

1393. Wilkins, A.S., Wrangham, R., and Fitch, W.T. (2014). The "domestication syndrome" in mammals: a unified explanation based on neural crest cell behavior and genetics. *Genetics* 197, 795–808.

1394. Williams, M.L.K. and Solnica-Krezel, L. (2017). Regulation of gastrulation movements by emergent cell and tissue interactions. *Curr. Opin. Cell Biol.* 48, 33–39.

1395. Williams, R.W., Hogan, D., and Garraghty, P.E. (1994). Target recognition and visual maps in the thalamus of achiasmatic dogs. *Nature* 367, 637–639.

1396. Winchester, G. (1996). The Morgan lineage. *Curr. Biol.* 6, 100–101.

1397. Winchester, G. (2004). Edward B. Lewis 1918–2004. *Curr. Biol.* 14, R740–R742.

1398. Winter, R.M. (1996). Analyzing human developmental abnormalities. *BioEssays* 18, 965–971.

1399. Wolff, G.L. (1955). The effects of environmental temperature on coat color in diverse genotypes of the guinea pig. *Genetics* 40, 90–106.

1400. Wolpert, L. (1969). Positional information and the spatial pattern of cellular differentiation. *J. Theor. Biol.* 25, 1–47.

1401. Wolpert, L. (1971). Positional information and pattern formation. *Curr. Top. Dev. Biol.* 6, 183–224.

1402. Wolpert, L. (1981). Positional information and pattern formation. *Philos. Trans. Roy. Soc. Lond. B* 295, 441–450.

1403. Wolpert, L. (1989). Positional information and prepattern in the development of pattern. In *Cell to Cell Signalling: From Experiments to Theoretical Models*, Goldbeter, A., editor. Academic Press, New York, pp. 133–143.

1404. Wolpert, L. (1996). One hundred years of positional information. *Trends Genet.* 12, 359–364.

1405. Wolpert, L., Tickle, C., Martinez Arias, A., Lawrence, P., Lumsden, A., Robertson, E., Meyerowitz, E., and Smith, J. (2015). *Principles of Development*, 5th ed. Oxford University Press, New York.

1406. Woltering, J.M. and Duboule, D. (2010). The origin of digits: expression patterns versus regulatory mechanisms. *Cell* 18, 526–532.

1407. Wong, R., Geyer, S., Weninger, W., Guimberteau, J.-C., and Wong, J.K. (2016). The dynamic anatomy and patterning of skin. *Exp. Dermatol.* 25, 92–98.

1408. Worley, M.I., Alexander, L.A., and Hariharan, I.K. (2018). CtBP impedes JNK- and Upd/STAT-driven cell fate misspecifications in regenerating *Drosophila* imaginal discs. *eLife* 7, e30391.

1409. Wright, S. (1915). The albino series of allelomorphs in guinea-pigs. *Am. Nat.* 49, 140–148.

1410. Wu, J., Greely, H.T., Jaenisch, R., Nakauchi, H., Rossant, J., and Izpisua Belmonte, J.C. (2016). Stem cells and interspecies chimaeras. *Nature* 540, 51–59.

1411. Wu, P., Hou, L., Plikus, M., Hughes, M., Scehnet, J., Suksaweang, S., Widelitz, R.B., Jiang, T.-X., and Chuong, C.-M. (2004). *Evo-Devo* of amniote integuments and appendages. *Int. J. Dev. Biol.* 48, 249–270.

1412. Wutke, S., Benecke, N., Sandoval-Castellanos, E., Döhle, H.-J., Friederich, S., Gonzalez, J., Hallsson, J.H., Hofreiter, M., Lõugas, L., Magnell, O., Morales-Muniz, A., Orlando, L., Pálsdóttir, A.H., Reissmann, M., Ruttkay, M., Trinks, A., and Ludwig, A. (2016). Spotted phenotypes in horses lost attractiveness in the Middle Ages. *Sci. Rep.* 6, 38548.

1413. Wyatt, T.P.J., Fouchard, J., Lisica, A., Khalilgharibi, N., Baum, B., Recho, P., Kabla, A.J., and Charras, G.T. (2020). Actomyosin controls planarity and folding of epithelia in response to compression. *Nat. Mater.* 19, 109–117.

1414. Xu, H.-J., Xue, J., Lu, B., Zhang, X.-C., Zhuo, J.-C., He, S.-F., Ma, X.-F., Jiang, Y.-Q., Fan, H.-W., Xu, J.-Y., Ye, Y.-X., Pan, P.-L., Li, Q., Bao, Y.-Y., Nijhout, H.F., and Zhang, C.-X. (2015). Two insulin receptors determine alternative wing morphs in planthoppers. *Nature* 519, 464–467.

1415. Yamamoto, D., Jallon, J.-M., and Komatsu, A. (1997). Genetic dissection of sexual behavior in *Drosophila melanogaster*. *Annu. Rev. Entomol.* 42, 551–585.

1416. Yan, B. (2010). Numb: from flies to humans. *Brain Dev.* 32, 293–298.

1417. Yang, L.Q., Zhang, K., Wu, Q.Y., Li, J., Lai, S.J., Song, T.Z., and Zhang, M. (2019). Identification of two novel single nucleotide polymorphism sites in the *myostatin* (*mstn*) gene and their association with carcass traits in meat-type rabbits (*Oryctolagus cuniculus*). *World Rabbit Sci.* 27, 249–256.

1418. Yang, Z.-Q., Zhang, H.-L., Duan, C.-C., Geng, S., Wang, K., Yu, H.-F., Yue, Z.-P., and Guo, B. (2017). IGF1 regulates RUNX1 expression via IRS1/2: implications for antler chondrocyte differentiation. *Cell Cycle* 16, 522–532.

1419. Ye, X.C., Pegado, V., Patel, M.S., and Wasserman, W.W. (2014). Strabismus genetics across a spectrum of eye misalignment disorders. *Clin. Genet.* 86, 103–111.

1420. Yip, R.K.H., Chan, D., and Cheah, K.S.E. (2019). Mechanistic insights into skeletal development gained from genetic disorders. *Curr. Top. Dev. Biol.* 133, 343–385.

1421. Yokoyama, H., Yonei-Tamura, S., Endo, T., Izpisúa Belmonte, J.C., Tamura, K., and Ide, H. (2000). Mesenchyme with *fgf–10* expression is responsible for regenerative capacity in *Xenopus* limb buds. *Dev. Biol.* 219, 18–29.

1422. Yuan, S. and Brueckner, M. (2018). Left–right asymmetry: myosin 1D at the center. *Curr. Biol.* 28, R567–R569.

1423. Zakany, J. and Duboule, D. (2007). The role of *Hox* genes during vertebrate limb development. *Curr. Opin. Genet. Dev.* 17, 359–366.

1424. Zalokar, M., Erk, I., and Santamaria, P. (1980). Distribution of ring-X chromosomes in the blastoderm of gynandromorphic *D. melanogaster*. *Cell* 19, 133–141.

1425. Zanella, M., Vitriolo, A., Andirko, A., Martins, P.T., Sturm, S., O'Rourke, T., Laugsch, M., Malerba, N., Skaros, A., Trattaro, S., Germain, P.-L., Mihailovic, M., Merla, G., Rada-Iglesias, A., Boeckx, C., and Testa, G. (2019). Dosage analysis of the 7q11.23 Williams region identifies *BAZ1B* as a major human gene patterning the modern human face and underlying self-domestication. *Sci. Adv.* 5, eaaw7908.

1426. Zanna, G., Docampo, M.J., Fondevila, D., Bardagi, M., Bassols, A., and Ferrer, L. (2009). Hereditary cutaneous mucinosis in shar pei dogs is associated with increased hyaluronan synthase-2 mRNA transcription by cultured dermal fibroblasts. *Vet. Dermatol.* 20, 377–382.

1427. Zanna, G., Fondevila, D., Bardagi, M., Docampo, M.J., Bassols, A., and Ferrer, L. (2008). Cutaneous mucinosis in shar-pei dogs is due to hyaluronic acid deposition and is associated with high levels of hyaluronic acid in serum. *Vet. Dermatol.* 19, 314–318.

1428. Zartman, J.J. and Shvartsman, S.Y. (2010). Unit operations of tissue development: epithelial folding. *Annu. Rev. Chem. Biomol. Eng.* 1, 231–246.

1429. Zhang, C., Wang, F., Gao, Z., Zhang, P., Gao, J., and Wu, X. (2020). Regulation of Hippo signaling by mechanical signals and the cytoskeleton. *DNA Cell Biol.* 39, 159–166.

1430. Zhang, H., Tang, C., Li, X., and Kong, A.W.K. (2014). A study of similarity between genetically identical body vein patterns. In *IEEE Symposium on Computational Intelligence in Biometrics and Identity Management (CIBIM)*. Orlando, FL, 151–159.

1431. Zhang, N., Guo, L., and Simpson, J.H. (2020). Spatial comparisons of mechanosensory information govern the grooming sequence in *Drosophila*. *Curr. Biol.* 30, 988–1001.

1432. Zhang, Y., Andl, T., Yang, S.H., Teta, M., Liu, F., Seykora, J.T., Tobias, J.W., Piccolo, S., Schmidt-Ullrich, R., Nagy, A., Taketo, M.M., Dlugosz, A.A., and Millar, S.E. (2008). Activation of β-catenin signaling programs embryonic epidermis to hair follicle fate. *Development* 135, 2161–2172.

1433. Zhang, Y., Tomann, P., Andl, T., Gallant, N.M., Huelsken, J., Jerchow, B., Birchmeier, W., Paus, R., Piccolo, S., Mikkola, M.L., Morrisey, E.E.,

Overbeek, P.A., Scheidereit, C., Millar, S.E., and Schmidt-Ullrich, R. (2009). Reciprocal requirements for EDA/EDAR/NF-κB and Wnt/β-catenin signaling pathways in hair follicle induction. *Dev. Cell* 17, 49–61.

1434. Zheng, Y. and Pan, D. (2019). The Hippo signaling pathway in development and disease. *Dev. Cell* 50, 264–282.

1435. Zhou, J.X. and Huang, S. (2010). Understanding gene circuits at cell-fate branch points for rational cell reprogramming. *Trends Genet.* 27, 55–62.

1436. Zhou, P., Byrne, C., Jacobs, J., and Fuchs, E. (1995). Lymphoid enhancer factor 1 directs hair follicle patterning and epithelial cell fate. *Genes Dev.* 9, 570–583.

1437. Zhu, J. and Mackem, S. (2017). John Saunders' ZPA, Sonic hedgehog and digit identity: how does it really all work? *Dev. Biol.* 429, 391–400.

1438. Zilles, K., Palomero-Gallagher, N., and Amunts, K. (2013). Development of cortical folding during evolution and ontogeny. *Trends Neurosci.* 36, 275–284.

1439. Zimova, M., Hackländer, K., Good, J.M., Melo-Ferreira, J., Alves, P.C., and Mills, L.S. (2018). Function and underlying mechanisms of seasonal colour moulting in mammals and birds: what keeps them changing in a warming world? *Biol. Rev.* 93, 1478–1498.

1440. Zuniga, A. and Zeller, R. (2014). In Turing's hands: the making of digits. *Science* 345, 516–517.

1441. Zylberberg, J. and Strowbridge, B.W. (2017). Mechanisms of persistent activity in cortical circuits: possible neural substrates for working memory. *Annu. Rev. Neurosci.* 40, 603–627.

Index